Public Concerns, Environmental Standards and Agricultural Trade

Public Concerns, Environmental Standards and Agricultural Trade

Edited by

Floor Brouwer

Agricultural Economics Research Institute (LEI)
The Hague, the Netherlands

and

David E. Ervin

Portland State University
Portland, Oregon, USA

CABI *Publishing*

CABI *Publishing* **is a division of CAB** *International*

CABI Publishing	CABI Publishing
CAB International	10 E 40th Street
Wallingford	Suite 3203
Oxon OX10 8DE	New York, NY 10016
UK	USA
Tel: +44 (0)1491 832111	Tel: +1 212 481 7018
Fax: +44 (0)1491 833508	Fax: +1 212 686 7993
Email: cabi@cabi.org	Email: cabi-nao@cabi.org
Web site: www.cabi-publishing.org	

A catalogue record for this book is available from the British Library, London, UK.

Library of Congress Cataloging-in-Publication data

Public concerns, environmental standards, and agricultural trade / edited by Floor Brouwer and David E. Ervin

    p. cm.

Includes bibliographical references.

ISBN 0-85199-586-1 (alk. paper)

    1. Agriculture--Environmental aspects. 2. International trade--Environmental aspects. 3. Agriculture and state. 4. Environmental policy. 5. Environmental protection--Standards. I. Brouwer, Floor. II. Ervin, David E.

TD 195.A34. P83 2002

363.73'1--dc21   2001043746

ISBN 0 85199 586 1

Printed and bound in the UK by Cromwell Press, Trowbridge, from copy supplied by the editors.

Contents

Contributors

Kym Anderson is Director of the Center for International Economic Studies, School of Economics, University of Adelaide, Adelaide, SA 5005, Australia (e-mail: kym.anderson@adelaide.edu.au).

David Baldock is Director of the Institute for European Environmental Policy, Dean Bradley House, 52 Horse Ferry Road, London SW1P 2AG, United Kingdom (e-mail: db@ieeplondon.org.uk).

Floor Brouwer is Head of the Research Unit Management of Natural Resources, Agricultural Economics Research Institute (LEI), P.O. Box 29703, 2502 LS The Hague, The Netherlands (e-mail: f.m.brouwer@lei.wag-ur.nl).

Chantal Line Carpentier is Program Manager, Environment, Economy and Trade Program, North American Commission on Environmental Cooperation, Montreal, Canada (e-mail: carpentier@ccemtl.org).

W. Bradnee Chambers is a Research Fellow and Coordinator, Multilateralism and Sustainable Development Programme of the Institute of Advanced Studies, The United Nations University, 53-67, Jingumae 5-chome, Shibuya-ku, Tokyo 150-8304, Japan (e-mail: chambers@ias.unu.edu).

Janet Dwyer is Senior Fellow, Institute for European Environmental Policy, Dean Bradley House, 52 Horse Ferry Road, London SW1P 2AG, United Kingdom (e-mail: jd@ieeplondon.org.uk).

David E. Ervin is Professor of Environmental Studies, Portland State University, 241 M Cramer, Portland, OR 97207-0751, USA, and Senior Policy Analyst with the Wallace Center for Agricultural and Environmental Policy, Winrock International, Arlington, VA (e-mail: ervin@pdx.edu).

Glenn Fox is Professor and Acting Chairman, Department of Agricultural Economics and Business, University of Guelph, Guelph, Ontario, Canada N1G 2W1 (e-mail: fox@agec.uoguelph.ca).

Ulrike Grote is Senior Research Fellow at the Department of Economics and Technological Change of the Center for Development Research (ZEF), Walter-Flex-Str.3, D-53113 Bonn, Germany (e-mail: u.grote@uni-bonn.de).

Theo H. Jonker is Researcher at the Agricultural Economics Research Institute (LEI), P.O. Box 29703, 2502 LS The Hague, The Netherlands (e-mail: t.h.jonker@lei.wag-ur.nl).

Jennifer Kidon is former graduate Research Assistant, Department of Agricultural Economics and Business, University of Guelph, Guelph, Ontario, Canada N1G 2W1 (e-mail: jkidon@hotmail.com).

Anton D. Meister is Professor, Resource and Environmental Economics, Department of Applied and International Economics, Massey University, Private Bag 11222, Palmerston North, New Zealand (e-mail: a.meister@massey.ac.nz).

Cindy van Rijswick is Economist in the Research Unit Agricultural Policy, Agricultural Economics Research Institute (LEI), P.O. Box 29703, 2502 LS The Hague, The Netherlands (e-mail: c.w.j.van.rijswick@hccnet.nl).

Huib Silvis is Head of the Research Unit Agricultural Policy, Agricultural Economics Research Institute (LEI), P.O. Box 29703, 2502 LS The Hague, The Netherlands (e-mail: h.j.silvis@lei.wag-ur.nl).

Randy Stringer is Deputy Director of the Center for International Economic Studies, School of Economics, University of Adelaide, Adelaide, SA 5005, Australia (e-mail: randy.stringer@adelaide.edu.au).

Ikuo Takahashi is Professor of Marketing, Faculty of Business and Commerce, Keio University, 2-15-45 Mita, Minato-ku, Tokyo 108-8345, Japan (e-mail: takahashi@fbc.keio.ac.jp).

List of Abbreviations

AAA	Agriculture - Advancing Australia (Australia)
AAFC	Agriculture and Agri-Food Canada (Canada)
AAP	Accepted agricultural practice (USA)
ACPNT	Animal confinement policy national task force (USA)
ACVM	Agricultural Compounds and Veterinary Medicines Act (New Zealand)
ADW	Agricultural drainage well (USA)
AFSC	Australian Food Standard Code (Australia)
AMS	Aggregate Measure of Support
ANZECC	Australia New Zealand Environment and Conservation Council
ANZFA	Australia New Zealand Food Authority
APEC	Asia-Pacific Economic Cooperation
APHIS	Animal and Plant Health Inspection Service (USA)
ARMCANZ	Agriculture and Resource Management Council of Australia and New Zealand
AU	Animal unit (USA)
AWB	Australian Wheat Board (Australia)
BA	Biosecurity Act (New Zealand)
BAT	Best available techniques (EU)
BMP	Best management practice (USA)
BSE	Bovine Spongiform Encephalopathy
BST	Bovine Somatotrophin
CAA	Clean Air Act (USA)
CAFO	Confined animal feeding operation (USA)
CAP	Common Agricultural Policy (EU)
CBD	Convention on Biological Diversity
CCC	Commodity Credit Corporation (USA)
CDC	Center for Disease Control and Prevention (USA)
CFIA	Canadian Food Inspection Agency (Canada)
CITES	Convention on International Trade in Endangered Species
COAG	Council of Australian Governments (Australia)
CODEX	Codex Alimentarius Commission
CREAMS	Chemical, Runoff and Erosion from Agricultural Management Systems model (Canada)

CRP	Conservation Reserve Program (USA)
CSE	Consumer Support Estimate
CURB	Clean Up Rural Beaches (Canada)
CWA	Clean Water Act (USA)
CWB	Canadian Wheat Board (Canada)
CZMA	Coastal Zone Management Act (USA)
DEIP	Dairy Export Incentive Program (USA)
DOC	Department of Conservation (New Zealand)
EARP	Environmental Assessment Review Process (Canada)
EEC	European Economic Community
EEP	Export Enhancement Program (USA)
EPA	US Environmental Protection Agency (USA)
EQIP	Environmental Quality Incentives Program (USA)
ERMA	Environmental Risk Management Authority (New Zealand)
ESA	Endangered Species Act (USA)
ESU	Economic Size Unit (EU)
EU	European Union
EUREP	European Retailer Produce Working Group
FAIR Act	Federal Agriculture Improvement and Reform Act (USA)
FAO	Food and Agriculture Organization of the United Nations
FARMER	Farmer Analysis of Research, Management and Environmental Resources (New Zealand)
FCCC	Framework Convention on Climate Change
FDA	US Food and Drug Administration (USA)
FFDCA	Federal Food, Drug, and Cosmetic Act (USA)
FIFRA	Federal Insecticide, Fungicide, and Rodenticide Act (USA)
FIPA	Farm Income Protection Act (Canada)
FOSHU	Food for Specified Health Use (Japan)
FQPA	Food Quality Protection Act (USA)
FSA	Farm Service Agency (USA)
GAMES	Guelph Model for Evaluating the Effects of Agricultural Management Systems on Erosion and Sedimentation (Canada)
GAP	Good Agricultural Practice (EU)
GATT	General Agreement on Tariffs and Trade
GDP	Gross Domestic Product
GMF	Genetically modified food
GMO	Genetically modified organism
GRIP	Gross Revenue Insurance Program (Canada)
GSSE	General Services Support Estimate
GTAP	Global Trade Analysis Project
HACCP	Hazard Analysis and Critical Control Point
HGP	Hormonal Growth Promotants Audit Program (Australia)
HSNO	Hazardous Substances and New Organism Act (New Zealand)
IGAE	InterGovernmental Agreement on the Environment (Australia)
ICM	Integrated Catchment Management (Australia)
IFCN	International Farm Comparison Network
IPM	Integrated Pest Management

IPPC	International Plant Protection Convention
IPPC	Integrated Pollution Prevention and Control (EU)
IPR	Intellectual Property Right
ISO	International Standard Organization
IUCN	International Union for the Conservation of Nature
JBIA	Japan Banana Importers Association (Japan)
KAWQ	Kentucky Agriculture Water Quality Act (USA)
LMO	Living Modified Organism
LU	Livestock Unit (EU)
MAC	Maximum Acceptable Concentration
MAF	Ministry of Agriculture and Forestry (New Zealand)
MAFF	Ministry of Agriculture, Forestry and Fisheries (Japan)
MAI	Multilateral Agreement on Investment
MCL	Maximum contaminant level
MDB	Murray Darling Basin (Australia)
MEA	Multilateral Environmental Agreement
MFE	Ministry for the Environment (New Zealand)
MPCA	Minnesota Pollution Control Agency (USA)
MRL	Maximum residue limit
NAC	Nominal Assistance Coefficient
NAFTA	North American Free Trade Agreement
NARM	National Antibacterial Residue Minimisation (Australia)
NGO	Non-governmental Organisation
NHT	Natural Heritage Trust (Australia)
NISA	Net Income Stabilisation Account (Canada)
NLP	National Landcare Programme (Australia)
NMP	Nutrient management plan (USA)
NORM	National Organochlorine Residue Management (Australia)
NPDES	National Pollutant Discharge Elimination System (USA)
NRA	National Registration Authority (Australia)
NRC	National Research Council (USA)
NRCS	Natural Resources Conservation Service (USA)
NRS	National Residue Survey (Australia)
NSESD	National Strategy for Ecologically Sustainable Development (Australia)
NTC	Non-trade concern
NVZ	Nitrate Vulnerable Zone (EU)
NZBS	New Zealand Biodiversity Strategy (New Zealand)
OECD	Organisation for Economic Co-operation and Development
OTA	Office of Technology Assessment (USA)
PCE	Parliamentary Commissioner for the Environment (New Zealand)
PFC	Production Flexibility Contract
PMS	Pest management strategy (New Zealand)
PPM	Process and Production Method
PSE	Producer Support Estimate
QSC	Queensland Sugar Corporation (Australia)
RDC	Research and Development Corporation (Australia)

RLMP	Rabbit and Land Management Program (New Zealand)
RMA	Resource Management Act (New Zealand)
SDT	Special and differential treatment
SDWA	Safe Drinking Water Act (USA)
SLM	Sustainable Land Management strategy (New Zealand)
SOILEC	Soil Conservation Economics model (Canada)
SPCA	Society for the Protection and Care of Animals (New Zealand)
SPS	Sanitary and Phytosanitary Measures
SSC	Special Safeguard Clause
STE	State Trading Enterprise
SWEEP	Soil and Water Environmental Enhancement Program (Canada)
SWCD	Soil and Water Conservation District (USA)
TBT	Technical Barriers to Trade
TED	Turtle Excluder Device
TMDL	Total maximum daily loads of nutrients (USA)
TRIM	Trade Related Investment Measure
TRIPS	Trade Related Intellectual Property Rights
TRQ	Tariff-rate quota
TSE	Total Support Estimate
UK	United Kingdom
UNCED	United Nations Conference on Environment and Development
URAA	Uruguay Round Agreement on Agriculture
US	United States
USA	United States of America
USDA	United States Department of Agriculture (USA)
USGS	United States Geological Survey (USA)
USLE	Universal Soil Loss Equation
WHO	World Health Organization
WIPO	World Intellectual Property Organization
WMD	Water management district (USA)
WMP	Waste management plan (USA)
WTO	World Trade Organization
WWF	World Wildlife Fund

Preface

Many environmental, animal welfare and human-health issues are of significant concern to the public and media in large parts of the world. Agriculture in the developed world increasingly responds to issues in the areas of environment, animal welfare and human-health. Governments have a wide range of standards, codes of good practice and other policy measures designed to control pollution, protect environmental resources, control animal welfare concerns, and avoid significant risks to human-health from modern agricultural systems. Farmers respond by adjusting their practices and subsequently internalise externalities they create or provide public benefits. Significant differences in the programmes and standards for the main environmental and human-health issues cause concern among producers who compete in world food markets. Such issues are also relevant in the context of multilateral trade agreements, to determine where differences in standards between countries mainly reflect specific natural resource, environmental and socio-cultural conditions, or where they appear likely to distort competitiveness.

This book deals with these public concerns and agricultural trade. We identify and compare the policy models used across developed agriculture to internalise external costs or provide public goods related to environment, animal welfare and human-health. By doing so, we will look at incentives given by policy regulations and changed market conditions in broad terms.

The book contains reviews of environmental, animal welfare and human-health related standards affecting agriculture in the EU, USA, Canada, Australia and New Zealand. These chapters build on a previous report 'Comparison of environmental and health-related standards influencing the relative competitiveness of EU agriculture vis-à-vis main competitors in the world market' contracted by the European Commission (Directorate-General Agriculture). The Agricultural Economics Research Institute (LEI) in the Netherlands, working with experts in the five countries undertook this original study. The original country reports were revised and updated. In addition, the editors identified a series of key related issues and identified key experts to prepare chapters on the topics.

We very much appreciate the substantial efforts made by Martin Scheele (European Commission, Directorate-General Agriculture) for conceiving and developing the European Commission study upon which this volume is built.

    Additional funds for preparing the book were obtained from the Ministry of
Agriculture, Nature Management and Fisheries in the Netherlands. This support is
appreciated.
    The editors are grateful to the authors for preparing excellent contributions
and for the secretarial assistance provided by Mrs. Urmila Koelfat from LEI. She
took responsibility for guiding the publication process, and prepared the several
drafts of the chapters. We also thank LEI and the Environmental Sciences and
Resources Program at Portland State University for their administrative support.
Without the assistance and support given by Tim Hardwick (CABI Publishing)
this volume would not have been in its present form.

Floor Brouwer and David E. Ervin

Introduction

Floor Brouwer and David E. Ervin

PUBLIC CONCERNS AND AGRICULTURAL TRADE

Freer trade and improved environmental[1] conditions have become robust public goals in most developed countries. International and national policies reduce government and industry trade barriers to foster more liberalised trade. Another set of policies treat market and government failures to improve environmental quality. Progress on each objective is necessary to increase a country's social welfare. Therein lies the potential for conflict. For example, programmes to reduce water pollution from animal agriculture may hinder the ability of farmers to compete in global food markets. Or, opening trade may induce growth in concentrated animal feeding operations that degrades the environment?

The challenge before policy makers is to design and implement trade and environmental policies that work in 'complementary' fashion. In short, actions to liberalise trade will be accompanied by programmes to meet public environmental objectives, and the environmental programs will use cost-effective approaches consistent with open trade rules (Ervin, 1999). The extent to which developed country policies for agriculture meet this standard of 'complementarity' is unclear.

This volume reports new analyses of the environmental concerns affecting agriculture in the European Union (EU), United States of America (USA), Canada, Australia and New Zealand, and the implications of related environmental policies for trade. Some overarching messages emerge from the findings to provide context for interpreting the book's contents. These messages can also inform national and international policy discussions and negotiations.

---

1 To simplify the exposition, 'environmental' is used to encompass the natural environment, animal welfare and human-health issues that exhibit public good and missing market (externality) problems.

Different countries may 'naturally' adopt different approaches and standards

Environmental, animal welfare and human-health issues generally take on more policy importance in developed countries with economic development and increasing income. However, the priorities and approaches may vary across nations due to differing public values (demands) and varying environmental conditions (supply conditions). Countries naturally respond to the environmental concerns within the context of their cultural, socio-economic and natural resource bases. For example, one nation may perceive water quality to be the highest priority for its citizens, while another views human-health risks as the overriding issue. Or, two countries may each focus on reducing water pollution, but select two very different quality levels because of differing aversions to the risks posed by the pollutants. Approaches may vary by issue. 'Command and control' regulations may be used for problems with documented human-health effects, such as pesticide use, whereas for issues such as fragile land conservation, public payments may be used. The preferred approach may also vary by country, with some preferring regulatory approaches and others emphasising private industry initiatives or private-public partnerships.

There is no single 'right' approach or standard across countries. The choices depend on the citizen's values for resolving the issues, and the costs and political constraints each country faces. The resulting mix of country approaches and standards are part of the forces that interact to shape international trade patterns, just as varying market conditions. Although different paths are taken, the public concerns will be addressed nonetheless.

When environmental effects cross country borders, such as air pollution or migratory species, will individual country actions not suffice. Unless such issues enter bilateral or international dispute resolution processes, the country experiencing the external effects cannot represent its interests. If negotiation occurs, the outcome may be a common approach with different country standards or a common standard but differing approaches. The extent to which the transboundary effect is treated depends on the strength of each country's bargaining powers.

Do regulations distort trade or internalise externalities? - It depends

Clear differences are observed in the broad patterns of environmental, animal welfare and human-health policies across the country analyses in this volume. However, the differences must be evaluated in the context of the varying environmental and human-health needs of the countries to determine where differences reflect specific natural resource, environmental and socio-cultural settings, and whether they appear likely to distort freer trade because of excessive cost. It is important to note that simple comparisons of national environmental programme provisions are insufficient to determine the nature of constraints on farmers. Information on the degree of enforcement at the local level is necessary to properly characterise the nature of any constraints and costs.

A major challenge to agriculture today is to properly internalise the social costs of production into output prices, in a way that least affects trade. The proper internalisation of such domestic concerns in farm management practices will improve national economic welfare and not 'distort' trade. The evidence presented in this volume suggests that public policies about environmental management, public health or animal welfare have not significantly altered agricultural trade to date.

Shifting perspectives on public concerns and agricultural trade

The traditional fear of farmers in addressing public concerns in agriculture has been reduced competitiveness and lost trade opportunities. Perhaps partly for this reason, mostly voluntary and compensatory programmes have been used to address public concerns about developed country agriculture until quite recently. That pattern appears to be gradually changing in two important ways. Due to the growth in demand for environmental quality, more direct controls are being used to assure the public of sufficient protection. The trend applies especially to the state, province and local levels of government, and to certain sectors, such as large-scale animal operations. The trend poses the threat of increased competitiveness effects by environmental controls, unless low-cost compliance measures are found for the targeted sectors. However, reduced competitiveness will only materialise if competing countries do not impose similar regulatory requirements.

Agricultural suppliers in some countries are trying to avoid the trend to more regulation. Instead, they are adopting 'clean and green' strategies supported by public and private efforts to appeal to buyers who desire certain environmental performance attributes. If such voluntary measures are successful, they may be the most economically sustainable approach to addressing public concerns.

Anticipate increasing market pressures for addressing public concerns

The recent rise in personal income has led to an increased demand for higher quality (safer) food, more convenience products and more variety of food products. Some economists have argued this trend requires a shift from the 'economics of quantity' to the 'economics of quality' for agriculture (Antle, 1999). It has also put pressure on parts of the agricultural industry to move voluntarily towards improved environmental performance.

Food processors and food retailers have established quality control systems throughout the whole supply chain. These trends are evident in many EU countries and the food processors and food retailers in the USA are setting requirements for sustainable cultivation practices to differentiate their products to reach new domestic food markets. Private certification systems are growing. Similar private sector actions and initiatives also play an important part in setting effective standards for farms, especially in Australia and New Zealand, to capture emerging export markets.

Country environmental policies have global repercussions

Just as one country's agricultural production and trade policies may affect other nations, the policies for agriculture's public concerns also may have spillover effects. A prime example is the establishment of permissible pesticide products and residues by a key global food buyer. Countries that desire to export to that country may have to adopt the same rules to maintain market access, unless domestic concerns outweigh the lost trade value. Such interactions suggest that international fora for discussing the spillover effects may be increasingly helpful in agricultural trade. Several institutions already perform such roles, such as the Codex Alimentarius Commission and the WTO's Sanitary and Phytosanitary Committee. However, their scope is limited and may not address all emerging concerns, such as biodiversity and biotechnology issues. The policy challenge is to devise a set of disciplines that permit the sovereignty of individual countries for environment and other public concerns, but permit cooperative international approaches where multiple countries will benefit.

THE PRINCIPAL ISSUES

A broad range of issues is identified in the five countries. The weights given by their societies to the issues and the policy measures used to address them are discussed. The pattern, legal status and range of application of these policy measures is assessed and placed in the context of the conditions applying in each country. To varying degrees these measures introduce standards applying to agriculture many of which give rise to constraints on farming activities.

The key issues addressed are:

- water quantity and quality;
- soil related issues;
- air quality;
- nature conservation, biodiversity and landscape protection and management;
- genetically modified organisms (GMOs);
- farm animal welfare;
- human-health issues.

THE ORGANISATION OF THE BOOK

The book focuses primarily on the key environmental and human-health issues affecting agriculture in the EU, USA, Canada, Australia and New Zealand. Also, it reviews the policy measures and standards established in these countries. The key aims are threefold:

- Comparison of main environmental, animal welfare and human-health issues causing concern to agriculture in the developed world. It includes a reasoned selection of the most important elements in the EU, USA, Canada, Australia and New Zealand.

- Comparison of the environmental, animal welfare and human-health standards as they apply at farm level. What are the key measures affecting the agricultural sectors in these countries exporting to the world market.
- What are implications of such standards for agricultural trade on the world market? This primarily qualitative analysis examines the trade implications in terms of competitiveness on the world market, linkages with developing countries and the importance of public concerns by developed countries importing food.

The introductory part of the book highlights changes taking place in agriculture on the world market. Public concerns related to environment, animal welfare and human-health increasingly affect the farming sector, including policy measures and changes in consumer preferences. Agricultural policy and the instruments applied, also respond to the demand for liberalisation of agricultural markets.

Chapter 2 examines agricultural policies in the context of trade liberalisation. Huib Silvis and Cindy van Rijswick offer an overview of agricultural policies in the EU, USA, Canada, Australia and New Zealand. In addition, an up-to-date and factual overview is given on the provision of support in these countries. The main categories of producer support, general services support and consumer support are distinguished. Agricultural support measures taken in these countries respond to the requirements of the Uruguay Round Agreement on Agriculture (URAA). The primary constraints that apply to agriculture policies include the demand for increasing market access and reducing export and domestic support. These items will likely be the central elements for the further liberalisation of agricultural trade. Meanwhile, policymakers also need to consider fair competition, as well as the control of environmental quality, safeguarding human-health and considering animal welfare.

Chapter 3 is on world trade and concerns for the human environment. Bradnee Chambers reviews the main areas of contention in the context of the WTO, the primary regime regulating world trade. The compatibility of measures to protect the environment with WTO rules have gained interests with environmentalists and free traders. The inability in the WTO to discriminate between products on the basis of how they were produced runs contrary to most environmental measures. A wide range of human-health related issues have caused much controversy in the trade and environment debate, including biosafety and the transboundary movement of GMOs. The author concludes that one of the greatest challenges for the trade and environment debate will be to strike a balance between issues that are of primary concern to developing countries (e.g. market access and agricultural subsidies) with environmental issues that are of major concern to the North. Opportunities for compromise will increase if the proposals better recognise and factor in the other's perspective.

The second part of the book identifies the leading environment and human-health issues in the EU, USA, Canada, Australia and New Zealand. These five regions play large roles in global food trade, and their relative competitiveness in the trade in major agricultural commodities may be affected by changes in standards. Five country case studies are presented, followed by a comparative synthesis chapter. Each country analysis reviews the available evidence on the incidence of environmental, animal welfare and human-health problems associated with agriculture in the country. These differences may support the imposition of different standards. Differences in standards may change the relative competitive position of agriculture in a country vis-à-vis its main competitors on the world market. In order to examine such issues, some examples are presented in the five case studies of compliance costs to meeting environmental, animal welfare and human-health standards.

A standard set of key issues was selected and grouped under seven main themes, and the case study chapters are structured to address each of these key issues.

- water quantity and quality (nutrient enrichment by nitrates and phosphates, water-borne sediments, pesticides and issues connected with irrigation);
- soil related issues (including soil erosion, contamination and salinisation);
- air quality (including odour, ammonia emission, pesticide drift, crop burning and noise);
- nature conservation, bio-diversity and landscape protection and management (including endangered species, habitat conservation and protected landscapes);
- genetically modified organisms;
- farm animal welfare (including housing for livestock, transport conditions and slaughter); and
- human-health issues (the use of hormones in livestock rearing, the ingredients used in animal feed, pesticide residues in food, safety of those applying agrochemical, hygiene rules for dairy farming and veterinary requirements).

Chapter 4 offers an overview of the main environmental and health-related standards applying to EU agriculture. Floor Brouwer and his co-authors, examine the mix of environmental and health policies applied in Europe that reflect a balance of regulatory approaches and voluntary approaches. Generally, regulatory approaches are more common in relation to the protection of physical resources and the prevention of irreversible losses of remaining valued landscape qualities, features or nature conservation sites. The most important environmental and health-related issues, which are of significant concern and with high priority in policy, are nutrient enrichment, pesticides, odour and nuisance, biodiversity and landscape, GMOs and animal welfare. Most policies have been reactive in the sense that measures were introduced in response to perceptions of emerging or newly documented problems. The EU seems to have made relatively widespread use of public funds to support agri-environmental programmes, which offer some financial compensation to producers in return for generating public policies. There is considerable diversity between Member States in the EU. At sub-national level legislative controls, guidelines and planning controls are evident in the EU.

Chantal Line Carpentier and David Ervin, in Chapter 5, identify four situations when human and environmental health are perceived as seriously threatened by agriculture that are regulated, including water pollution from large confined animal feeding operations, pesticides applications, wetlands alterations, and endangered species habitat alterations. They argue that serious environmental problems caused by agriculture are real and prevalent in the USA, but not universal. The regulation of GMOs and animal welfare issues are limited. The available evidence indicates that compliance requirements in agriculture have not imposed large cost burdens. There are several reasons. Voluntary-payment programmes instead of mandatory measures are applied widely, and most mandatory programmes are targeted to limited regions. Also, mandatory programmes are targeted to industries with the most intensive pollution problems. Finally, operations that are most likely to be regulated often capture economies of scale in their production and compliance costs. Water quality policies are the most constraining, especially for confined animal feeding operations (CAFOs). CAFOs are a small proportion of the total animal feeding operations, but produce a disproportionately large share of output.

Chapter 6 examines the relationship between agricultural production and the protection of environmental values in Canada. Glenn Fox and Jennifer Kidon argue that surface water quality and air quality degradation from livestock odour have been the most important agriculturally related environmental problems in Canada. The recent Walterton incident, however, has raised the profile of bacterial contamination of groundwater. The overall incidence of agriculturally related environmental problems in Canada seems low relative to the USA or the EU, largely due to the lower intensity of agricultural production in Canada. Environmental policy in Canada is a shared responsibility by federal, provincial and local governments. This underscores the need to consider the full range of policy instruments employed by all levels of government in international comparisons of environmental policies, especially for comparisons of the impacts of regulations in competitiveness. This chapter also reviews the available evidence on environmental compliance costs in primary agriculture in Canada. This review indicates that these costs are not high. When farmers are allowed flexibility in selecting the means of attaining environmental targets, compliance costs are generally less than 3% of gross revenue.

Chapter 7, by Randy Stringer and Kym Anderson, examines the integration of environmental, economic and social considerations in agriculture in Australia. A main policy strategy adopted for agriculture is to enhance the country's reputation as 'clean and green'. This is seen as an important strategy to gain an edge in markets that are increasingly concerned about food quality and human health. The major agriculture-related environmental concerns are water quantity, salinity and soil erosion, and loss of biodiversity. The main policy instruments adopted to control land and water degradation problems are information, education, community participation, dissemination of research and the development of codes of good agricultural practice. Market-oriented approaches are adopted to reduce water use inefficiencies and the removal of water subsidies. Landcare programmes have been developed across the country. Farmers are promoted to work together with local communities and governments to address environmental problems too large for individuals to handle. The integration of natural resource management with land, water and vegetation is advocated under such programmes. Finally, the loss of biodiversity is also recognised as a threat to the main market advantage of Australia as a producer of 'clean and green' goods. A range of quality assurance programmes is available to maintain the reputation of the country as a high-quality and safe supplier, including dairy, wheat, red meat, pork, fruit and vegetables.

In Chapter 8, Anton Meister offers an overview on New Zealand. Farmers continually need to respond to changing societal values and attitudes. Society increasingly demands higher environmental quality levels, and more preserved natural environmental resources. The image of 'clean and green' has helped the country in the international marketplace, as strong measures have been taken to curb pollution, manage the landscape and protect flora and fauna. Most changes are achieved through voluntary means. The food industry is under continuing pressure to achieve standards that are set by the major trading partners, such as pesticide residue levels, and to satisfy the desires of domestic consumers. The Landcare Trust has been established to develop a network of trained, landcare or community group facilitators, and to provide support for community-based projects. In addition, there is a range of industry-led initiatives to promote sustainable agriculture.

Chapter 9 offers a synthesis of the principal environmental and health related standards applying to agriculture in the EU, USA, Canada, Australia and New Zealand. Floor Brouwer and David Ervin identify areas of clear differences in standards between the countries, at the farm level. These differences need to be evaluated in the context of the different environmental and health needs of the countries. This contextual analysis is considered crucial to draw conclusions about trade distortion from information about standards. The analysis offers the basis for a preliminary assessment of the implications of these differences in standards for the relative competitiveness of EU agriculture, vis-à-vis its main competitors on the world market. The chapter concentrates on the economic implications of compliance with on-farm constraints arising from environmental, animal welfare or human-health related standards. In crop production, the analysis did not find evidence to suggest that compliance with environmental regulations have been or will be a driving force determining the location of production. The story may be different with respect to livestock production. Here, the costs of compliance with nutrient regulation and measures to control odour and nuisance from intensive livestock production units are increasing in several parts of the world (e.g. EU, USA and Canada). The compliance costs of pigs and poultry have increased in these countries, and may have a significant effect on the location of production in the future.

The third part of the book examines the importance of public concerns in the context of international trade and multilateral trade agreements (e.g. WTO). Linkages are made with the integration of public concerns in farming practice, both in developed and developing countries. The authors assess the importance of public concerns in countries importing food, and the trade implications of public concerns in the context of WTO.

Chapter 10, by Ulrike Grote, examines the importance of environmental standards from the perspective of developing countries. She compares environmental legislation on three trade products (vegetable oils, grain and broilers) for three countries (Brazil, Germany and Indonesia) that play a significant role on the international market, and future liberalisation of agricultural trade may increase competition between these countries on the world-market. Environmental compliance costs tend to have a share of up to 4% of total production costs. However, total production costs in Germany exceed those in Brazil and Indonesia, and the costs of environmental compliance are marginal relative to other factors like the wage level, prices of land, buildings and machinery. The analysis offers evidence that environmental regulation appears to be adjusted to national conditions (availability of resources, population density and welfare). She recommends that enforcement of legislation is an important factor. Eco-labelling would give incentives to farmers to use environmentally-friendly production practices.

Chapter 11 examines public concerns from the perspective of a major importer of food. Theo Jonker and Ikuo Takahashi examine the role of public concerns in Japan. They stress that more than half of the total calories needed to feed the country are imported. The country depends, among others, on the import of beef, pork and poultry. This chapter examines the role of public concerns in Japan. Public health issues play a more significant role in consumer behaviour relative to environmental and animal welfare concerns. Good taste and high quality of food are first priority when consumers choose food products. Issues of great significance to Japanese consumers in their purchasing behaviour are the product's freshness, appearance and place of origin. The agrifood industry is well aware of these preferences and target them in their promotion of food products.

Chapter 12 examines several environmental proposals for reform of the WTO. The track record of the WTO, and before it the GATT, on environmentally related trade disputes has attracted considerable attention from environmental groups, NGOs and academics. Glenn Fox evaluates several of these proposals and also revisits the basis of environmental criticism of the WTO. One widely held perception is that domestic environmental policy measures generally face a precarious existence when subjected to a WTO challenge. This chapter concludes with some practical guidance on how domestic environmental policies could be developed that face a reduced risk of conflict with member obligations under the WTO.

REFERENCES

Antle, J. (1999) The new economics of agriculture. *Americal Journal of Agricultural Economics*, 81(5), 993-1010.
Ervin, D.E. (1999) Towards GATT-proofing environmental programmes for agriculture. *Journal of World Trade*, 33(2), 63-82.

Agricultural Policies and Trade Liberalisation

Huib Silvis and Cindy van Rijswick

INTRODUCTION

With diverse objectives and in different ways, governments are influencing the structure and development of agricultural production, agricultural trade relations and agricultural price and income formation. Typically, agricultural policies are aimed at domestic objectives, though they usually involve repercussions for other countries. Therefore, international co-ordination of agricultural policies is valued highly, but only achieved with great difficulty.

As a matter of fact, the formulation and successive reforms of the Common Agricultural Policy (CAP) of the European Union (EU) offer an interesting example of international co-ordination. The Treaty of Rome, by which six countries set up the European Economic Community (EEC), and which entered into force in 1958, formed the starting point. Article 38 of the Treaty stated the direction for agriculture: free trade within the Common Market. The policy making process that followed not only addressed the trade concerns of the EU Member States, but also their environmental and health-related concerns. In the beginning of the CAP the concerns of other countries were not in the forefront, but these would become of increasing importance, as this chapter will show.

The chapter starts with an overview of the agricultural policies in the EU, the United States of America (USA), Canada, Australia, and New Zealand, focusing on the general policy objectives and the types of instruments applied (Van Rijswick and Silvis, 2000). Next we deal with the quantitative level of agricultural support in the countries concerned, based on measurements by the Organisation for Economic Co-operation and Development (OECD). In the following section, the Uruguay Round Agreement on Agriculture (URAA) is summarised and the main constraints applying to agricultural policies are identified. It is to be expected that these constraints, which refer to market access, export support and domestic support, will also be central elements for the further liberalisation of agricultural trade. Like in the development of the CAP, in the liberalisation process at the global level, policymakers will have to tackle not only matters of fair competition, but issues of environmental protection, food safety, and animal welfare as well.

AGRICULTURAL POLICIES

Agricultural policies in the EU

According to Article 39 of the Treaty of Rome, the objectives of the CAP are to increase agricultural productivity, to ensure a fair standard of living for the agricultural community, to stabilise markets, to assure the availability of supplies, and to ensure that supplies reach consumers at reasonable prices. Similar objectives are found in other countries. As explained in the introduction of this chapter, the CAP was developed to allow agriculture to take part in the Common Market. As a consequence, a great deal of attention was devoted to market and price policy. In practice the central objective was to foster reasonable income formation in agriculture.

The Common Market for agricultural products was perceived to be established by specific rules. After a transition period of some years, common guarantee prices for the main agricultural products of the EEC-6 were introduced in 1968. These prices were to be realised by a combination of policy instruments: variable levies on imports, intervention in the domestic market, control of stocks (bought at minimum prices) and variable export subsidies (restitution or refunds). Some products were not included in the system of levies on imports and subsidies on exports, especially oilseeds, as was agreed with the partners in the General Agreement on Tariffs and Trade (GATT), mainly the USA.

Initially, the CAP could be operated with rather low budgetary costs. Over the years however, the CAP became increasingly linked with negative effects such as uncontrollable government expenditure, surpluses of production, market disruption for other countries, and pressures on the environment. In response to this, in recent decades reforms have taken place, among which the introduction of supply control measures such as the quota system for dairy and set-aside in arable farming.

Also trading partners of the EU called for changes in the CAP. After long and difficult negotiations, the URAA on liberalising agricultural trade was concluded in 1994. The disciplines on domestic support, market access and export support of this agreement have implied rather strict constraints for the operation of the CAP (see the section on the URAA). Consequently a tendency has developed to adjust the market organisations of the CAP and to provide income support by direct payments related to area planted and number of animals. This process started with the Mac Sharry reform of 1992, and was followed by the decisions on the Agenda 2000 package made in 1999 (Berlin agreement). The result of the Mac Sharry reform and the Berlin agreement is a policy system with (lower) guarantee prices combined with various direct payments linked to the volume of production.

The main initiatives regarding agri-environmental and rural policies have been combined in a new framework regulation (Regulation 1257/1999), which includes various support measures. Generally, the level of aid from the EU amounts to 50-75% and the remaining part is covered from Member States. Some of the measures for which farmers can receive aid are the setting up of new farms, early retirement, compensation for farming in less favoured areas, support for forestry, biodiversity measures, management of water resources, promotion of tourism and crafts, and land improvement.

In relation to the Agenda 2000 reforms, the objectives of the CAP have been reformulated as follows (EC, 1999):

- To promote a competitive agricultural sector which is capable of exploiting the opportunities existing on world markets without excessive subsidy, while at the same time ensuring a fair standard of living for the agricultural community.
- Promoting production methods which are safe, capable of supplying quality products that meet consumer demand.
- Promoting diversity, reflecting the rich tradition of European food production.
- To maintain vibrant rural communities capable of generating employment opportunities for the rural population.
- Promoting an agricultural sector that is sustainable in environmental terms, contributes to the preservation of natural resources and the natural heritage and maintains the visual amenity of the countryside.
- A simpler, more comprehensible policy which establishes clear dividing lines between the decisions that have to be taken jointly at the EU level and those which should remain in the hands of the Member States.
- To promote an agricultural policy that establishes a clear connection between public support and the range of services which society as a whole receives from the farming community.

From this listing one may conclude that agriculture in the EU is to perform multiple functions. In relation to agricultural trade, the EU does not strive for a model of autarchy, but on the contrary wants to play an active and acceptable role in future world markets. Reform is considered to be vital in achieving that.

Agricultural policies in the USA

With the roots of their agricultural policies in the 1930s, the USA applies a wide range of agricultural assistance measures to provide a safety net for farmers, support agricultural prices and incomes, and to help farmers manage their risks. In terms of the gross value of total assistance by the US Department of Agriculture (USDA), food and nutrition programmes form the most important element of the USA agricultural policies, whereas commodity support programmes are placed second. Hence, the composition of agricultural assistance in the USA differs quite strongly from that in the EU and other OECD countries. In most developed countries, needy people receive financial assistance instead of help in kind; such assistance is usually included in the social security system.

In 1996 the USA passed a major reform of its agricultural policies with the Federal Agriculture Improvement and Reform Act of 1996 (1996 FAIR Act). One of the intentions of the Farm Bill was to create more market-oriented government programmes (USDA, 1996). The most significant aspect of the Farm Bill, which became law in April 1996, was the introduction of farm support decoupled from farm production, and abolishment of supply management through acreage reduction. Furthermore, the Bill aimed to put resources into rural development and extend conservation and environmental programmes and to make them simpler and more workable for agriculture. Besides these more or less new aspects, many of the earlier programmes were maintained.

The commodity programmes historically consisted of agricultural price support and direct income payments. Price support programmes are still handled primarily through loan programmes. Also import restrictions and purchase programmes are used to keep domestic prices at certain levels, in particular for sugar and dairy.

The mechanism of the loan programmes is that producers may hand their produce over to the Commodity Credit Corporation (CCC) at the loan rate. Commodity loan rates are based on a moving average of past market prices. According to the Farm Bill of 1996 the loan rates reach their maximum level in 1996. The value of the loan is the product of the announced loan rate and the quantity placed under loan. Non-recourse loans allow producers to pay their bills and other loan payments when they come due, without having to sell crops at a time of year when prices are at their lowest. Farmers can reclaim the produce later if prices improve. Then, when conditions are more favourable, farmers can sell the crop and pay off the loan and a certain fee for administration costs and interest. If the prevailing price of the crop remains below the loan level set by USDA, farmers can keep the loan proceeds and give the crop to the CCC instead. Hence, these loans are called non-recourse: producers can forfeit or deliver the commodity to the CCC to discharge the loan in full.

Contrary to the intent of the 1996 FAIR Act, commodity expenditures have been increased sharply since 1998 to compensate for the effects of natural disasters and depressed world market prices. Since the mid-1990s prices on the world market showed a steady downward trend. According to the OECD the economic slowdown in Far East Asia and Russia and an increase in world production of major agricultural commodities have been the main factors responsible for this development in recent years. In response to the stronger global economy and lower stock levels, international market prices for agricultural commodities, expressed in US$, increased during 2000. For many products, this was the first price rise since the mid-1990s. For example, world cereal and butter prices remained 50% and 40% lower respectively than in 1995 (OECD, 2001a).

If the FAIR Act was aimed at a gradual abolishment of market price support, in practice the opposite has happened. Because of depressed market prices, the maximised loan rates (level 1996) for cereals and soybeans regained significance. Since the 1996 FAIR Act, Congress has increased direct payments such as the production flexibility contract payments with 50 to 100%. Additional payments have also been made as natural disaster payments and for conservation programmes. Farmers, who have suffered a loss due to a natural disaster, may be eligible for assistance under one of the natural disaster assistance programmes of the Farm Service Agency (FSA). In 1998 and 1999 this type of ad hoc and ex-post income payments became important. Hence, total payments, based on output, area planted and support provided in the past, have been increased recently to $28 billion in 2000.

With the intended shift towards a policy aimed at less government intervention in agriculture, the need for a better understanding of farm risk and risk management was recognised. Natural and economic events in 1998 and 1999 reinforced the need to provide farms with tools to successfully manage their own risks. Hence, agricultural policymakers are addressing the issue of risk management, in order to develop a longer-term risk management policy. Most of the insurance programmes are 'yield based'. This means that the historical actual production history of the farmer is used to calculate the guarantee level. Revenue insurance programmes combine the production guarantee component of yield insurance with a price guarantee.

Next, there is a wide variety of programmes available to support agricultural exports, including export credit guarantees and export subsidies. In addition to this, foreign food aid plays an important role. Finally, several general services programmes are in operation, such as rural development support, credit and farm loans, and research, education and extension programmes. The activities of the Foreign Agricultural Service, such as concessional sales, payments, direct credits and other supporting activities, account for a significant share (7%) of total assistance to agriculture. Marketing and promotion is the largest category of 'general service support' in the USA. The other OECD countries do not devote such a large percentage of general service support to export marketing and promotion as the USA.

With respect to conservation two programmes should be mentioned here. The Conservation Reserve Program (CRP) is a voluntary programme for which participants agree long term contracts (usually 10 to 15 years) in exchange for annual rental payments and cost share assistance for carrying out certain conservation practices. The Environmental Quality Incentives Program (EQIP) helps farmers (through incentive payments, cost-share assistance, technical support, and education) to improve their properties to protect the environment and to conserve soil and water resources. CRP payments amount to nearly $2 billion per year, whereas EQIP payments are some $150 million per annum. Other main categories of green box support in the USA are natural disaster assistance, food aid, domestic nutrition assistance, research and development, and rural development support. Disaster assistance has become increasingly important since 1998, not only under influence of natural disasters but also due to adverse price developments. Supplemental disaster payments have increased from approximately $3 billion in 1998 to about $10 billion in 2000 (The Commission on 21st Century Production Agriculture, 2001).

Agricultural policies in Canada

In Canada four groups of products account for some 90% of total gross farm sales: the red meats sector; grains and oilseeds; dairy, poultry, and eggs; and the horticultural sector (Barichello, 1996). An equal share of the budgetary support payments relates to these groups. Payments are provided in the context of income support and stabilisation policies. Like the other countries considered in this chapter, Canada provides export support and general support such as credit to farmers, agri-environmental programmes, rural support measures and research, education and extension initiatives. Federal and provincial governments are jointly responsible for the implementation of Canada's agricultural policies. In 1999, about 46% of the Canadian expenditures for the agri-food sector (C\$4.2 billion) was provided by the provinces (AAFC, 1999). Total agri-food support expenditures have shown a declining trend in the 1990s.

Canadian agricultural support policies fall into two major groups: market regulations and income stabilisation. Also input subsidies have been used quite heavily. However, the most important of these related to grain transportation and are gradually phased in response to GATT-commitments.

The marketing regulations are aimed at stabilising agricultural markets and improving or stabilising producer prices. The existence of marketing boards is one of the characteristic features of the market regulations. As early as in the 1920s there were marketing boards in Canada. The purpose of these boards was the 'orderly marketing' of agricultural commodities (Ash, 1998). There was a major development in the evolution of these boards in the late 1960s and early 1970s when some of them gained powers to restrict domestic supply and some of them received import protection. This occurred in the dairy, poultry and egg boards. These boards form the so-called supply management system. The other application of marketing boards is the Canadian Wheat Board (CWB). This board does not have the powers of supply control, but has other powers in the grains and oilseeds sector that has made it the largest marketing board in Canada. Other elements of Canadian market regulations are the Price Pooling Programme and the Advance Payments Programme. The Price Pooling Programme assists and encourages co-operative marketing of products. The programme provides a price guarantee to marketing agencies that protects them against unanticipated declines in the market price of their products. The Advance Payment Programme improves cash flow at or after harvest. It operates through the provision of a cash advance to the farmer. This allows a farmer to store the crop and sell it throughout the crop year to achieve higher returns.

The Canadian income stabilisation policies are generally directed towards the agricultural commodities that are not covered by the supply management system. Stabilisation policies include crop income insurance programmes. In 1991, an umbrella statute (the Farm Income Protection Act, FIPA) was introduced to provide a general framework for income stabilisation and safety net programmes for virtually all commodities. All the programmes under this framework are guided by the principles of market neutrality, equity among commodities, social, economic and environmental sustainability, and consistency with international obligations (Ash, 1998). The main programmes of this safety net framework, developed to address the different needs of different sectors of the industry, are crop insurance, the Net Income Stabilisation Account (NISA), and province-specific companion programmes.

Crop insurance provides production risk protection to producers by minimising the economic effects of crop losses caused by natural hazards like drought, hail, flood, frost, wind, fire, excessive rain, heat, snow, unpreventable disease, insect infestation and wildlife. The Crop Insurance Programme is a provincially delivered programme whereby federal financial contributions are made to provincial crop insurance schemes. Crop insurance is a voluntary programme extended to a wide variety of crops. The programme can provide coverage for virtually all farmers. Coverage varies according to crops grown in a certain province.

NISA is a voluntary programme that assists farmers in stabilising incomes for the long term. Essentially, establishing a fund that receives contributions from farmers during good years in order to provide withdrawals during poor years does this. The contributions are matched by the government, which also provides interest subsidies. NISA is available in all ten Canadian provinces.

Next to the above-mentioned policy instruments, there are policies that may be important for particular commodities or in certain regions, such as agri-environmental policy, and rural policy and research, education and extension initiatives. The number of policy programmes on agri-environmental issues has been growing in the last years. Most of the spending on these programmes has come through federal-provincial programmes or directly from provincial initiatives.

Agricultural policies in Australia

Governmental support to the agricultural sector of Australia is fundamental but rather low. In the nineties, the Australian Federal Government developed an overall strategy in order to make farming more competitive, sustainable and profitable (AFFA, 1998). The Government has provided funds for a rural R&D programme, provided support for sustainable agriculture, supported exports through the 'Supermarket to Asia initiative' and invested more resources in quarantine. The programmes under the Agriculture - Advancing Australia (AAA) package have to deliver benefits across the entire farm sector. The major objectives of the AAA-package are to help individual farm businesses profit from change, to ensure that the farm sector has access to an adequate welfare safety net, to provide incentives for ongoing farm adjustment, and to encourage social and economic development in rural areas (AFFA, 1998). The other main package of Australia's agricultural policy is the Natural Heritage Trust (NHT) initiative. NHT is a large environmental policy effort. The programmes of NHT play a role in the development of sustainable agriculture and natural resource management, as well as the protection of biodiversity (NHT, 1996). One of the major programmes under the NHT is the National Landcare Programme (NLP). Landcare is a locally based approach to protect the environment and Australia's resources. The NLP supports projects that contribute to an integrated programme of sustainable management of land, water, vegetation and biological diversity.

An important element of the strategy concerns 'industry control of industry matters'. The Government has been working closely with individual industries such as wool, meat and wheat to remove undesirable statutory controls, particularly in marketing, and to allow each industry to take direct responsibility for determining its operations (AFFA, 1998). As in Canada, producer boards are a well-known phenomenon in Australia. The most important ones are the Australian Wheat Board (AWB) and the Queensland Sugar Corporation (QSC). The AWB is the sole exporter of wheat from Australia and the major seller on the deregulated domestic market. Besides export control, the AWB uses a price pooling system for different grades and classes of wheat. As of 1 July 1999 the AWB has been converted into a private corporation. Hence, it needs to seek funding in international financial markets, while maintaining its exclusive export authority (AWB, 1999). The QSC is a state-level marketing board. It procures Queensland's total produce of raw sugar, which it markets to refineries. Domestic price support is arranged by pool pricing for raw sugar by grade. The government does not underwrite the pool losses (USDA, 1997). QSC has the exclusive right to export Queenland's raw sugar.

Agricultural policies in New Zealand

The New Zealand government follows a market led policy approach. Sector specific policies are rare since the removal of virtually all agricultural subsidies and concessions following the election of the Labour Party to power in 1984. Few shifts in New Zealand's economic policy approach have occurred following the national elections in 1996. As in Canada and Australia, producer boards are an important feature of agricultural policy in New Zealand. General services account for more than 55% of total agricultural support. The primary general services are basic research and the control of pests and diseases. In addition to this, there is some support for agri-environmental policies and compensation for adverse events. The role of the government in promoting a sustainable agriculture is primarily that of encouraging market-led adjustment to sustainable practices.

Most of New Zealand's agricultural and horticultural exports are in some way controlled or influenced by producer boards or licensing authorities. These organisations are creatures of statute but the New Zealand government has indicated that it intends to remove the statutory backing of agriculture's producer boards (OECD, 1999). There are three different types of producer boards in New Zealand (MAF, 1996). The first type is the export trading board. Boards under this type have export powers. Examples of these boards are the New Zealand Dairy Board, the New Zealand Apple and Pear Marketing Board and the New Zealand Kiwifruit Marketing Board. The second type of producer boards are those with both export and domestic market (monopoly) powers, such as the New Zealand Hop Marketing Board and the New Zealand Raspberry Marketing Council. The non-trading boards form the third type of boards. These boards are not directly involved in the purchase or sale of commodities. Their roles are mainly directed at the areas of licensing exporters, promotion and market development, research, industry support, and quality control. The five non-trading boards in New Zealand are the New Zealand Wool Board, the New Zealand Meat Producers Board, the Game Industry Board, the New Zealand Pork Industry Board and the New Zealand Horticulture Export Authority.

Box 2.1 Policy instruments in the EU, USA, Canada, Australia and New Zealand

	EU	USA	Canada	Australia	New Zealand
Price support	X	X	X		
Marketing quota	X	X	X		X
Area- or livestock-based direct income payments	X				
Decoupled income payments		X			
Natural disaster assistance	Member States	X		X	X
Funds for farmers' savings			X	X	
Yield-based income insurance	Member States	X	X		
Revenue-based income insurance		X	X		
Other types of risk management programmes	Member States	X	X	X	
Direct (producer) retirement assistance	X			X	
Loan and credit provisions	Member States	X	X	X	
Marketing boards with export powers			X	X	X
Export credit guarantee or insurance	Member States	X		X	
Export subsidies	X	X			
Export promotion and assistance	X	X	X	X	X
Food aid	X	X	X		
Marketing and promotion support	X	X	X	X	X
Domestic nutrition assistance		X			
Direct payments subject to environmental commitments	X	X			
Agri-environmental programmes	X	X	X	X	X
Rural development support	X	X	X	X	X
Research, extension, education	X	X	X	X	X

Overview

This section has shown that the agricultural policies of the EU, USA, Canada, Australia and New Zealand contain important similarities and differences. Traditionally, the markets of agricultural products and the income formation of the producers have attracted a lot of policy attention in most countries. The objectives of the agricultural policies have now been broadened and relate to the sustainable development of the agricultural sector and the rural areas. The new objectives are pursued with a whole range of different instruments (Box 2.1). Before going into the trade policy aspects of these instruments, the next section provides measurements of the level and composition of agricultural support.

THE PROVISION OF AGRICULTURAL SUPPORT

Based on measurements published by the OECD in its annual Monitoring and Evaluation Report (OECD, 2001b), this section focuses on trends in the overall level and composition of agricultural support in the EU, USA, Canada, Australia and New Zealand. The overall transfers associated with agricultural support is measured by the TSE (Total Support Estimate). Together these five countries account for 64% of total agricultural support in OECD member countries, as measured by the TSE (Table 2.1). In the group of other OECD countries, accounting for 36% of support, among others, Japan, Korea, Switzerland and Norway maintain high levels of support.

The OECD indicators measure support arising from agricultural policies relative to a situation without such policies, i.e. when producers and consumers are subject only to general policies (including economic, social, environmental and tax policies) of the country. It should be stressed that the support measures have a static character and do not gauge dynamic effects on income or trade if the agricultural policy measures were to be abolished.

Table 2.1   Agricultural production value and TSE in selected OECD countries, average 1998-2000

Country	Production value (Million Euro)	TSE		
		(Million Euro)	(% share of OECD total)	(Euro per capita)
Australia	18,076	1,570	0.5	84
Canada	19,042	4,930	1.5	162
EU	213,454	115,072	34.8	307
New Zealand	5,582	122	0.0	32
USA	183,268	90,741	27.5	333
Other OECD countries	192,529	118,109	35.7	-
OECD	631,952	330,544	100.0	313

Source: OECD, PSE/CSE database.

The TSE contains the agricultural support financed by the consumers (transfers from consumers) and taxpayers (transfers from taxpayers) net of import receipts (budget revenues). The classification of total transfers associated with agricultural policies groups the policy measures into three main categories:

- PSE (Producer Support Estimate): transfers from consumers and taxpayers to producers individually. These transfers include the broad range of market price and income policies. Market price support arises from policy measures (such as stock intervention, import restrictions, production quota) that create a gap between domestic market prices and border prices of a specific agricultural commodity. It is measured at the farm gate level. Direct income support are transfers from taxpayers to agricultural producers arising from policy measures based on current output of a specific commodity, the historical entitlements, the area planted etc.
- GSSE (General Services Support Estimate): transfers from general services provided to agriculture collectively. These measures include taxpayer's transfers to improve agricultural production (research), to improve education, to control the quality and safety of food, to improve environmental conditions, to improve off-farm collective infrastructures, to support marketing and promotion, and other general services.
- CSE (Consumer Support Estimate): transfers to (from) consumers of agricultural commodities individually. A positive CSE is associated with market price support to producers. When the value of CSE is negative, transfers from consumers measure the implicit tax on consumption associated with policies to the agricultural sector.

The TSE is the sum of the PSE, the GSSE and the transfers from taxpayers to consumers (in CSE). The values of the different indicators are expressed in monetary terms (PSE, GSSE, CSE, and TSE), but also in ratios. An example is the percentage PSE, the PSE as a ratio to the value of total gross farm receipts, measured by the value of total production (at farm gate prices) plus budgetary support. A percentage PSE of 60% means that some 40% of gross farm receipt is derived from the market without any support. Another example is the producer Nominal Assistance Coefficient (producer NAC), which expresses the PSE as a ratio to the value of total gross farm receipts valued at world market prices, without budgetary support. A producer NAC of 2.50 means that the gross value of the producer's receipts is 250% of (or 150% higher than) what the producer would obtain if he received the world price for his commodities without any support. The percentage GSSE is defined as the share of general service support in the total support to agriculture (TSE). The percentage TSE is defined as the share of total support in the total Gross Domestic Product (GDP), as transfers from consumers and taxpayers are both included in the GDP.

Trends in the overall level of agricultural support

The overall level of support to agriculture in the OECD area is calculated at 1.4% of GDP in 1998-2000, down from 2.3% in 1986-1988 (Table 2.2).

Table 2.2   Estimates of support to agriculture in the OECD area

	1986-1988	1998-2000	2000p
PSE (billion Euro)	215	250	266
Market price support (billion Euro)	166	165	172
Percentage PSE (%)	39	35	34
Producer NAC (ratio)	1.63	1.54	1.52
GSSE (billion Euro)	38	55	60
GSSE as a share of TSE (%)	13	17	17
CSE (billion Euro)	-151	-154	-160
Percentage CSE (%)	-33	-28	-26
Consumer NAC (ratio)	1.49	1.38	1.35
TSE (billion Euro)	271	331	354
TSE as a share of GDP (%)	2.3	1.4	1.4

Note: p = provisional.
Source: OECD.

However, significant differences in the sources of financing as well as in the level and composition of support to agriculture persist across countries (and commodities). Among the selected countries, the TSE per capita ranged from about 333 euro in the USA to 32 euro in New Zealand in 2000 (Table 2.1). In 1998-2000, the share of agricultural support in a country's GDP, as measured by the percentage TSE, ranged from below 0.25% in Australia and New Zealand to almost 1.5% in the EU.

The percentage PSE, which measures the level of support to agricultural producers, has been on a slow downward trend in the OECD area, declining from 39% in 1986-1988 to 35% in 1998-2000. In 1998 and 1999, the percentage PSE has been rising again, in reaction to a sharp fall in world market prices.

The composition of support to producers has also changed over the last decade. The share of market price support fell from 77% in 1986-1988 to 66% in 1998-2000. Simultaneously, the share of payments based on output decreased initially. However, since 1997 this share is growing again. The share of payments based on area or animal numbers amounted to more than 11% in 1999, as opposed to almost 7% in 1986-88 (OECD, 2001b). This increase is closely linked with the introduction of hectare and animal premiums in the EU. On the other hand, the level of payments based on input use (for example, interest concessions, capital grants) has been fairly stable around 8%. Although payments based on constraints to the use of fixed and variable inputs (including environmental constraints) have increased nearly threefold, they represent only about 3% of total support. The share of payments based on overall farming income remains very low, representing less than 1% of support. Such support, which has some significance in Australia and has risen in Canada, is a kind of social security system (safety net) for farmers with incomes below a certain minimum.

Trends in the level of support by country

There are wide variations in the level and composition of support for individual countries (and commodities) among OECD countries, as there are also wide variations in farm structures, natural, social and economic conditions, and trade positions. The decrease of total support to agriculture as a percentage of GDP in the nineties is clearly visible in Figure 2.1. This is partly due to the declining share of agriculture in GDP, which has grown steadily in OECD countries over the last years.

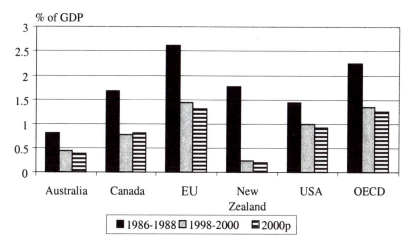

Figure 2.1   TSE in selected OECD countries (percentage of GDP)
Notes: p = provisional; EU-12 for 1986-1988, EU-15 for 1998; OECD excludes most recent Member countries for 1986-1988.
Source: OECD.

The share of TSE provided for general services to agriculture, as measured by the percentage GSSE in 2000, ranged from 9% in the EU to 79% in New Zealand (Figure 2.2). Over the last decade, Australia, Canada, New Zealand, and the USA have had lower levels of support than the OECD average, as measured by producer NAC (Figure 2.3).

In New Zealand, the percentage GSSE has shown a sharp increase between 1986-1988 and 1998-2000. This increase was largely due to the major reduction of the TSE during the 1990s, as the GSSE was rather stable during the period. In absolute terms, the GSSE of the USA is the highest of all countries, the bulk of which consists of marketing and promotion.

In countries with a low level of support, domestic prices are in general closely aligned with world market prices. In contrast to the other countries, support to agricultural producers has been above the OECD average in the EU during the last decade. The EU grants market price support for a number of commodities (especially dairy and sugar), and has the highest share of market

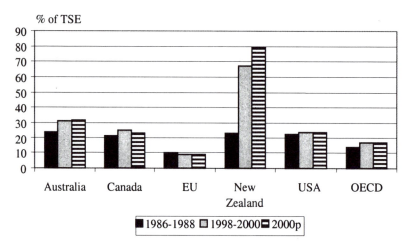

Figure 2.2   GSSE in selected OECD countries (percentage of TSE)
Notes: p = provisional; EU-12 for 1986-1988, EU-15 for 1998.
Source: OECD.

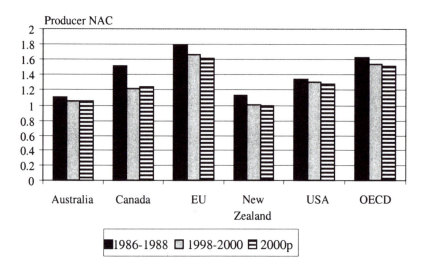

Figure 2.3   Producer NAC in selected OECD countries
Notes: p = provisional. EU-12 for 1986-1988, EU-15 for 1998.
Source: OECD, PSE/CSE database.

price support in overall support. The relative importance of market price support
is mirrored in the levels and changes in the consumer NAC (Figure 2.4).

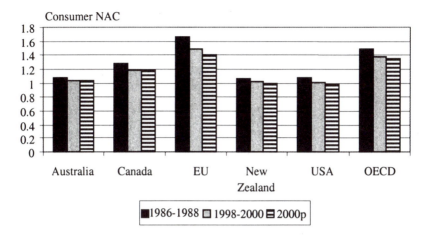

Figure 2.4 Consumer NAC in selected OECD countries
Notes: p = provisional. EU-12 for 1986-1988, EU-15 for 1998.
Source: OECD, PSE/CSE database.

Conclusions

The major trends in support since 1986-1988 can be summarised as follows:

- The share of total agricultural support in GDP has been on the decline. However, in 1998 and 1999 support rose again, due to the sharp fall in world market prices in those years.
- The sources of agricultural support showed some substitution, with the contribution by the consumer decreasing and that by the taxpayer increasing. This trend was especially observed in the EU.
- Support to general services provided to agriculture remained stable; the importance of general services in agricultural support is relatively low in the EU, but relatively high in the USA, Canada, Australia, and New Zealand.
- Although there have been reductions in the share of market price support, it remains the main source of support for the OECD in total. That is because the share of price support is relatively large in countries that mark the TSE, like Japan and the EU. The reductions in price support in the EU have largely been replaced by an increase in support based on area planted or animal numbers; payments in the USA based on area planted have been replaced by payments based on historic entitlements.
- The share of support based on input use is rather stable, with an increase in the share of support based on input constraints, including environmental constraints; despite significant increases in some countries, this share has remained rather low.

- The share of support based on overall farm income is low, but is still significant in Australia and has risen sharply in Canada; this form of support is least coupled to production of commodities.
- In general, the majority of support to producers still comes from support based on output, area planted or animal numbers. However, some payments have limits attached to the levels of output, area or animal numbers that attract the support, or are associated with environmental constraints.

THE URAA

Countries agreed under the URAA to substantially reduce agricultural support and protection by establishing disciplines in the areas of market access, export subsidies, and domestic support. In addition, an Agreement on the Application of Sanitary and Phytosanitary Measures (SPS Agreement) was established to prevent countries from using arbitrary and unjustified health and environmental regulations as barriers to trade. Finally, a new Agreement on Technical Barriers to Trade (TBT) was concluded to ensure that regulations, standards, testing and certification procedures do not create unnecessary obstacles.

The URAA has been implemented over a 6-year period, from 1995 to 2000. For the developing countries, the implementation periods generally are 10 years. Least-developed countries are not required to undertake across the board reduction commitments, but tariffs and domestic support are bound at base levels. The World Trade Organization (WTO) disciplines make up the latitude of a member state's current and future agricultural policy. Hence, this section gives a broad outline of the framework of WTO provisions (WTO, 2001).

Increasing market access

The market access provisions of the URAA consist of two parts: 1) tariffication and 2) minimum and current access provisions. Tariffication implied that non-tariff barriers had to be converted into tariff equivalents equal to the difference between internal and external prices existing in the base period, which was set at 1986-1988. All tariffs had to be bound, which means that they could not be increased above the maximum level without notification and compensation. Subsequently, tariffs had to be reduced by 36% on average (with a minimum of 15%), in six equal steps. Developing countries' reduction commitments are generally two-thirds of those for developed countries and implementation periods are longer: 10 versus 6 years. And in case of products deemed of importance to food security, exemptions in reduction commitments are provided. After the tariffication and the following reduction in tariffs, tariffs remained high on average. The base period chosen was a period of very high protection levels, contributing further to the retention of high tariffs under tariffication. The requirement for reductions of 36% (on average) had limited significance too. By making rather large tariff cuts for commodities that do not compete with domestic production or large percentage cuts in tariffs that already were low, the 36% reduction could be achieved with minimal cuts in politically sensitive tariffs.

To ensure that historical trade levels were still maintained and to create some new trade opportunities, minimum and current access provisions were introduced. Minimum access provisions implied that minimum access opportunities had to be provided for products subject to tariffication with imports below 5% of domestic consumption in the base period. Furthermore, countries had to maintain current access opportunities equivalent to those existing in the base period. To ensure that these access opportunities were offered, countries have established tariff-rate quotas (TRQs), subject to a low duty of imports. Imports above that amount are subject to the tariff established through tariffication. Minimum access quotas had to be increased from 3% to 5% of domestic consumption during the implementation period. In allocating TRQs, special consideration can be given to developing country exporters.

For products subject to tariffication, countries can put a special temporary agricultural safeguard mechanism (SSC: Special Safeguard Clause) in place. This mechanism, applying additional levies, may be used when an increase in imports or a drop in price of imports exceeds certain trigger levels. In this way, market prices and import quantities are prevented from large fluctuations. The special safeguard provisions have been invoked for numerous products since the sharp decrease in world market prices.

Reducing export support

Countries that employed export subsidies agreed in the URAA to reduce this kind of support. Export subsidy commitments implied a reduction of a country's volume of subsidised exports by 21% (14% for developing countries) and a reduction of the value of export subsidies by 36% (24% for developing countries) during the implementation period (see Box 2.2 for the definition of export subsidies). Although with some exceptions, the base period has been set at 1986-1990.

For each country, reduction commitments were specified in a schedule, which contains 22 groups of agricultural products, including a group of processed goods. Countries are not allowed to initiate export subsidies for commodities that are not included in their schedule. If a country under-utilises its commitment in any year, it can add the excess commitment quantity or value to the next year's commitment. However, this does not apply to the last year of implementation. On certain conditions, food aid has been exempted from the URAA disciplines on export support. Donors of international food aid shall ensure that the provision of food aid is not tied to commercial exports of agricultural products. Furthermore, food aid transactions must be in accordance with the FAO 'Principles of Surplus Disposal and Consultative Obligations' and Article IV of the Food Aid Convention 1986.

Of the 25 countries that have export commitments in their WTO schedules, the EU is by far the largest export subsidiser. The EU accounted for almost 90% of the export subsidies notified to the WTO in 1995-1998. The EU relies on export subsidies to bridge the gap between domestic prices and lower world prices. The EU is most reliant on subsidies for dairy products, poultry, bovine meat and processed products. The USA employ export subsidies through their

Export Enhancement Programme (EEP) and their Dairy Export Incentive Programme (DEIP). New Zealand, Australia and Canada do not make use of export subsidies. Although the level of export subsidies of the USA used to be relatively small, it has increased since world market prices were depressed.

Box 2.2   The definition of export subsidies in the URAA (based on Article 9)

According to the URAA, export subsidies are defined as follows:

- The provision by governments of direct subsidies, including payments-in-kind, to (groups of) producers, contingent on export performance.
- The sale or disposal for export by governments of non-commercial stocks of agricultural products at a price lower than the comparable price charged for the like product to buyers in the domestic market.
- Payments on the export of an agricultural product that are financed by virtue of governmental action, including payments that are financed from the proceeds of a levy imposed on the agricultural product.
- The provision of subsidies to reduce the costs of marketing exports of agricultural products (other than widely available export promotion and advisory services) including costs of transport and processing.
- Internal transport and freight charges on export shipments, provided or mandated by governments, on terms more favourable than for domestic shipments.
- Subsidies on agricultural products contingent on their incorporation in exported products.

Export credit programmes

The USA and Australia guarantee commercial credit for sales of agricultural products, and, in some cases insure sales on special terms if these are viewed to be in their 'national interest'. Export credit guarantees expand importers' demand when they have difficulties obtaining foreign exchange. In this way, credit guarantees can help stabilise exports in times of (for example) economic crises. Since export subsidies are reduced under the URAA, the competitive aspects of credit guarantees have come under growing criticism. During the Uruguay Round, the WTO partners agreed to continue talks on the establishment of disciplines on export credit guarantees. However, no agreement on this issue has yet been achieved.

State trading

In Canada, Australia and New Zealand State Trading Enterprises (STEs) are a well-known phenomenon. As early as in 1947, the contracting parties of the GATT recognised that STEs could distort global trade. GATT Article XVII recognises STEs as legal enterprises, but requires that they do not discriminate among importers or exporters when they make purchases or sales and that STEs act 'in accordance with commercial considerations'.

Countries must report information about their STEs to the WTO. The WTO defines STEs as 'governmental and non-governmental enterprises, including marketing boards, which have been granted exclusive or special rights or privileges, including statutory or constitutional powers, in the exercise of which they influence through their purchase or sales the level or direction of imports or exports'. Several WTO partners have criticised STEs, as such entities have been granted exclusive or special rights or privileges that contribute to distortions in international agricultural trade. Critics of state trading argue that STEs' lack of transparency could be used to mask export subsidies and import tariffs. Furthermore, it is argued that statutory authorities provide STEs with opportunities unavailable to commercial firms that compete against them. STEs may have exclusive rights to purchase or sell particular commodities destined for the domestic or export market. This statutory power might be used to act as a monopsonist or monopolist, offering lower prices than prices prevailing in the world market or charging consumers higher than those in international markets. Other exclusive rights granted to STEs may restrict trade or competition. STEs may control standards of imported products. Consequently, this may lead to discriminatory treatment against goods of certain national origin. Furthermore, governments may give STEs preferential exchange rates for imports. And occasionally, STEs are allowed to keep over-quota tariff revenues or resale price differential. These revenues can be used to subsidise other aspects of the operations of STEs to the disadvantage of private entrepreneurs. Some of the various other facilities that can be provided to STEs and may not be available to private firms, are tax breaks, transport subsidies, and preferential rates on utilities.

Reducing domestic support

In addition to the agreement on increasing market access and reducing export support, WTO member countries agreed to limit domestic support, as some of such measures could have a distorting effect on trade. In the discussions leading up to the URAA, domestic support policies were segregated into three main categories, to indicate the relative acceptability of the policies: the amber box, the blue box, and the green box (Box 2.3).

Amber box

In the final agreement, only the domestic policies that deemed to have the largest effect on production and trade, the so-called amber box policies, were subjected to limitations. In general, these policies provide economic incentives to producers to increase current resource use or current production ('coupled' incentives). In accordance with the 'de minimis clause', product-specific or non-product-specific domestic support 'which does not exceed 5% of a Member's total value of production of a basic agricultural product or otherwise the Member's total agricultural production during the relevant year' is excluded from reduction commitments. The support category that had to be reduced is quantified by the

Aggregate Measure of Support (AMS). The AMS had to be reduced by 20% from 1995 to 2000, in comparison with the base period level (1986-1988). Least developed countries are granted additional exemptions, including delayed applications of the provisions, a *de minimis* exclusion of 10% and exemption of certain domestic support (for example for encouraging diversification).

Examples of amber box policies are price support, input subsidies, and direct payments related to production. According to the URAA definition, New Zealand's agricultural policies do not count as amber box support. Because of its modest level, most of Australia's farm support measures (for example the safety nets, crop insurance, and exceptional circumstance assistance) are qualified as 'de minimis' and not subject to reduction commitments. Although Canada has shifted away from commodity-linked market price support over the last decade, there is still some amber box support provided, especially for dairy. Canada's NISA is included in the AMS calculations as well.

In the USA, price support is mainly provided through border measures, export support and minimum-price provisions operating through administered prices, nonrecourse loans, loan deficiency payments and marketing loans. Policies for dairy, sugar, peanuts, poultry, cotton, oilseeds and grains are the main amber box policies. The crop and revenue insurance subsidies have also become part of the amber box, as a result of increased subsidy levels. Reforms under the two major Farm Acts in 1990 and 1996 reduced amber box support considerably. However, price support through loans was during 1998-2000 on the rise again, due to the unfavourable world market situation.

On the other side of the spectrum, EU agricultural policies are still primarily based on market prices support, particularly in cereals, oilseeds, cotton, tobacco, wine and grassland based livestock products (dairy, beef, sheep meat). Market price support is provided through administered prices, export subsidies and tariffs. Price support is often combined with supply management tools and compensatory payments (which are blue box policies). The policy reforms (Mac Sharry reforms) of the EU, prior to the URAA, reduced support prices and increased its reliance on direct payments. The Agenda 2000 reforms, which will be implemented during the period between 2000 and 2006, will further reduce price support.

Blue box

The second category of domestic support measures is the blue box. To accommodate the USA and the EU and to bring the negotiations to a conclusion, countries agreed to redefine some amber box policies. Support measures placed in the blue box are in fact amber box payments related to supply management programmes. The blue box is viewed as a special temporary exemption category for the 1995-2000 period. Payments can be placed in the blue box if the amount of payments is based on a fixed area and fixed yields, or a fixed number of livestock, or if they are based on no more than 85% of the base level of production. Examples of blue box policies are the compensatory payments of the EU (hectare payments and animal premiums) and the former deficiency payments of the USA. Canada, Australia, and New Zealand do not provide any agricultural support of the blue box category.

Box 2.3   The structure of domestic support commitments in the URAA

Total support	**Amber box** Expressed in AMS, are domestic policies presumed to have the largest potential effects on production and trade.	*Reduction commitment*
	Product-specific *de minimis* support When product-specific support expenditures are below a certain threshold (5% of the production value for developed countries and 10% for developing countries).	*Ceiling*
	Non-product-specific *de minimis* support Applies when total support expenditures are below a certain threshold (5% of the production value for developed countries and 10% for developing countries).	
	Blue box Temporary exemption category for amber box payments related to production limiting programmes.	*Exempt criteria (No ceiling)*
	Green box Policies considered having the smallest potential effects on production and trade.	

Green box

The third category of domestic support is the green box. These policies have been considered to have the smallest effects on production and trade. Hence, these policies were fully exempted from support reduction commitments. Broadly speaking, there are four categories of green box policies provided for by the URAA (Swinbank, 1999):

• income support payments that are totally decoupled from production;
• public stockholding for food security services and domestic food aid;
• a variety of support measures for the farm sector, such as the provision of general services; and
• payments under environmental and regional assistance programmes.

More specific criteria for the green box are presented in Box 2.4.

Box 2.4   URAA criteria for the green box

Green box policies shall meet the fundamental requirement that they have no, or at most minimal, (distorting) effects on trade and production. Furthermore, the support in question shall be provided through a publicly-funded government programme, not involving transfers from consumers; and the support in question shall not have the effect of providing price support to producers. Such (green box) policies include:

- General services (research; pest and disease control; training services; extension and advisory services; inspection services; marketing and promotion, excluding expenditure for unspecified purposes that could be used by sellers to reduce their selling price or confer a direct economic benefit to purchasers; infrastructural services).
- Public stockholding for food security purposes.
- Domestic food aid.
- Direct payments to producers (based on the green box criteria).
- Decoupled income support. Eligibility for such payments shall be determined by clearly-defined criteria such as income, status as a producer or landowner, factor use, and production level in a defined and fixed base period. The amount of such payments shall not be related to or based on the type or volume of production, prices, or factors of production employed in any year after the base period.
- Government financial participation in income insurance and income safety-net programmes, that are based on various criteria (Annexe 2 of the URAA).
- Payments (made either directly or by way of government financial participation in crop insurance schemes) for relief from natural disasters and like disasters (including disease outbreaks, pest infestations, nuclear accidents and wars). To be classified as green box policies, these payments have to meet some specific criteria (Annexe 2 of the URAA).
- Structural adjustment assistance based on various criteria (Annexe 2 of the URAA).
- Payments under environmental programmes: eligibility for such payments shall be determined as part of a clearly-defined government environmental or conservation programme and be dependent on the fulfilment of specific conditions under the government programme, including conditions related to production methods or inputs; and the amount of payment shall be limited to the extra costs or loss of income involved in complying with the government programme.
- Payments under regional assistance programmes: eligibility for such payments shall be limited to producers in disadvantaged regions on basis of neutral and objective criteria and not on the basis of contemporary circumstances.

In absolute terms, the USA is the largest provider of green box support. An important category of green box support are the Production Flexibility Contracts (PFCs), a form of decoupled income support. The two largest conservation and environmental programmes are the CRP and the EQIP. The major share of green box policies in the EU, aimed at supporting structural adjustment, rural development, marketing and promotion, research and extension and environmental benefits, is co-financed or entirely financed by Member States.

FURTHER LIBERALISATION OF TRADE

In Article 20, the URAA mandated new negotiations before the end of 1999, in order to continue the agricultural liberalisation process. The WTO Ministerial meeting in Seattle in December 1999 did not succeed in launching a new broad round of trade liberalisation. The lack of time available for serious negotiations, difficulties in the management and organisation of the conference, and the exclusion of many developing countries from the discussions, made it too difficult to bridge differences. Instead of a broad-based comprehensive round, separate negotiations on Agriculture have started in 2000. During the first phase of the negotiations approximately forty proposals have been submitted, covering all major areas of the negotiations. Some proposals, such as those from the EU and the USA cover a wide range of subjects. To give an idea of the contents of the proposals, some of the main elements are summarised below.

There is a general consensus about the approach to be pursued in the agricultural negotiations in the context of the three pillars established under the URAA: market access, export support and domestic support. Some countries argue that an important objective of the new negotiations should be to bring agricultural trade under the same rules and disciplines as trade in other sectors. Others, however, argue that agriculture is different and that negotiations should be based on five pillars. As well as the existing three pillars, the negotiations should include non-trade concerns (NTCs) and special and differential treatment (SDT) for developing countries (DCs) as separate issues in their own right. Concern has been expressed to ensure that DCs, in particular those that are net food importers, benefit evenly from trade liberalisation. DCs are active in the negotiations and have tabled a number of proposals reflecting a diverse range of interests. Here we will not go further into the SDT-pillar, but focus on the other topics. The presented information is drawn in part from the OECD (2001b) and WTO (WTO, 2001).

Issues of market access

Most of the WTO members are favourably disposed towards a further reduction of tariffs. Nevertheless, it is questionable whether the EU, Canada and the USA are willing to significantly reduce tariffs of 'sensitive' products, such as sugar and dairy. It is likely that the EU is willing to lower tariffs modestly but probably only through a sector by sector approach. The USA principally wants to reduce the highest tariffs, but also through a sector by sector approach. The Canadian view on market access is unclear. Canada will probably resist too drastic a reduction in tariffs, as these are particularly high for Canada (Josling and Tangermann, 1999). Conversely, Australia and New Zealand want to lower *all* the tariff peaks.

For commodities subject to TRQs, increasing the tariff quota volumes might have a larger impact on trade than tariff reduction. Furthermore, the administration of the TRQs could be improved, as the administration of the current TRQs has become a major problem in agricultural markets. After all, licensing procedures of TRQs are involved with a great deal of governmental interference. TRQs have provided a playground for rent-seeking traders, who in turn have acquired an incentive to lobby for the continuation of high above-quota tariffs (Josling and Tangermann, 1999). Most countries seem to be willing to ensure stricter disciplines and improve transparency in the administration of tariff rate quotas.

Issues of export competition

So far, the negotiations have focused mainly on how much and how quickly export subsidies should be reduced and on the extension of the provisions to cover all measures affecting export competition. On export subsidies, New Zealand and Australia try to negotiate an abolishment. Canada and the USA are slightly more reluctant to abolish them. The EU, by far the largest export subsidiser, seems to be willing to further reduce export subsidies, but argues for continuation of the provision to be enable to carry over unused export subsidies and to accumulate them with their annual commitments in subsequent years.

What concerns the coverage and rules of other measures that affect export competition, important issues are the operation of STEs, abuse of food aid, price pooling and subsidised export credits. Export credits are among other tools used by the USA. Like the EU, New Zealand and Australia want to abolish the use of subsidised export credits, whereas the USA are willing to make them somewhat more transparent. Since Canada, Australia and New Zealand use STEs, these countries would have difficulties agreeing with an abolition of export monopolies. The USA and the EU on the other hand, would like to bring STEs up for discussion.

Issues of domestic support

The negotiations on domestic support have focused mainly on how much further trade and production distorting support should be reduced, the status of the blue box and a review of the green box criteria. The USA as well as New Zealand and Australia have pushed for the elimination of the blue box. This should take place because the box contains measures, which are only partly de-linked from production. The EU is defending the maintenance of the blue box, and Japan, Norway, Korea and Switzerland have also taken this position. If this would prove to be unrealistic, the EU could change its own compensation policies without too much inconvenience to make them compatible with the green box provisions, which would imply decoupling. However, the size and composition of the green box will also be an issue during the negotiations, because some important countries want to sharpen the criteria for the green box. Canada has proposed to impose ceilings on all types of support: amber, blue box and green box. On the issue of further amber box reductions, the USA could differ from New Zealand and Australia. Recent sharp drops in prices and natural disasters in the USA have led the Congress to raise support to farmers. The EU has argued for a strengthening of rules concerning non-product specific domestic support and reductions of the *de minimis* clause for developed countries.

Non-trade concerns

Most countries recognise that agriculture plays an important role in achieving societal objectives such as the sustainable use of natural resources, environmental protection, development of rural areas, poverty alleviation and food security. The main question under debate is whether 'trade-distorting' support, or support outside the green box, is needed in order to help agriculture to perform its multiple roles. According to the EU NTC measures should be well targeted, transparent and implemented in minimally trade-distorting ways and could be addressed inside the green box.

In the same broad context, issues related to animal welfare has led to a proposal to compensate farmers for the extra costs they incur when they are required to meet higher standards of animal welfare. Under the proposal these payments would be in the green box. Another proposal concerns food safety and quality. To meet consumer concerns on these issues, several countries have proposed to apply precautionary principles and labelling schemes. Since these issues have not specifically been covered in the URAA, some other countries argue that they should be handled in other WTO bodies such as the Council on Trade Related Intellectual Property Rights (TRIPs) and the Technical Barriers to Trade (TBT) Committee.

CONCLUDING REMARKS

The process of agricultural policy reform in OECD countries have seen mixed results (OECD, 2001b). For the USA, greater reliance on market outcomes intended under the 1996 FAIR Act has subsequently been wound back by measures taken to buffer producers from market realities. The EU has embarked on further reforms for some crop sectors and started it for others under the Berlin Agreement on the Agenda 2000 proposals, but has postponed decisions on more fundamental changes on some of the more heavily regulated and supported agricultural industries. In the near future many programmes will be reviewed and a new multilateral agreement on agriculture in WTO is anticipated. At the same time governments are being confronted with a broader range of emerging issues on their agenda (market power of large companies, consumer concerns, food safety, environmental concerns, human and animal health issues and animal welfare).

It is inappropriate here to try and provide ready answers to the question which policies should be pursued. Such recommendations would anticipate considerations, which will have to be made in and among the countries. The argument is often heard that market participants would arrive at acceptable solutions without government intervention, i.e. with a radical liberalisation of agricultural policies. Partly with reference to the recent developments and experiences in the USA, which saw the failure of major elements of the agreed liberalisation of the agricultural programmes in 1996, some scepticism is due with regard to the satisfactory functioning of free market forces in agriculture. In order to direct the development process of the agricultural sector in a socially desirable direction, both market and public policies, including income support, environmental, and rural development elements, may be necessary as supporting conditions. It is a challenge for the countries to reconcile domestic and international policy interests and design policies that support a sustainable development of agriculture.

REFERENCES

AAFC (Agriculture and Agri-Food Canada) (1999) *Farm Income, Financial Conditions and Government Assistance Data book 1999* (Update October). Agriculture and Agri-Food Canada, Ottawa.

AFFA (Ministry of Agriculture, Fisheries, Forestry Australia) (1998) Agriculture-Advancing Australia: Overview and Background. <www.affa.gov.au> (printed on 13-10-1999).

AWB (Australian Wheat Board limited) (1999) Our Business. <www.awb.com.au> (printed on 6-10-1999).

Ash, K. (1998) *Agri-food Policy in Canada.* AAFC Economic and Policy Analysis Directorate Policy Branch, Ottowa.

Barichello, R. (1996) Overview of the Canadian Agricultural Policy Systems. <www.library.ubc.ca/ereserve/agec420/overv.htm>.

EC (European Commission) (1999) Fact-Sheet CAP reform: A policy for the future. <www.europa.eu.int/comm/dg06/ag2000/index_en.htm>.

Josling, T. and Tangermann, S. (1999) Implementation of the WTO agreement on agriculture and developments for the next round of negotiations. *European Review of Agricultural Economics,* 26 (3), 371-388.

MAF (Ministry of Agriculture and Fisheries New Zealand) (1996) Agriculture Industry Governance: Post Election Briefing. <www.maf.govt.nz> (printed on 11-11-1999).

NHT (Natural Heritage Trust Australia) (1996) Natural Heritage Trust-Introduction: Helping Communities Helping Australia. <www.nht.gov.au> (printed on 07-10-1999).

OECD (1999) *Agricultural Policies in OECD Countries - Monitoring and Evaluation 1999.* Organisation for Economic Co-operation and Development, Paris.

OECD (2001a) *Agricultural Outlook 2001-2006.* Organisation for Economic Co-operation and Development, Paris.

OECD (2001b) *Agricultural Policies in OECD Countries - Monitoring and Evaluation 2001.* Organisation for Economic Co-operation and Development, Paris.

Rijswick, C.W.J. van and Silvis, H.J. (2000) *Alternative Instruments for Agricultural Support: A Survey of Measures Applied by Competitors of the EU.* Report 5.00.04, Agricultural Economics Research Institute (LEI), The Hague.

Swinbank, A. (1999) CAP reform and the WTO: compatibility and developments. *European Review of Agricultural Economics,* 26 (3), 389-407.

The Commission on 21st Century Production Agriculture (2001) *Directions for Future Farm Policy: the Role of Government in Support of Production Agriculture.* Report to the President and Congress, Washington, DC.

USDA (1996) Glickman pledges swift implementation of new farm law. In: *Press Release No. 0173.96*, United States Department of Agriculture, Washington, DC.

USDA (1997) *Agricultural Situation Annual.* Attaché Query Detail: AGR Number AS 7067, United States Department of Agriculture, Washington, DC.

WTO (2001) http://www.wto.org/english/tratop_e/agric_e/agric_e.htm.

World Trade and Concerns for the Human Environment

W. Bradnee Chambers

INTRODUCTION

In 1998, the Second WTO Ministerial Meeting held in Geneva, Switzerland, put in motion a process to begin the ninth round of global trade talks held since 1947. The last trade negotiation of this kind, the Uruguay Round, was the most ambitious trade negotiation ever held. This round had taken on a negotiating agenda that covered virtually every imaginable trade issue and lasted an unprecedented seven and a half years, earning it the historical title of 'the round to end all rounds'. To the participants of the 1998 Geneva meeting, it seemed fitting to kick off a new round of negotiations two years later, on the eve on a new millenium, in the port city of Seattle which stood as a model city of the new global economy. At stake were many of the issues which had been left unresolved in the Uruguay Round or which had been built into future negotiating round agendas. Included were issues such as agriculture and services, the review of agreements reached under the WTO such as Trade Related Intellectual Property Rights (TRIPS), Trade Related Investment Measures (TRIMs) and the Textile and Clothing Agreement, as well as the so called 'new issues' that had been tabled in Singapore at the First WTO Ministerial Meeting in 1996.

Trade and environment was categorised as one of the 'new issues', yet there was really nothing new about this old debate that had long held a high profile in World Trade Organization (WTO) activities. It was, in fact, the continued failure to resolve the issue and the tendency to regard it as an ancillary concern to increasing trade liberalisation that had so angered many of the world's environmental activists. In the opening days of the 3rd WTO Ministerial in 1999, this anger spilled over into the Seattle streets.

The question is, why had the trade and environment issue been left unresolved for so long and such a high level of frustration been allowed to develop. So much work had already been carried out in an effort to resolve some of the key issues, and some would have argued that the negotiating environment was ripe for a decision, yet an actual consensus had continually eluded negotiators. The answer is not an easy one. Disagreements among countries remain prevalent. On one side of the debate are the countries arguing that present WTO provisions already deal with trade and environment issues adequately. On the other side, are those countries arguing that as environmental agreements and issues are growing both in number and economic significance, preemptive action is urgently needed in order to ensure the level of policy coherence required to avoid potential conflicts.

Perpetuating this debate are forces outside the WTO, such as non-governmental organisations (NGOs) and parts of civil society, which claim that WTO rules and trade liberalisation threaten the natural environment and challenge the sustainable utilisation of resources of future generations. These environmental advocates argue that the exercise of WTO rules have the effect of putting health, development and environmental policy decisions in the hands of trade policy makers and international jurists working behind closed doors. They widely perceive the WTO as an instrument of globalisation that is non-transparent and unaccountable to any national citizenry. This criticism stems from a history of dispute settlements in either the GATT or the WTO whereby domestic measures taken to protect amiable species such as dolphins, or endangered animals such as sea turtles, have been struck down as being incompatible with WTO rules. As a result, the green community cautions against the further liberalisation of trade at the WTO without adequate regard to the environment and sustainable development.

Trade proponents argue that the accrued benefits from progressive trade liberalisation have been significant. Today more than six trillion dollars of goods and services are traded on international markets and regulated by WTO rules. Tariff reductions have steadily declined to an average 4% on industrial products in 1999. International trade has increased, marking tremendous figures of 8% per year in the 1950s and 1960s. Generally the increase in trade has created economic growth and stability that has led to higher standards of living, fuller employment, and the eradication of poverty in many places.

In 1995 these contrasting views were placed under one roof when the WTO was created as a permanent organisation to house the GATT and other trade agreements. The WTO preamble juxtaposed trade liberalisation and sustainable development and the need to protect and preserve the environment. Since this preambular recognition, the WTO's 134 members have struggled to determine exactly how this objective should be integrated into WTO agreements.

At the first WTO Ministerial Meeting in Singapore the new Committee on Trade and Environment presented its first report. Yet, at this meeting members failed to reach any substantive conclusions and instead only agreed, in recognition of the 'breadth and complexity' of the issue, that more work was needed. Similarly, very little substantive progress was made at the second Ministerial Meeting. Nonetheless, the new millennium has brought with it fresh hope that somehow the differences between what seems to be a clash of cultures can be settled and a mutually reinforcing set of multilateral regimes - promoting both trade liberalisation and environmental protection - can be created.

Perhaps the greatest obstacle to solving the trade and environment conundrum is the polarised political divides that exist between developed and developing countries. Generally, while developed countries do not share an exact consensus in regard to the specifics of how to resolve the trade and environment debate, they agree that some sort of policy coherence is a minimal requirement.

In recent years political support for environmental issues at the WTO has grown significantly among developed countries. United States President, Bill Clinton, in his address to the 1998 WTO Ministerial Meeting underscored the idea that enhancing trade should also enhance the environment. In a similar message to the 1999 High Level Symposium on Trade and Environment, Clinton emphasised the need to strengthen environmental protection; ensure trade rules support national policies providing for high levels of environmental protection and effective enforcement; and achieve greater inclusiveness and transparency in WTO proceedings. He reiterated these sentiments in addresses in Seattle. In its final communiqué from the 1999 meeting in Köln (Germany), the G8 urged WTO members to 'fully take account of environmental considerations in the next round' and also that there was an urgent need for clarification of the relationship between multilateral environmental agreements, key environmental principles, and WTO rules. Similarly, at the 2000 G8 Summit in Okinawa, leaders emphasised the need to balance trade with environment. Yet this apparent political will represents only one half of the story.

Contrary to the growing consensus in the north, there still exists a clear lack of political will to take up trade and environment issues in developing countries. In fact, there is outright opposition that has resulted in a determination to block environmental issues from the global trade agenda. Environmental issues in the WTO are viewed with a deep suspicion that verges on contempt by developing countries. They are widely regarded as a façade to green protectionism or as clandestine efforts to restrict developing country access to northern markets. Developing countries are also concerned that the introduction of environmental standards, particularly those that would include production and process methods, will create a legal precedence for the negotiation of labour or human rights standards. This is a path that developing countries would rather not go down.

Trade and environment issues are by no means as straightforward as portrayed by North-South, or free trade vs. environmentalist, perspectives. The debate is multifaceted and key questions are woven around complex nuances in country positions vis-à-vis WTO and GATT agreements. This chapter is aimed at examining some of the main areas of contention in regard to the WTO as the primary regime regulating global trade.

TRADE AND ENVIRONMENTAL MEASURES[1]

One of the contentious areas where environmentalists and free traders have been at loggerheads relates to the measures put in place to protect the environment and their compatibility with WTO rules. Most of this controversy centers upon the key WTO discipline of 'like products'. According to WTO and GATT rules, the term 'like products' refers to the state of the product when it arrives at the border

---

[1] This section draws from Chambers (2001).

of a country[2]. 'Alikeness' of products should be decided upon on a case-by-case basis, and should take account of 'the product's end uses in a given market, consumers' tastes and habits, which range from country to country, the product's properties, nature and quality'[3]. In other words, under a strict interpretation of WTO clauses, which are designed to prevent unfair constraints on commerce, the environmental impact of a production process is irrelevant to a WTO decision. It would be in contravention of WTO and GATT rules to discriminate against 'like products' because they were produced using environmentally damaging processes. This said, the inability of the WTO to discriminate between products on the basis of how they were produced runs contrary to the objectives of environmentalism and most environmental measures.

Understanding WTO rules on exceptions for the environment

The WTO does provide exceptions under Article XX that could potentially confront such problems of incompatibility between environmental measures and the WTO. The two subparagraphs that relate to the environment under Article XX are (b) and (g), which state the following:

> Subject to the requirements that such measures are not applied in a manner which constitute a means of arbitrary or unjustifiable discrimination between countries where the same conditions prevail, or a disguised restriction on international trade, nothing in this Agreement shall be construed to prevent the adoption or enforcement by any contracting party of measures:
>
> (b) necessary to protect human, animal or plant life or health;
>
> (g) relating to the conservation of exhaustible natural resources if such measures are made effective in conjunction with restrictions on production or consumption......

Article XX of the GATT is a typical provision often found in trade agreements. Generally it allows for the main disciplines or obligations under the agreement to be waived if there are justifiable grounds. Normally the grounds are for reasons of national security, morality, human health or environmental protection[4]. Article XX subparagraphs (b) and (g) have been at the centre of much of the trade and environment debate and at the heart of most environment related disputes among members.

One of the main sources of contention has been the manner in which Article XX (b) has been interpreted by GATT and WTO panels and the Appellate Body. Subparagraph (b) requires that any WTO illegal measure that is enacted

---

[2] The WTO has referred to the way in which a product is produced as the 'process and production method' (PPM).

[3] Report of the Working party on Border Tax Adjustments, BISD 18S/97 (1972) par. 18.

[4] For a similar provision in the EU see Article 36 of the Treaty of Rome or Article 104 of the North American Free Trade Agreement (NAFTA).

domestically in an effort to protect human, animal or plant life or health must be the least inconsistent with WTO rules in comparison to other possible measures. This requirement is sometimes referred to as the 'necessary test.' This test was examined extensively within a different context in the *Section 337* case. This case involved a different issue in that it related to Article XX subparagraph (d), which protect patents, trademarks and copyrights and the prevention of deceptive practices. The case established that 'a contracting Party cannot justify a measure inconsistent with other GATT provisions as 'necessary' in terms of Article XX (d) if an alternative measure which could be reasonably be expected to be employed and which is not inconsistent with other GATT provisions is available.'[5] It further stated that if an alternative measure is not available, the measure selected must be the least inconsistent of the available measures, with the GATT. Subsequently, the environmental context of the 'necessary test' has been upheld for subparagraph (b) in the *Thai Cigarettes* case, [6] and thereafter in the two *Tuna Dolphin* decisions[7].

Many countries and other stakeholders concerned with environmental protection consider the threshold that a domestic environmental measure must first pass before it can be considered a legitimate exception to WTO rules is too stringent. In practice, Article XX (b) gives predominance to trade rules over optimal environmental solutions. What is required of Article XX (b) is a more balanced proportionality test between the environmental measure and the level of inconsistency with the GATT rules. The article requires a 'rule of reason' that renders the measure 'necessary' in order to protect human, animal or plant life or health while being proportional to the rules and principles under the WTO. Such a rule is essentially a test of reasonableness, i.e. could an alternative measure have achieved the same level of protection to human, animal, or plant life and less inconsistent with WTO disciplines? If so, would the measure have been feasible to implement, would it be cost-effective, could it be monitored, would it be consistent with national legislative practices etc. If such reasonableness is taken into account, a level of equity will be reached between the environment and trade.

To a certain degree Article XX (g) provides a better balance between the legitimacy of domestic environmental measures with the necessity of being the least trade inconsistent measures. For this reason, Article XX (g) has become a more probable defense for discriminatory domestic policies aimed at environmental protection. The 1998 *Shrimp Turtle* Appellate Body decision clarified past WTO interpretations of the (g) provision and put in place a series of tests that appears to strike a more appropriate balance between the trade and environment regimes. The Shrimp Turtle Case established that the environmental measure must be related to protecting the natural resource in question[8]. To establish the intent of the measure the Body stated that a substantial relationship must exist between the measure's effect and the objective of protecting an exhaustible resource. The Body referred to the method it set in the *U.S. -*

---

[5] *Section 337* case, 7 November 1990, BISD 36S/345.
[6] *Thai Cigarettes* case, 7 November 1990, BISD 37S/200.
[7] *Tuna Dolphin* case Panel 1, 1992 BISD 29S/9 and *Tuna Dolphin* case Panel 2, 1993 39S/155.
[8] What was deemed a natural resource was broadly interpreted to include both living and non-living resources.

Reformulated Gasoline case, that a 'close and genuine relationship of ends and means' must exist in order to establish the 'substantial relationship'. Once this requirement was met, then its consistency with whether the measure was the least trade inconsistent came into question. In interpreting this requirement the Appellate Body recognised that a balance must be struck between the right to invoke the exception to protect the natural resource in question and the rights of Members to the main provisions contained in the WTO.

Unilateralism vs. multilateralism

The *Shrimp Turtle* ruling was viewed as a success in many ways, but it did leave open an important question of whether a WTO member has the right to unilaterally protect the environment that is outside its national territory. Until the *Shrimp Turtle* case it had been assumed by the majority of observers that exceptions could only be justified on the basis of some form of multilateralism. This would include a multilateral agreement, widely practised and documented standards, or accepted principles of international law.

In one of the earliest environmental cases of the GATT, the landmark *Tuna Dolphin* case, the Panel noted 'that a country can effectively control the production or consumption of an exhaustible natural resource only to the extent the production or consumption is under its jurisdiction.' The case suggested that Parties could not invoke Article XX to protect the environment outside its own jurisdiction, such as the global commons. This for many environmental advocates raised direct concerns that the GATT rules could restrict the use of measures or polices pursuant to certain Multilateral Environmental Agreements (MEAs). In determining whether the exception under consideration was a necessary one, however, the report of the same panel also used arguments to the effect that the US action was unjustified because it had not exhausted other efforts under international law to protect dolphins: 'The United States had not demonstrated to the Panel....that it had exhausted all options reasonably available....in particular through the negotiation of international cooperative agreements...' Although not substantiated thus far in other WTO panels, this ruling implies that internationally adopted standards such as those pursuant to MEAs could be grounds for justifying an exception.

Thus, the Panel left open the jurisdictional question. It was reasonable, however, to interpret these results as indicating that a unilateral action which has an extraterritorial effect has more weight as a legitimate exception to international trade law if it is carried out pursuant to recognised enabling provisions, or endowed by an MEA, than if it is carried out without any explicit reference to, or authorisation under, an MEA. Such an interpretation would be in keeping with the way certain other exceptions contained in Article XX have been treated as well as other parts of the WTO such as the Sanitary and Phytosanitary (SPS) and Technical Barriers to Trade (TBT) agreements.

The *Shrimp Turtle* Appellate ruling, however, has brought this rationale into question. The case in which the US lost, placed a ban on trade from countries that did not install a mechanism that would allow sea turtles, if inadvertently caught, to escape through a trap door in the shrimp net[9]. Basically, the ban was enacted to protect turtles not inside US national territory. There was also no international environmental or conservation treaty in place to protect them as a justification. In other words, the action taken by the USA, was a unilateral measure to protect sea turtles. Yet, the Appellate Body did not rule against the Americans because of unilateralism. They deemed the US ban as arbitrary and unjustifiable as it was applied without taking into consideration different conditions, which may occur in territories outside of its own[10]. Moreover the Appellate Body actually said that the measure was viewed as legitimate.

> What we *have* decided in this appeal is simply this: although the measure of the United States in dispute in this appeal serves an environmental objective that is recognized as legitimate under paragraph (g) of Article XX of the GATT 1994 (Shrimp Turtle case, WT/DS58/AB/R, 12 October 1998, par. 186).

The ruling, thus, raises strong questions and uncertainty as to the scope of Article XX (g) and whether discrimination can be justified for exhaustible resources outside national jurisdictions. With this ruling it will undoubtedly beckon attempts by governments to justify unilateralism and projectionist domestic measures in the future.

BIOSAFETY AND THE TRANSBOUNDARY MOVEMENT OF GENETICALLY MODIFIED ORGANISMS (GMOs)

Few issues have caused as much controversy and created as much uncertainty in such a short time span as biosafety. As biotechnology progressively develops, these types of issues will increasingly require the attention of the global community. Pressures from these advances in technology will also require coherent and swift responses from governments and multilateral institutions as the stresses on natural and societal structures increase. Biotechnology itself is, of course, not a new issue. It has been around since Gregor Mendel's pioneering work on plant breeding, gene dominance, and inheritance over a century ago.

The development of modern biotechnology, particularly recombinant gene technology (where genes from a different species have been introduced into host organisms), has recently generated deep concern because of its potential impacts on the environment and human health. Very little of the concern on the human health side has been based on conclusive scientific evidence. Some research has shown that GM genes can jump species in oilseed rape through the pollen ingested by honeybees and which has been transferred to bacteria living inside the bee's digestive track. This has led to speculation that the GM foods could in fact jump species in bacteria contained in human digestive systems and alter the

---

[9] The *Appellees* were India, Malaysia, Pakistan, and Thailand.
[10] Appellate Body Report, para 164.

important role it plays in blood clotting, digestion, and the immune system[11]. Concerns have also been voiced ·in various fora and in the media over recombinant DNA from pork or meat being added to non-meat foods, raising ethical concerns for vegetarians and followers of certain religious faiths. The strongest concern over biosafety, is the potential impact of GM crops and seeds on the environment. General risks relate to biodiversity particularly agrodiversity, the impact on non-targeted organisms or related plants, and the potential for increased weediness and invasiveness (Zakri, 2001).

The main agreement under the WTO umbrella that relates to the regulation of trade in GM foods and GMOs, is the SPS agreement. This agreement emerged out of the Uruguay Round of negotiations as a means of harmonising domestic sanitary and phytosanitary measures. Such measures include those required to protect animal, or plant life or health from risks arising from the entry, establishment or spread of pests or diseases, disease causing or disease carrying organisms. The SPS agreement also covers safety regulations for food additives, toxins and contaminants in foods. The agreement does not create independent rules on SPS but, rather, regulates the way in which the laws, decrees, regulations, requirements, procedures etc. are set-up at the domestic level. The SPS agreement does this by encouraging harmonisation on the basis of rules developed by international organisations, including; the Codex Alimentarius, the International Office of Epizootics, and the International Plant Protection Convention (IPPC).

If a domestic measure is adopted according to the standards set by these international instruments then it is presumed to be in compliance with them. A lower domestic standard does not present any difficulties because it will not pose any additional barriers to trade. If a country adopts a higher standard of protection than those of the international instruments listed, then this must be justified and must satisfy several criteria including; science-based risk assessment, transparency, and minimum trade restrictiveness.

Recent attention in the trade and environment debate has focused on the compatibility of the SPS agreement with a new international agreement, under the Convention on Biological Diversity (CBD), on the transboundary movement of GMOs. The agreement, the Cartagena Protocol on Biosafety, was adopted in 2000 after gruelling negotiations that spanned several years. Its main objective is to ensure the 'safe transfer, handling and use of living modified organisms LMOs[12] that may have adverse effects on the conservation and the sustainable use of biological diversity, taking into account human health'[13]. Generally, the Protocol sets out the guidelines for introducing LMOs on to the market and an internet notification procedure, the Biosafety Clearing House, that provides information on national laws and regulations, risk assessments, emergency response procedures and identification of LMOs[14]. The Protocol also sets out very

---

[11] Antony Barnett, "GM Genes Jump Species Barrier", *Dawn/The Observer News Service*, 29 May 2000. Note the research was conducted by Han-Hinrich Kaatz.
[12] GMO and LMO have virtually the same meaning, the Cartegena Protocol defines LMO as 'any living organism that possesses a novel combination of genetic material obtained through the use of modern technology.'
[13] Cartagena Protocol on Biosafety to the Convention on biological Diversity, Article 1.
[14] Ibid., Article 20.

specific details on the elements and methodology that a risk assessment should be based upon[15] and what has been hailed as an authoritative interpretation on the precautionary approach[16].

While concerns were raised about the compatibility of the Biosafety Protocol with the WTO during the negotiation of the Protocol, the issue was essentially left unresolved. The ambiguity between the two international instruments was reflected in the Protocol's final preambular text. One line states that the Protocol does not change the rights or obligations of any party under other international agreements, while the following line makes a seemingly contradictory statement that the Protocol should not be subordinate to any other international agreement[17].

It is difficult to ascertain the compatibility of the two agreements in the absence of a concrete case or authoritative interpretation and, as such, analysts are divided. It is clear, however, from an analysis of the SPS agreement and the Biosafety Protocol of the CBD, that possible inconsistencies at the level of overall approach exist in regard to risk assessment methods and the use of the precautionary principle.

Generally the Biosafety Protocol is more flexible to the establishment of individual country protection levels. The SPS, with its provisions that work towards harmonisation, lists certain international standards as the benchmark[18]. In biotechnology, however, there is strong divergence concerning levels of protection. International standards are also a long way from being adopted although the Codex Alimentarius has formed a task force aimed at adopting guidelines on risk assessment and risk management by 2004. One of the deepest concerns with the Codex process, however, is that for developing countries there has been a lack of participation in international standard setting.

Risk assessment priorities in both the SPS and Biosafety Protocol include different objectives and methods. For instance while the WTO prioritises trade liberalisation and economic considerations by stipulating that risk assessments must be the least trade restrictive, the Biosafety Protocol looks to many other factors and puts environmental protection above all else[19]. In terms of the precautionary principle there are elements of similarity between the SPS and the Biosafety Protocol in that they both acknowledge the necessity to take decisions in the face of scientific uncertainty. But SPS (Art. 5.7) is far more restrictive as such decisions can only be used provisionally, with qualified time periods (case by case). The thresholds for invoking the precaution SPS is also much higher as it requires proof that the pertinent scientific information is lacking[20]. This edge cuts both ways, as it may decrease the abuse of the principle but it may also limit its application.

In regard to their other provisions, the two regimes diverge. The SPS, for example, requires that the party adopting a higher level of protection, other than

---

[15] Ibid., Annexe 3.
[16] Ibid., Article 11.8.
[17] Ibid., Preamble.
[18] Codex Alimentarius, International Office of Epizootics, and the IPPC.
[19] See, for example, Cartagena Protocol on Biosafety to the Convention on Biological Diversity, Article 26 which sets out socio-economic considerations to be taken into account.
[20] See Beef Hormones Case EC Measures Concerning Meat and Meat Products (Hormones), WT/DS26/R/USA, 18 August 1997.

an international standard, must prove that the higher level is necessary and justifiable. This involves a test to show that no less trade restrictive measure is available. There is also much divergence in regard to the concept of equivalence. The Biosafety Protocol has no provision for recognising the standards of protection of countries that may be similar to their own domestic measures. The SPS has a strong provision in this direction.

These are just a few preliminary examples of the potential incompatibilities between the trade and environment regimes in regard to GMOs. Right now, most of them are located at the hypothetical level, although they may be played out in the real world very soon. Conflict becomes increasingly likely as both the environment and trade related regimes expand in size and maturity. If such a conflict were to occur now, it would most likely be played out in the context of the WTO, as it has the only binding dispute settlement system. If this were the case it is likely that full consideration to the environment might not be offered in such a WTO setting.

TRANSPARENCY AND PARTICIPATION

Anti-globalisation protests have shocked the world and have caused millions of dollars of property damage to quiet cities such as Seattle, Quebec and Genoa, thousands of injuries, and even the death of a 23-year-old protester at the 2001 G8 meeting in Italy. These protests, from the average person's point of view, depict the WTO as an organisation that is unresponsive to growing concerns about the environment, transparency and the need for greater participation. This perspective is not a balanced one. Even before the first anti-globalisation protest in Seattle in 1999, many constructive NGOs were working hard to put participation and transparency on the WTO agenda and they had some successes in this regard.

These successes mainly came in the form of opening up WTO documentation to public scrutiny. Initially, the bulk of the organisation's documentation would remain unnecessarily restricted for long periods of time. It is only in response to increased external pressure that the WTO has become much better at declassifying its documents in recent years. Many WTO decisions can now be found easily on the organisation's website and not much later than when they are made available to members.

The real transparency and participation issues revolve around two main areas; the submission of *amicus curiae* briefs and the participation of NGOs in committee meetings and panel and appellate body deliberations (see also Esty, 1998).

The Latin term, *Amicus curiae*, literally means 'friend of the court.' *Amicus curiae* briefs are reports that may assist a court in examining a particular case. They are used in various national legal systems, including the Australia, Canada and the USA. In 1998, four environmental NGOs submitted amicus briefs to the WTO in the Shrimp and Turtle case. These supported the American right to ban imported shrimp from countries that did not install turtle excluder devices (TED) in merchant fishing fleets in order to avoid the unnecessary taking of endangered turtles. The WTO secretariat immediately rejected the submissions, but three briefs were subsequently annexed to the American arguments to the Appellate Body and the Appellate body later ruled that panels have the right to accept non-requested submissions from non-parties[21]. For many NGOs this ruling marked a progressive step forward in terms of opening up the WTO dispute settlement proceeding to civil society. However, two years later the WTO Appellate Body established additional criteria governing the submission of amicus briefs stating, *inter alia,* that the organisations submitting the briefs had to be of relevance to the actual dispute at hand[22]. In the first case where these new rules were enforced, the Asbestos Case between Canada and the EU[23], all the submissions were rejected on the grounds that they were not relevant to the dispute[24]. This has infuriated NGOs, re-raised the question of their participation in disputes, and returned pressure for the WTO to loosen its rules in relation to these types of submissions.

The other main area of concern is the participation of NGOs in actual WTO committee, council, and dispute settlement proceedings. NGOs argue that their participation would greatly increase the transparency and legitimacy of the WTO. Commentators such as Daniel Esty claim that 'legitimacy, authoritativeness, and a commitment to fairness' of the WTO would be greatly enhanced if NGOS were allowed to participate more fully in WTO proceedings (Esty, 1999). Although some countries such as the US, under the Clinton Administration, have advocated the opening up of closed meetings to NGOs, most countries have rejected these claims outright. They argue that the WTO is an intergovernmental process and that lobbies such as NGOs already enjoy the right to raise their concerns at the national level before issues come before the WTO. Furthermore, questions relating to the legitimacy and representation of NGOs and their constituencies have also been raised by member states at various times.

It is true that the WTO is an intergovernmental organisation, and decisions in the final analysis must stem from agreement between governments. Yet, if one were to examine other intergovernmental processes such as the United Nations Commission on Sustainable Development or the Climate Change negotiations it would become readily apparent that many NGOs can play a very positive role. For developing countries with limited resources to cover all the technical issues and the many committee meetings they can be a source of technical advice and can raise issues that would not otherwise be raised. If NGOs are allowed to follow the meeting closely they often make very constructive submissions. Similarly, their research often serves as useful background material even when it is in opposition to the viewpoints of certain governments and can often help provide the foundation for arguments by other member countries.

When deliberations enter an advanced stage of negotiations, or when sensitive matters are under discussion, most NGOs recognise that their participation would not be possible. Most NGOs recognise that WTO decisions are between member states, what they ask is that they are able to contribute effectively to the frameworks in which these decisions are made. Perhaps the

---

[21] Shrimp Turtle case, Appellate Body Report, WT/DS58/AB/R, 6 November 1998, paras. 99-110.

[22] WTO document WT/DS135/9, 8 November 2000.

[23] Canada European Asbestos case WT/DS135/AB/R 12 March 2001.

[24] There were 13 submissions that were returned to the senders.

relatively peaceful nature of climate change and other Multilateral Environmental Agreement (MEA) related meetings serves as proof positive that NGOs can be constructively engaged in inter-governmental negotiations.

BIODIVERSITY, GENETIC RESOURCES, TRADITIONAL KNOWLEDGE AND INTELLECTUAL PROPERTY RIGHTS (IPR)

The TRIPS agreement, like the SPS agreement, was a new addition to the group of agreements arising out of the Uruguay Round. Unlike the SPS, however, there has been widespread debate as to whether the TRIPS agreement actually belongs at the WTO as a multilateral legal regime that is aimed at trade liberalisation. Protecting intellectual property rights through patents, copyrights, and other means is an outright intervention in the free market. Countries such as the USA and Switzerland have long been strong advocates of the inclusion of TRIPS in the WTO. They argue that the TRIPS agreement is necessary to adequately protect the rights of innovators and to create positive incentives for research and development and the advancement of science and technology.

The goal of the TRIPS agreement is to create a minimum standard of global protection for intellectual property. TRIPS seeks the harmonisation of national laws, it does not represent a uniform law in itself. Instead, the agreement attempts to achieve its goals by laying out a list of basic requirements that national IPR regimes should follow. These standards are set out in the second part of the agreement and contain specific requirements on trademarks, copyrights, appellation rights, industrial design, patents, integrated circuit board design, trade secrets and licensing. The standards are loose and do not stipulate that WTO members must create IPR regimes that are modelled after predominate western systems, such as the USA or the EU. A member wishing to create its own IPR regime in order to protect its special interests can do so with a higher standard, so long as the system created is an 'adequate and effective' one.

The last noteworthy characteristic of the TRIPS agreement is its dispute settlement provision. One of the aims of bringing the TRIPS agreement into the WTO is to place the agreement under the organisation's strong compliance mechanisms. The prospect of an effective dispute settlement body and enforcement measures against non-compliant countries, made the WTO, at the time of formulising general rules on trade related intellectual property rights, a much more attractive home for TRIPS than the World Intellectual Property Organization (WIPO). However, the WTO dispute settlement procedures have been used sparingly for TRIPS disputes. This could be due to the fact that IPR is a relatively new area for countries to associate with a trade agreement such as the WTO. In addition, a transitional period has been agreed upon for phasing in the agreement for developing and least developed countries.

The TRIPS has several important implications to the environment. These include broad arguments on the sustainability of patenting life forms, which is possible under TRIPS, its specific implications for the CBD, access and benefit sharing of genetic resources, the transfer of environmental technologies, and the protection of indigenous knowledge.

Article 27.3 (b) states that members may not patent plants and animals biological processes, but this does include the patents on micro-organisms and non-biological processes for producing plants and animals. Now on the eve of a possible new round of trade talks there is concern that that a re-negotiation of 27.3 (b), which has a special clause for review on this subparagraph (after four years) might open greater possibilities for the patenting of life forms. Civil society believes that industries in large developed countries working in life sciences are intent on this direction and would like to see widespread patenting of life forms moving towards a second green revolution, a so called 'genetech revolution' (Friends of the Earth, 2001). With greater opportunities for the patenting of life forms, civil society argues that companies will then be able to create agricultural systems dominated by their patented crops (Friend's of the Earth, 2001). Concerns are already being raised over the linking of certain plant varieties with agro-chemicals. Civil society argues that such developments are creating far too much power over life-vital sectors such as agriculture for a handful of multinationals (Shiva, 2001).

A great deal of concern has also been focused on monoculture practices. Of particular worry is the way in which IPR promotes the reduction of plant variety by creating market incentives to use varieties that may have higher yields or certain qualities that make them more attractive to farmers (i.e. weed resistance, durability etc.). In the last several decades, the intensification of the use of modern biotechnology has caused agro-diversity to sharply decline. Many countries that once planted numerous plant varieties for food or as cash crops now use only very limited varieties[25]. Environmentalists claim that this dependence is increasing the vulnerability of crops to disease and pests and providing the wrong incentives to clear rich biodiverse land in order to plant single or less agro-diverse commercial crops. However, as Graham Dutfield points out, the real question is whether IPRs are the actual culprit behind the reduction of agro-diversity. He argues that several commentators, such as Walter Reid, have made this link, stating that IPRs are responsible for creating research incentives leading towards "centralized crop breeding and the creation of a uniform environmental condition, and discourages agro-ecological research or local breeding tailored to local conditions." this may have biodiversity-erosive effects and reduce environmental heterogeneity (Dutfield, 2000).

Yet another environmental issue related to IPRs and TRIPS is the question of access and benefit sharing and traditional knowledge. A primary objective of the CBD is to create equitable arrangements by which countries can protect their genetic resources that may have potential commercial or human value, while at the same not creating overly restrictive regimes that would inhibit the development of those resources by research and commercial interests.

---

[25] Lorin Ann Thrupp points out that several examples and estimates that of "7000 species of plants, only 150 species are commercially important...103 species account for 90% of the world's food crops".See L.A. Thrupp, "Linking Biodiversity and Agriculture: Challenges and Opportunities for Sustainable Food Security", *World Resources Institute*, Washington DC.

Many undiscovered plants and microbials are likely located in developing countries and their potential value as medicines or health products make them very attractive to illegal prospectors. For environmentalists these 'biopirates' are stealing the future legacies and birthrights of some of the poorest countries in the world. The vast majority of cases of biopiracy also involve the exploitation of indigenous knowledge[26].

Estimated losses from the exploitation of genetic resources in developing countries are enormous, with estimates in the billions of dollars US per year[27]. Environmentalists have placed the blame for the loss of revenues on the patenting system that promotes the rights of the commercial interests but does not adequately protect the rights of lesser-developed and poor countries and their indigenous groups.

PERVERSE SUBSIDIES AND THE ENVIRONMENT

Another crucial environmental issue that has strong links to world trade is subsidies. Of greatest concern to environmentalists are subsidies on fishing, which are depleting the world fisheries and marine diversity and on fossil fuels, which are a major cause of climate change. The Europeans, and to limited extent the Japanese, have also linked agriculture subsidies to the environment, by arguing that their heavy subsidisation has the multifunction of promoting environmental protection. This link to the environment has come under tremendous controversy from Canada, the USA and others, which have to compete with the subsidies on global markets. The Europeans are under pressure to lower these subsidies which are sensitive domestic issues that, to a certain extent, have been linked to self-perceptions regarding the European way of life. If these countries wish to maintain these subsidy levels through the next round of global trade negotiations, they would be expected to make major concessions in other areas.

Many environmentalists point to fishing subsidies as being a prime target for some form of action. As David Shorr argues, most of the world's fisheries are exceeding their sustainable capacity and there are 'too many fishing boats chasing too few fish' (Shorr, 1999, p. 144). Many of the large fishing fleets are maintained by a level of subsidization which has caused, like any subsidy, a distorted market price that is much lower than the actual value and which does not adequately reflect the scarcity of the natural resource. Actual figures vary, but most have put subsidies as high as 20-25% of the overall fishing revenue worldwide[28].

Even with such perverse subsidies as these figures suggest the scope for challenge under WTO rules is quite limited. According to WTO rules, a

---

[26] Often a local medicine man or shaman knows how to use plants or extracts for medicinal purposes this facilitates finding applications for genetic discoveries.

[27] World Resources Institute website, *Questions and Answer on Bioprospecting*, http://www.wri.org/biodiv/bp-facts.html, visited July 2001.

[28] WTO Committee on Trade and Environment, *Environmental Benefits of Removing Trade Restrictions and Distortions (Note by the Secretariat)*, WTO Doc. No. WT/CTE/W/67 (7 November 1997), para. 93.

prohibited subsidy is linked to export performance (export bonuses, favourable shipment charges), or which gives preference to the use of a domestic good over an imported good. Most fishing subsides would not fall under this definition. Another category of subsidy that is considered illegal under the WTO is 'actionable subsidies.' This category focuses on the result of the subsidy and whether it has caused a WTO member 'adverse effects' of a 'serious prejudice' or nature. What is defined as 'serious prejudice' under Article 6.1 of the WTO Subsidies Agreement is if a 'products total subsidisation exceeds 5% of the value of sales', 'if the subsidy covers an operating loss of an industry' other than a one time measure to give the industry time to find a long term solution, and direct debt forgiveness. Article 6.1 does not apply to developing countries.

At first glance, the definition of an actionable subsidy seems to suggest that it could be applicable to fishing subsides that are much higher, at least globally, to the 5% of the value of sales. Unfortunately, this clause had a 5-year sunset provision and it lapsed at the end of 1999. If the clause were to be extended (Shorr, 1999, p. 154), it may provide an avenue to bring a case before the WTO dispute settlement panel. One drawback of this approach, however, is that it may not address many of the indirect or hidden subsidies that exist and, in which case, it may only partially address the real problem. Perhaps what is really needed, instead of trying to find a solution through a WTO panel ruling, is to address the question of perverse fishing and energy subsidies more fully within the new round of trade talks.

CONCLUSION: BRIDGING THE NORTH SOUTH DIVIDE

One of the greatest challenges for resolving the trade and environment debate will be striking a balance between the issues that are of primary concern to developing countries and those of interest to developed countries during the next round of global trade negotiations. Developing countries are concerned with such questions as market access, dumping, and agriculture subsidies. Developed countries, on the other hand, are placing a much greater emphasis on the environment, particularly in the face of growing public and political pressure. In the past, the developing countries have succeeded in stalling any alterations to GATT or WTO rules concerning the environment through the use of blocking tactics and rhetorical arguments. This strategy is unlikely to work much longer as the complexity of the issues is increasing, and developed countries have initiated their own strategies to mainstream the environment and sustainable development throughout WTO agreements.

For developing countries, preparing for a round of negotiations that will include environment and sustainable development issues will be essential. In the past, the scale of resources available to developed country delegations has put developing countries at a disadvantage in negotiations. In regard to the trade and environment debate, the position of developing countries is now being worsened by the massive information generating and disseminating capacities of northern environmental NGOs and by the sheer voracity of civil society activists.

One of the main impediments to a consensus between the developed and developing worlds is the surprising lack of recognition and understanding of the other's perspective. Many of the proposals that have been put forward from developed countries relating to trade and environment, have lacked sensitivity to the needs of developing countries. Often the proposals have not had due regard to principles such as common but differentiated responsibility, the right to development, or even the right for basic human needs that developed countries take for granted such as food, health and education. Greater opportunities for compromise will exist if such proposals recognise and factor in these principles more constructively.

Equally true is that developing countries have not fully engaged in the debate. To gain authority means formulating positions, making trade-offs, and leveraging for issues that are of developing countries' concern. With growing pressure from civil society and the political constituents of developed countries on the trade and environment issues, this could be a natural bargaining chip for developing countries. Hopefully, for the sake of the environment, developing countries will realise this opportunity and engage in a frank discussion on what is really required to balance trade with environment. They may also take greater advantage of the value of their natural environments by bargaining on rules for compensation if they preserve these resources.

In conclusion, the debate on trade and environment is extremely complex and interwoven between issues and agreements under both WTO and environmental regimes, and between the polarities of environmentalists and free traders. This chapter has given an overview of some the primary concerns in the trade and environment debate and will serve as a background and preliminary analysis for the more detailed national case studies that appear later in this volume.

REFERENCES

Chambers, W.B. (2001) International trade law and the Kyoto Protocol: potential incompatibilities. In: Chambers, W.B. (ed.) *Inter-linkages: The Kyoto Protocol and the International Trade and Investment Regimes.* UNU Press, Tokyo, pp. 92-99.

Dutfield, G. (2000) *Intellectual Property Rights, Trade and Biodiversity.* Earthscan, London.

Esty, D.C. (1998) Non-governmental organizations at the World Trade Organization: cooperation, competition, or exclusion. *Journal of International Economic Law,* 1, 123-147.

Esty, D.C. (1999) Environmental governance at the WTO: outreach to civil society. In: Sampson, G.P. and Chambers, W.B. (Eds) *Trade, Environment and the Millennium.* United Nations University Press, Tokyo, pp. 97-117.

Friend's of the Earth (2001) The citizens' guide to trade environment and sustainability: TRIPs, the environment and the review of Article 27.3 (b). < http://www.foei.org/activist_guide/tradeweb/tripenv.htm> 2 July 2001.

Shiva, V. (2001) TRIPs and the environment. *Third World Network,* <http://www.twinside.org.sg/titletrips-ch.htm> 2 July 2001.

Shorr, D. (1999) Fishery subsidies and the WTO. In: Sampson, G.P. and Chambers, W.B. (Eds) *Trade, Environment and the Millennium.* United Nations University Press, Tokyo, pp. 143-170.

Zakri, A.H. (2001) International standards for risk assessment and risk management. Paper presented at the workshop on Biotechnology, Biosafety and Trade: Issues for Developing Countries. International Centre for Trade and Sustainable Development, Geneva, Switzerland, 18-20 July.

European Union

Floor Brouwer, Janet Dwyer and David Baldock

INTRODUCTION

Agriculture is not a major economic sector in the European Union (EU). The agricultural sector contributes a limited share of Gross Domestic Product (GDP) in most Member States. In the United Kingdom (UK), the figure is particularly low, at around 1%, although the total agro-industrial complex has a much bigger share in national income. The agrifood sector (primary production, processing and deliveries to these sectors) has a share of around 6% of total gross value added in the EU. Agriculture, however, is a dominant user of land in most European countries. More than three-quarters of the territory of the EU is agricultural or wooded land. Forests cover about a third of the total land area in the EU. In marked contrast to the situation in other parts of the world, a large proportion of the land area of Europe has been farmed for several millennia. Only the extreme north and the most mountainous areas, including Sweden, Finland, the northern tip of the UK and the Alpine regions of France, Italy, Austria, Switzerland and southern Germany, remained almost completely covered by forest or other natural vegetation, as recently as 100 years ago.

Similar to the USA, the EU is a key producer of food in the world market. The EU has a share of more than 10% in global production of potatoes, sugar beet, citrus fruit and primary fruit. Fruit and vegetables production is concentrated in the Mediterranean part of Europe. The main production areas for citrus fruit are Spain, Italy and Greece. The gross production value of agriculture in 2000 is estimated at around € 265 billion, including crop products (€ 150 billion) and livestock products (€ 115 billion). This includes a wide diversity of crop and livestock products, such as:

- Fresh milk, which has a share of 14% of total production value, and is the source of a wide diversity of products, including butter, cheeses and milk powder. About 25% of the total export of dairy products is to non-EU countries, including the Russian Federation, the USA and Saudi Arabia.

- Cereals, with a share of slightly less than 14% and representing an important source of livestock feed.
- Beef and veal, with a share of 10% of total agricultural production. A structural imbalance of supply and demand during the 1980s gave incentives to reduce intervention prices and introduce headage premia for bulls and steers, subject to an upper limit on the number of eligible male animals. The outbreak in the UK of Bovine Spongiform Encephalopathy (BSE) also known as 'mad cow disease' is of particular importance since beef production was reduced between 1995 and 1997 by around 30%. Exports from the UK of all bovine animals, their meat and all derived products were prohibited in 1996. EU exports of beef to third countries remained fairly stable during that period.
- Pigmeat, with a share of 9% of agricultural production. The sector has faced some severe animal disease outbreaks in recent years, including classical swine fever in the Netherlands in 1997 and the foot-and-mouth disease crisis during the first half of 2001. More than 3 million pigs were killed in the Netherlands during the swine fever of 1997, including 2 million animals that were not affected but were kept in regions where movements were strictly forbidden. While movements of animals were forbidden, animal welfare conditions became too difficult for animals to be kept indoors. More than 1,500 cases of the foot-and-mouth disease were observed in the UK during the first half of 2001, as well as outbreaks in France, the Netherlands and Ireland. Intra-EU trade of pigmeat covers around 80% of total EU trade in pigmeat. More than half of the exports of swine meat to non-EU countries is to Japan. These exports are dominated by Denmark, because this country meets the strict veterinary requirements for the Japanese market.
- Vegetables, with a share of 8% of total agricultural production.
- Other products with a considerable share in production value include fruit, flowers and plants, and wine, with each of these three categories having a share of 6% of total agricultural production. The EU accounts for about 60% of world production of wine, and is the leading exporter on the world market. Recently, there has been a growth in export from countries like Chile, South Africa, USA and Australia. The export of vegetables, fruit and flowers is mainly as intra-community trade. The export of tomatoes to third countries accounts for less than 20% of total export value, mainly to the USA and the Russian Federation.

This chapter offers an overview of the main environmental and health-related standards apply to EU agriculture. Where possible the implications for standards at farm level are drawn out, essentially where they give rise to direct, operational constraints upon farming practices.

AGRICULTURE AND THE COUNTRYSIDE

As a result of its longstanding management of the land, farming in Europe has co-evolved with its ecology, landscapes and other environmental resources. Today, many of Europe's species and their characteristic habitats are dependent upon continued management to sustain their diversity and their range. European landscapes are primarily cultural, heavily influenced by centuries of farm and woodland management. However, this largely positive relationship between management and environmental quality has depended upon farming practices that were relatively low-input, in terms of their capital use and nutrient levels, while labour inputs were relatively high, by comparison to the current situation.

Today, Europe's agriculture remains a highly diverse industry, with a broad range of outputs and structural and environmental characteristics. In particular, there are many surviving examples of characteristic, low-intensity, high nature-value farming systems in different parts of Europe, which demonstrate a positive and largely symbiotic relationship with the natural environment. Greece, Spain, France, Ireland, Italy, Portugal and the United Kingdom include well over 50 million ha of farmland under low-intensity farming systems, representing around 40% of agricultural area (Bignal and McCracken, 1996).

In marked contrast to the other principal exporting nations in the global agro-economy, average farm size in Europe is small - less than 20 ha - and mixed farming remains an important feature of European landscapes. By 1997, the EU had almost 7 million holdings (Table 4.1). Farm size differs markedly between Member States, and on average ranges between 4 ha (Greece) and 69 ha (UK). Around 70% of the holdings are smaller than 5 ha in Greece, Italy and Portugal.

Stocking density (in terms of number of livestock units per ha of land) broadly reflects the intensity of livestock production. It is therefore highest in regions with an emphasis on intensive livestock production (Figure 4.1). Animal density at regional level in the EU exceeds 2 livestock units (LU) per ha of utilised agricultural area in the Netherlands, parts of Germany (some regions in Niedersachsen and Nordrhein-Westfalen), part of France (Bretagne), the northern part of Italy (Lombardy) and some parts of Spain (regions of Galicia and Cataluna). A stocking density of 2 LU ha^{-1} is considered to be close to the maximum amount of nitrogen from livestock manure that should be applied to land, according to the rules of the EU Nitrates Directive.

Among all Member States today an increasing range of environmental issues relate to two dominant trends in current farming change in Europe - intensification, concentration and specialisation in some areas; and marginalisation and abandonment in others. Both these processes involve a move away from the traditional forms of low-input, labour-intensive crop and livestock production which have characterised most of rural Europe for many centuries, and upon which its landscapes, its biodiversity and cultural heritage, and the quality of its soils, water and air have depended.

Intensification and specialisation

This involves the development of capital-intensive and geographically specialised farming, leading to problems for landscape and biodiversity, but also for soil, water and air (CEC, 1999). Such farming systems mainly apply to large-scale arable or horticultural production on the most fertile or accessible land and very large numbers of stock are concentrated on relatively small areas of land or are kept in large buildings for all or most of the year. Intensive cropping systems often involve significant modification of water resources - increased irrigation in arid areas and for horticulture, and widespread land drainage in wetter areas - and the application of fertilisers and pesticides to arable, horticultural and fodder crops, including grass. Intensive livestock systems also produce large quantities of manure and other wastes and a range of veterinary products are used. Large areas of land, which are drained, ploughed and managed by large machinery can be prone to soil erosion by wind and water.

Table 4.1 Structure of agricultural holdings in the EU in 1997, classified according to ESU[a)]

Country	Farm size (ha)			All holdings	
	< 4 ESU	4 - 40 ESU	> 40 ESU	Total number (1,000)	Average size (ha)
Belgium	2.0	11.7	36.4	67.2	20.6
Denmark	6.0	19.6	78.2	63.2	42.6
Germany	3.5	18.1	99.1	536.1	32.1
Greece	1.6	7.3	30.7	821.4	4.3
Spain	6.7	23.5	166.8	1,208.3	21.2
France	3.7	30.7	90.3	679.9	41.7
Ireland	10.4	28.2	72.8	147.8	29.4
Italy	1.8	10.3	60.2	2,315.2	6.4
Luxembourg	4.4	26.5	76.4	3.0	42.3
Netherlands	2.4	7.4	26.0	107.9	18.6
Austria	10.9	17.4	48.3	210.1	16.3
Portugal	2.5	12.8	156.7	416.7	9.2
Finland	7.8	22.2	45.7	91.4	23.7
Sweden	7.6	31.6	97.0	89.6	34.7
UK	18.4	46.2	149.1	233.1	69.4
EU-15	3.6	18.1	90.5	6,990.8	18.4

a) Agricultural holdings are classified according to the economic size and type of farming. Coefficients are determined in terms of €, per hectare or per animal. The economic size of a holding is converted into Economic Size Unit (ESU) (1 ESU = € 1,200).
Source: Eurostat (Eurofarm), Luxembourg; adaptation LEI.

Intensification of production is mainly observed in regions where agriculture is most productive but it is also a phenomenon in some marginal areas, particularly in southern Europe. Some European regions have a competitive advantage compared to others through better biophysical conditions, more rationalised farm structures and the integration of primary production with food processing industries or through well-equipped farm extension services. Pig production, for example, is largely concentrated in regions with the available infrastructure for production and processing industries, with easy access to the main consumption regions (e.g. Denmark, the Flanders region in Belgium, the Netherlands, Bretagne in France and the Po Valley in Italy). These regions also have highest stocking densities (see also Figure 4.1).

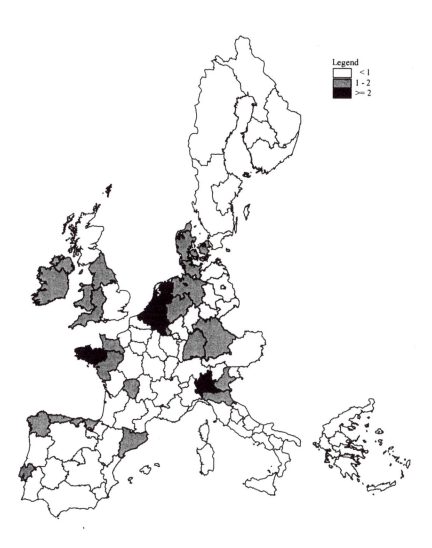

Figure 4.1 Livestock units per hectare of utilised agricultural area in the EU (LU per ha)

Source: Eurostat (Farm Structure Survey); adaptation LEI.

Marginalisation and abandonment

This tends to occur in remote areas or on less fertile land where traditional extensive agriculture is threatened by its inability to compete effectively with intensive production in other regions. In these areas, farm incomes are low and there are few incentives for young people to take on farms from the previous generation. As older farmers retire, land may be abandoned, leading to the loss of traditionally managed, semi-natural habitats and an increased risk of disasters such as fires, particularly in arid regions. Alternatively, land may be consolidated into larger holdings which are managed with much less labour so features and habitats become degraded - a style of farming which has been termed 'ranching'. Among northern Member States, significant areas of farmland in Finland, Ireland, the mountainous or hilly parts of Germany and the UK suffer from this phenomenon and it is also recognised as an increasing threat to the mainly small, mixed farms of Austria. In the south, marginalisation and abandonment are significant problems across much of the interior of southern France, the Iberian Peninsula and Greece, and in many parts of Italy.

Generally speaking, intensive agriculture has increased in Europe over the past few decades at the expense of more traditional systems. Whilst in some northern regions this trend has now slowed as most farms have restructured, it continues in southern Europe where many farms remain small, diverse and heavily dependent upon labour. Austria is unusual, in that despite having many small farms and high labour use, farm sizes and types have remained fairly stable over the past 10 - 15 years. Austria appears to have experienced a lower incidence of both these trends than other Member States, perhaps because of the willingness of its consumers and taxpayers to support the maintenance of Austria's particular style of farming, to a greater extent than is evident elsewhere. For example, 8% of Austrian farms are organic.

Pressures on the environment are observed with intensification and abandonment

The spread of intensive methods on crop and livestock farms has led to a loss of biodiversity and increased pollution in many Member States. It has also increased the energy used in the sector and its contribution to major problems such as global warming due to greenhouse gases, the degradation of river, sea and ground water, soil erosion and contamination, and acid rain. In areas including central Germany, southern England, northern France, Denmark and Flanders in Belgium, intensification within the last 20 years has also involved a significant loss of permanent pasture land to cropping, which has increased its vulnerability to many of these other problems.

In many southern Member States, the most significant change in land use in recent years has been associated with the development of irrigated and highly intensive horticulture on flat areas of land (often along coastal strips), while the traditional, small-scale irrigated polyculture, which was practised further inland and often on sloping ground, has declined.

At the same time, the area of marginal land in Europe that is threatened by abandonment and inadequate management has increased, due to increased competition within the single European market and in the wider global economy. Abandonment, degradation and economic decline now threaten both the extreme north and south of Europe and especially those areas where harsh natural conditions, poor soils and remote locations increase the costs of agricultural production and rural populations are falling. These include central and northern Finland and Sweden, the German Alps, western Ireland and the extreme north of the UK, as well as much of inland Spain, Portugal and Greece, and parts of Italy and southern France.

The European model of agriculture

European agriculture is characterised by a broad heterogeneity of production systems with wide-ranging geographical features. Intensive production systems tend to put pressure on the environment, whereas traditional farming practices may jointly provide agricultural commodities and environmental goods and services. The multi-functional nature of European agriculture and its role in conserving the countryside is vital to understanding agriculture's role in society, and its importance for the economy and the environment. This was acknowledged with the reform of Europe's agricultural policy under Agenda 2000, which highlighted the importance of the multifunctional 'European model of agriculture' as a key feature for agricultural development in the near future. In this context, emphasis has been placed upon establishing an economic sector that is versatile, sustainable, competitive and dispersed throughout Europe, which should be capable of maintaining the countryside, conserving nature and making a key contribution to the vitality of rural life.

The European Council at Cardiff in June 1998 endorsed the principle that major policy proposals made by the European Commission should be accompanied by an appraisal of their environmental impact. A report from the European Commission also proposed a strategy for environmental integration and sustainable development within different policy areas, including agriculture (CEC, 1998b). Sectoral practices are now identified as being the origin of most environmental problems in Europe and therefore they are increasingly seen as the source of the solutions, as well.

ENVIRONMENT AND HUMAN-HEALTH LEGISLATION

A range of policy instruments is used in Europe to arrest environmental damage, reverse environmental damage and improve human-health. Particularly within the past 20 to 30 years, policies have been devised both at European and at Member State level, with varying degrees of success. Some Member States have seen improvements in recent years while in others, negative environmental impacts are increasing. Overall, current agricultural practices and trends still negatively affect the state of the environment.

The mix of environmental and health policies applied in Europe reflects a balance of regulatory approaches - following the polluter pays principle - and voluntary approaches (such as grant schemes and free advice). This reflects the relationship between natural resources and the tradition of active but extensive management, which has maintained and enhanced them over many centuries.

Generally speaking, regulatory approaches are more common in relation to the protection of physical resources from significant and dramatic pollution or contamination (water, soil and air) and the prevention of dramatic irreversible losses of remaining valued landscape qualities, features or nature conservation sites. Measures to protect human health and animal welfare also tend to be of a regulatory nature. Site safeguard by public purchase or by grant-aiding the purchase of land by voluntary conservation groups has been of relatively minor overall impact, although in some countries such as the Netherlands and the UK it has been a significant factor in protecting some fragmented habitats. Taxes and charges have been used in some particular sectors in a few Member States, particularly in relation to water use and energy consumption, and in more recent years taxes have been applied to fertiliser or pesticide usage in a few countries including Denmark and Sweden. Planning controls, particularly to regulate the development of intensive indoor livestock units, are widely applied within the Member States in order to protect landscape quality in certain areas, or to prevent nuisance to nearby residential settlements, including noise and odour. In some countries these controls also extend to other land-use changes. For example, in the Netherlands there are controls to prevent the ploughing of meadows and in the UK there are some controls on the removal of hedgerows. In some States a range of such controls can apply at the regional or more local level, as determined by local government.

Regulatory approaches are applied to agriculture in the EU to protect human-health and animal welfare. Standards are set in relation to hygiene, disease control and safety and can affect many aspects of farming practice. Many controls apply to different kinds of building, including grain stores and livestock housing as well as waste storage and treatment facilities and dairy parlours. In general, most of these areas are now covered by EU legislation, but Member States may retain some discretion in how they achieve and enforce EU standards (e.g. whether the farmer or the state must pay for enforcement).

However, particularly within the last 10 years or so, policies which offer voluntary incentives to fund specific land management practices have become one of the most common policies for achieving wider landscape, biodiversity and general environmental benefits from land management. The most notable expansion of these kinds of policy has been under the EC's Agri-environment Regulation 2078/92. This regulation was introduced as an accompanying measure to reform of the Common Agricultural Policy (CAP), in 1992 and is strengthened and enlarged following the reform of the CAP under the Commission's Agenda 2000 package, as a single chapter within Regulation 1257/1999 on Rural Development. More than 1.3 million agreements have been made in the context of Regulation 2078/92. The agreements cover around 17% of agricultural holdings (CEC, 1997).

A common feature of almost all EU environmental policy approaches has been the provision of advice and information to farmers and other businesses. This is often linked to time-limited aids to encourage environmentally-beneficial structural adjustment on farms, such as the installation of new waste storage and treatment facilities or the restoration of valued habitats including wetlands and native woodlands. Such advice is frequently, but not always, paid for partly or fully by the State on the grounds that it promotes public benefits that go beyond what it would be in the farmer's own interest to pay for.

At the same time, there has been an increasing range of commercially driven initiatives linking environmental quality to product marketing. These have been applied both to large-volume, commodity outputs (grains and vegetables) where standards are increasingly established through 'producer protocols' and farm assurance schemes; as well as to small-volume, speciality outputs (regional cheeses and meats, unusual horticultural and other crops and organic foods) where the standards are more closely linked to individual product branding. Voluntary agreements, among others, are developed between growers and retailers. Some quality assurance schemes, for example, are based on protocols of Good Agricultural Practice (GAP) of the European Retailer Produce Working Group (EUREP). A reduction in the use of agrochemicals is a major goal in the EUREP GAP initiative. Retailers and food processors are also demanding better and audited farming systems in response to changing consumer 'demands. The adoption of Integrated Crop Management or Integrated Pest Management techniques on farms qualifies for the formulation of retailers' GAP standards (EUREP, 1999). Co-operative agreements also exist between farmers and water supply companies, in some countries. These agreements prevail in Germany, where compensatory payments may be provided to farmers for taking preventive measures to reduce pollution. Such agreements include contracts between drinking water suppliers and farmers to change farming practices, based on self-regulation and voluntary participation (Heinz, 2002).

The government in the Netherlands has taken a particularly active stance in supporting these kinds of private-sector, producer-led environmental initiative in recent years and promoting their potential benefits over state-led policy measures. The phenomenon seems likely to increase throughout the EU in future, which could potentially affect future environmental policy developments. A Communication of the European Commission concluded that environmental agreements 'can offer cost-effective solutions when implementing environmental objectives and can bring about effective measures in advance of and in supplement to legislation' (CEC, 1996).

Environmental, animal welfare and health legislation

Within the EU, the standards affecting farmers are generally embodied in legislation or in land use planning procedures, rather than in voluntary agreements between the agriculture sector and public authorities. Nonetheless, the legislation, which forms the main subject of this chapter, is accompanied by supporting policy measures and a growing number of more informal arrangements, as discussed above. Voluntary produce assurance schemes may embody standards for different foodstuffs marketed under particular labels. In some cases, the standards may apply to a significant proportion of the total output of a commodity or a region. There are also instances where farmers are obliged to meet certain standards set by purchasers with a powerful place in the market. There are many cases where pesticide residue standards in foodstuffs or hygiene standards in food production and processing are required by major supermarket groups, which create more stringent requirements for farmers than those laid down in national or EU legislation.

The EU's stance towards many health issues, most notably genetically modified organisms (GMOs), has tended to be increasingly sensitive to consumer concerns, now commonly adopting a precautionary stance to the control of these issues at source. However the response of different Member States varies, with the nordic countries generally more precautionary than the southern and central ones, where some kinds of disease or welfare problems associated with modern methods are seen more as acceptable hazards. For example, in some Member States it is now accepted that salmonella is more or less endemic to intensive livestock rearing and therefore policies concentrate upon the containment of any epidemics, while in Sweden, a large-scale salmonella eradication programme has been instituted. Likewise in relation to welfare issues, Denmark has banned battery egg production and the UK has banned pig production using tethers and stalls, while these practices remain widespread in other States. Sweden also banned the use of all antibiotics as growth enhancers in livestock production before it joined the EC. This country has also played a strong role in achieving recognition of this issue within the Union, such that all Member States agreed in January 1999 to impose an EU-wide ban on four such substances.

The growing use of a precautionary approach in EU health and welfare standards applied to agriculture no doubt relates to the serious and lasting effects upon consumer confidence and agricultural markets of the problem of BSE. This is a prime example of an issue where previously lax policies encouraged an undue confidence in livestock production safety up to a point where the disease was able to become a serious epidemic in a few Member States and consumer confidence was very badly affected across the whole EU. The EC policy developments in relation to tighter standards for antibiotic use, for animal welfare and towards GMOs may be interpreted as moves made in order to avoid undermining consumer confidence in food still further, following the severe problems of the BSE crisis. Another issue where the same principle has applied is that of the use of growth hormones in beef production, which is banned in the EU in relation to both domestic production and imports, pending the findings of further research.

The precautionary principle is important in the context of environmental policy in the EU, primarily when scientific evidence regarding the health effects and/or ecological impacts of certain substances is inconclusive. According to this principle, such substances may not be introduced unless scientific evidence regards them to be safe.

EU legislation and the implementation of national standards

This chapter is concerned primarily with EU legislation. However, it must be emphasised that the standards faced by farmers in any Member State are likely to arise from a combination of EU and national or regional standards since the EU does not have unique competence with regard to legislation in all the relevant fields. Many environmental and health standards are laid down at EU level but a significant proportion, such as the standards applying to irrigation, to protected areas and to various aspects of odour and nuisance, are determined by national or regional authorities. In several cases, such as pesticide law, authorities at a variety of different levels may be involved in setting the standards that apply to farmers. Furthermore, legislation at Community level, where it exists, does not necessarily result in the application of identical standards at farm level. EU legislation takes a variety of different forms, the two most relevant being Regulations and Directives:

- *Regulations* are the dominant form of legislation in agricultural and health policy but they are much less prevalent in environmental policy. About 10% of environmental law takes the form of regulations (CEC, 1998a). They are directly applicable as law in the Member States and allow rather limited discretion for national authorities to depart from the obligations set out in the text. Council Regulations are legal mechanisms which establish uniform rules, and are enforceable throughout the Union. Normally, they are used when a unified system is required (e.g. eco-label schemes or trade regulations regarding endangered species).
- *Directives* are the primary forms of legislation for most EU environmental policies for the achievement of environmental quality objectives. They are binding as to the results to be achieved, but they leave to the Member States the choice of form or method to be used. They must be sufficiently flexible to take into account the various legal and administrative procedures and traditions that are applicable to Member States. Also, they may contain different requirements, which may take into account the different environmental and economic conditions in Member States. States may choose to adapt their existing legislation, introduce new legislation or pursue the required goals through administrative procedures. Frequently, Directives allow Member States two years in which to meet the requirements set out in the text. As a result, there can be significant differences in the legislation in place in different parts of the EU. Furthermore, implementation is not always satisfactory and may be subject to delays, as in the case of the Nitrates Directive, 91/676.
- *Decisions* differ from Regulations and Directives in that they tend to be quite specific in nature and targeting. For example, the Commission has agreed a Decision to consent to the placing on the market of products, containing or consisting of, GMOs. However, Decisions alone are not very common in the context of environmental legislation. The Commission may occasionally implement environmental Regulations and Directives through Decisions.

Since environmental policy relies heavily on Directives, leaving Member States with considerable scope for choosing the most appropriate form of implementation, there is a greater possibility of variations in standards than would be expected if the legislation took the form of Regulations. For example, the Commission has reported that in 1995 only 91% of the then total number of Community environmental Directives had been the subject of transposition notifications from the Member States. Such notifications indicate how the EU measure has been implemented at national level. Consequently, around 22 Directives appear not to have been transposed into national measures in some Member States. A total of 265 suspected breaches of Community environmental law were registered by the Commission in the same year. Indeed, the Commission has over 600 environmental infraction cases outstanding against the Member States at any one time (CEC, 1996).

ENVIRONMENTAL, ANIMAL WELFARE AND HUMAN-HEALTH CONSTRAINTS

Seven themes were selected for inclusion in the analysis, covering a broad range of individual issues:

- water quality (notably nutrient enrichment by nitrates and phosphate, sediments, pesticides) and quantity (irrigation);
- soil quality (notably salinisation and soil contamination) and soil erosion;
- air quality (notably odour and noise, ammonia, pesticide drift and crop burning);
- biodiversity and landscape;
- GMOs;
- animal welfare (notably housing, transport and slaughter); and
- human-health (notably hormones and animal feed ingredients, pesticide residues in food, applicator safety, hygiene rules for dairy farming and veterinary requirements).

An overview of the main environmental and health-related issues of concern is provided in the following text. Wherever possible, farm level operational constraints are identified.

Water quality and quantity

Nutrient enrichment by nitrates and phosphates

Nutrient enrichment by nitrates and phosphates is a high priority issue for the EU, since contamination of both ground and surface waters and soils is a serious problem in parts of the EU. Linkages between farm management practices and leaching of nitrates are highly complex and subject to large variations, depending inter alia on soil types, climatic conditions and farming practices. Model calculations indicate that the maximum admissible concentration of nitrates for drinking water - 50 mg l^{-1}, following the standards of the World Health Organisation (WHO) - are exceeded on about 20% of agricultural land (Stanners and Bourdeau, 1995). Contamination occurs particularly in regions where there are concentrations of intensive livestock production (mainly pigs and poultry) or large areas of specialised crop farms (including intensive horticulture). Affected regions include parts of Belgium, Denmark, France, Germany, northern Italy, the Netherlands, coastal Spain, and the UK. In these areas, heavy nitrate and phosphate loading of soils is common and this can lead to significant water pollution problems. The main causes include high production levels and use of manure and chemical fertilisers. Nitrate pollution of surface waters is highest in the northern part of Western Europe, where much intensive agriculture is concentrated. Excess phosphate from manure is also leading to eutrophication problems in some Member States, including pollution along the Baltic coast and part of the Adriatic Sea. The Nitrate Directive provides an EU framework for controlling pollution, with potentially significant constraints on farmers in Nitrate Vulnerable Zones (NVZ), which cover sizeable land areas. A wide range of on-farm constraints applies to farming to meet the requirements of policies to control nutrient enrichment by nitrates and phosphates and associated pollution from livestock wastes and fertilisers. Many of them are multi-purpose. However, implementation of the Nitrates Directive by many Member States is currently not complete.

- *Manure application requirements and limits on nutrient surplus*

The application of manure to agricultural land is one of the most important and direct ways of creating potential pollution problems from nutrients. There is an increasing trend to seek more control over manure applications so as to limit the application of nutrients and control leaching of nutrients to the environment. Some manure application requirements at EU level result from the Nitrates Directive and these are applicable to NVZs, but more general limits not otherwise set at EU level. In NVZs, Member States must devise and implement Action Programmes affecting all farms, and manure may not be applied in amounts exceeding 170 kg N ha^{-1}. Some Member States or regions, particularly in Northern Europe, put additional restrictions on the maximum amount of animal manure which may be applied, restrictions on fertiliser applications, and restrictions on spreading of animal manure. In Germany, for example, the application of livestock manure must not exceed 170 and 200 kg N ha^{-1} on arable land or grassland, respectively. Several EU Member States have also put limits on the application of manure, for example in the form of stocking densities (e.g. Belgium, Denmark, Germany, the Netherlands, France and Italy), resulting in a range of different standards. On-farm constraints show wide regional variation due to the diversity of pressures on the environment (with highest potential for leaching of nitrates on the sandy soils) and the intensity of livestock production.

- *Production permits for intensive livestock units*

Permits are applied to control the environmental impact of certain categories of livestock production and impose standards on producers. Such permits are often required for constructing new facilities only, but may also be needed for continuing an existing operation. More rarely, they impose constraints on current production units. At EU level, the IPPC (Integrated Pollution Prevention and Control) Directive requires Member States to impose their own emission limits in environmental permits which are mandatory for potentially polluting plants of a given scale. The Directive is applicable to installations for the intensive rearing of poultry and pigs with more than (a) 40,000 places for poultry; (b) 2,000 places for production pigs (over 30 kg) or (c) 750 places for sows. It is now coming into force over the next five years for new buildings and it will gradually also incorporate all existing installations. These producers require investments so that they are able to apply best available techniques (BAT), to control several forms of pollution in an integrated way. Water pollution and emissions of ammonia are both included.

Nearly all Member States have additional or more local controls on the construction of intensive livestock units, which may include specific permitting systems and/or procedures established under land use planning legislation. This legislation mainly limits the expansion of current production levels. In the Netherlands, where there is a particular concentration of livestock production, there is a kind of moratorium on new units and increases in livestock production units are limited to the equivalent of 125 kg of phosphates from livestock manure per ha.

Manure storage requirements exist in most Member States, often associated with permits for livestock farms, mainly in regions with a concentration of intensive livestock production (to control nuisance from odour and meet restrictions on manure application during part of the year).

- *On-farm nutrient budgets*

Farm nutrient budgets are a tool used primarily to improve management of the input and output flows of nutrients by agricultural holdings. Such accountancy systems may guide the work of, and influence the nature of the advice given by, extension services. Alternatively, they may also be used as a basis for compulsory measures in cases where nutrient budgets exceed certain thresholds. Hence, there is a relationship between nutrient budgets and on-farm constraints.

There are no mandatory on-farm nutrient budgets required at EU level. However, they are now compulsory in some EU Member States, usually for holdings meeting certain livestock density or farm size criteria. An example is the Netherlands, where, as of 1998, a 'mineral declaration system' is required as a tool to identify mineral balances for all livestock holdings with livestock densities exceeding 2.5 LU ha^{-1}. This system gradually developed into an accounting system including all livestock and crop production in the Netherlands. Farms with excess amounts of manure need to dispose of surplus and many intensive livestock producers have to meet subsequent transport and disposal costs. Mineral balances also need to be developed on holdings in Germany of a size exceeding 10 ha, and also wherever the supply of nitrogen from livestock manure exceeds 80 kg of nitrogen per ha. Sanctions apply to holdings where the spreading or disposal of manure may cause damage to the environment.

- *Manure storage requirements*

Storage of livestock manure in solid or liquid form is required in many Member States because of constraints on the application of manure on the field (in some countries this is forbidden during part of the year due to the high leaching potential) or to control odour and emission problems. Manure storage requirements commonly are needed in countries where climatic conditions or environmental constraints are a limiting factor for the application of livestock manure during part of the year (e.g. nordic countries where the ground may be frozen for significant periods). Manure application tends to be restricted during part of the year when the uptake of nutrients by crops would be limited or when rainfall is highest (during part of the autumn and winter period), or when land is frozen and would be vulnerable to the leaching of nitrates. Requirements may include a mandatory level of storage (expressed in months of production for example) or building and equipment standards.

Manure storage requirements are common in the EU and determined at the Member State level. Where there are explicit requirements typically they range between 4 and 10 months. Requirements on manure storage tend to be longest in regions with concentrations of intensive livestock production (mainly pigs and poultry), and tend to be in the range between 6 and 10 months. There, rules that restrict manure application during part of the year require storage of livestock manure. However, in some Member States, there are no explicit manure storage requirements. Storage facilities are rather long in the northern part of the EU because of climatic conditions, which limit the application of livestock manure during a large part of the year. Manure storage regulations may include technical measures to reduce emissions of ammonia, which adds to the investment costs as well.

- *Buffer strips and cover crop requirements*

Buffer strips are a constraint on agricultural land use designed to control nutrient leaching to surface water or in groundwater protection zones. Cropping/grazing may also be constrained or prohibited.

There are varying requirements between Member States regarding the use of buffer strips and crop cover requirements. Neither is specified in EU legislation. Where they are used, buffer strips are normally applied in water catchment areas close to wells, mainly to limit the application of nutrients from fertilisers and livestock manure. For example, farmers may be strictly limited in their use of nutrients inside a zone of 3 to 5 m along banks of rivers and streams. Such buffer zones are normally required along watercourses in certain Member States, leaving the land uncultivated. Some northern Member States have introduced relatively stringent buffer zone standards in recent years. However, there are other Member States where the use of these measures remain rare.

Cover crops are required in some parts of the EU to control nitrate leaching during the autumn or winter period. In Denmark, for example, the proportion of winter crops must exceed 65% on all farms, while in Sweden, at least half of the winter-grown land needs to have green cover during the winter period in the southern and central part of the country. Requirements are less stringent or absent in most other countries, although payments for buffer strips and cover crops are available under several agri-environment schemes, for example in Germany and Finland.

Sediments in water

Sediments are an issue at the regional scale particularly in some regions of southern Europe where the most severe cases of water erosion in recent years have led to water pollution by sediments, leading to flooding and damage to houses and infrastructure. However, no comprehensive rules on sediments apply to farming in the EU.

Pesticides

The health and environmental impact of pesticide use is a continuing concern and a range of measures at EU and national scale reflect the high priority it is given. Pesticides may cause contamination of groundwater resources and surface water across the EU by leaching, run-off and spray drift. Utilisation of pesticides varies greatly within the EU. High levels of pesticides are observed mainly in areas with intensive horticulture, permanent crops and arable farming - including forage maize cropping on livestock farms. Monitoring frequently reveals levels exceeding the very strict standard of 0.1 μg l^{-1} (the EU standard) in between 5 and 25% of the samples taken in regions with intensive arable production and horticulture, including part of northern France and southern England. As a result, drinking water samples may exceed EU limits for commonly used pesticides such as atrazine.

Pesticide drift is a particular problem at the regional and local scale. Impacts of pesticide use on biodiversity and the quality of watercourses are widespread and probably are a significant factor in the decline of many species. In addition, health and safety aspects of pesticide application are a concern for farmers and workers applying pesticides in some regions. As in other countries, there are limits on residue levels in food.

Interest in organic farming is growing rapidly in the EU, partly because of pesticide concern. In 1998, nearly 2% of all agricultural land, on 1.4% of all holdings, was devoted to organic farming (Eurostat, 2001), and organic farming practices were adopted on 2.3 million ha of land. Four countries (Italy, Germany, Austria and Spain) account for almost 70% of the total organic area in the EU.

The use of pesticides on farms is subject to a variety of on-farm constraints, among others relating to the products available, permissible doses, target crops, timing, application techniques, applicator safety, product disposal, contamination of water, soil and vegetation, and residues in food. From the farmers' perspective the restrictions on the use of a particular substance, generally printed on the label and varying considerably between substances, is likely to be of particular significance.

- *Constraints on the use of pesticides*
Within the EU, the majority of on-farm constraints are set at Member State level. Certain policy targets are determined at EU level including maximum acceptable

concentrations of pesticides in drinking water, maximum recommended levels of pesticide residues in food and animal feed and outright bans or restrictions on certain highly toxic ingredients, principally mercury and organochlorine compounds under Directive 79/117. A common authorisation procedure of pesticide active ingredients applies at the EU level, with a common system for evaluating their efficacy and impact on the environment and human health. The 'authorisation Directive' 91/414 provides a legislative framework for this process for agricultural pesticides. More than 100 tests of a product's environmental and health impact must be completed before registration is granted. On average this takes a period of 10 years and costs involved are about € 125 million. However, since the Directive came into force, only about 20 active substances have yet been approved through this route and thus the majority of authorised substances in the EU remain approved under the prior legislative procedure of the individual Member States.

A measure of strategic importance in the EU is the Drinking Water Directive (98/83), as amended, which sets strict standards for maximum concentrations of pesticides, both individually and for all individual pesticides detected and quantified during the specified monitoring procedure. This measure also affects the authorisation procedure under Directive 91/414 which, inter alia, seeks to protect water from contamination. In addition, it has influenced the pattern of legislation at Member State level where a range of different policies is in place. All Member States have legislation and labelling procedures for pesticides but some have more extensive policy measures, including water protection zones where pesticide use is restricted, blanket controls on certain application systems, pesticide reduction programmes, and, in certain cases, taxes on pesticide sales.

Standards for pesticide residues in food (MRLs) are set at EU level and these will influence pesticide use on farms, mainly via the instructions for use printed on labels.

- *Number of active ingredients authorised*

The EU authorisation procedure may reduce the number of active ingredients available for agricultural use, over time. Products may be banned for specific use following a procedure to approve or reject their active ingredients. Depending on the result, this may limit the availability of pesticides for farming, and put a constraint for agricultural production. Any on-farm constraints that might result from the number of active ingredients authorised remain difficult to assess. Some active ingredients on the list of authorised compounds, for example, may hardly be used in agriculture, and the use of authorised active ingredients may be highly restricted. Around 850 active substances (organic and inorganic) are currently authorised for use in the EU. As approval procedures are tightened and authorisations are reviewed under the EU Directive, there will be a downward trend in the number of authorised substances. This may act as a critical limit upon use in some countries (e.g. Germany), especially where particular chemicals are used for growing crops that are limited and thus the market for these pesticides is small.

Atrazine is a commonly used pesticide, which is increasingly restricted at Member State level, following the detection of significant levels of this pesticide in water, in many parts of Europe. The use of atrazine is banned in several Member States (e.g. Denmark and Germany), requiring farmers to search for more expensive alternatives to treat maize. Specific controls also apply to the use of atrazine in some Member States, and this may increase farm costs where alternatives may be limited.

- *Targeted measures affecting pesticide uses*

The great majority of restrictions on the use of pesticides in the EU apply at national or sub-national levels. Under EU legislation, pesticides must be used in accordance with the principles of good plant protection practice, etc.

Spray drift to surface water is considered a very important diffuse source of water pollution from pesticides. Different measures are required to take mitigation action, including the use of buffer or no-spray zones, and low drift application technology. In the EU, such restrictions on the use of pesticides in watercourses are mainly determined at national level or sub-national level. Mandatory inspection of spraying equipment is applied in several Member States. However, major variations exist between Member States concerning restrictions on the use of pesticides. The use of pesticides may be prohibited or severely restricted in environmentally sensitive areas (e.g. tightly-defined water catchment areas) in some parts of the EU, as well as along streams and lakes in some countries. Such zones commonly increase labour costs or reduce revenues from agriculture, but they tend to be relatively small areas by comparison with the total cropped area. Cost-sharing programmes exist in several countries to assist farmers in meeting the requirements of buffer zones along watercourses. Several Member States control spraying of certain pesticides within a certain distance of any watercourse, or prohibit cultivation of the land (e.g. Denmark) in a zone of 2 m along a watercourse. Restrictions on use may also apply to groundwater protection zones (e.g. some products are banned).

Aerial spraying of pesticides is mainly controlled to limit spray drift and reduce pollution effects of pesticides use on air and water. The periodic inspection of pesticide spraying equipment may also be required in some countries. Aerial spraying is prohibited in some countries (e.g. Austria, Greece, Sweden) and tightly controlled in most other Member States. Some Member States require permits with constraints on spraying (e.g. minimum surface to be treated, under certain climatic conditions only) or limit aerial spraying to cases where the direct application of pesticides is otherwise not possible. In these cases, notices have to warn members of the public about spraying activities, before they take place.

- *Safety rules on application, storage and disposal of pesticides*

Safety rules on the application of pesticides are common since occupational exposure is most likely to occur while handling these chemicals during preparation, loading, cleaning and application. Mandatory training and certification is required in some Member States of the EU, by all those who apply pesticides. Periodic approval of the types of spraying equipment used is also required in some Member States. Certification procedures for the authorised use of pesticides have been developed in some Member States.

Strict controls on storage and disposal of pesticides are established in some Member States, where they are mainly applied to protect occupational health but may also fulfill a water protection function. Disposal of pesticide containers may pose serious risks for the environment. In the EU, controls may apply to the disposal and rinsing of packaging. Take-back programmes exist for active ingredients, whose authorisation has expired. Contaminated packaging may need to be delivered to chemical waste sites in many Member States. Containers may also need to be cleaned, under controlled procedures, before they can be collected by a household waste system.

Irrigation

EU countries, on average, abstract around 21% of their renewable water resources, which is regarded as a sustainable position (EEA, 1999). A growing area of farmland in Europe is irrigated, accounting for only about a quarter of the water abstracted, but agriculture is the single most significant user of water in some southern Member States, particularly where horticulture has intensified. Water availability problems thus occur mainly at local and regional scales. The majority of irrigated farmland is in Southern Europe, with over 3 million ha in Spain and around 1.5 million ha in Greece. Water stress occurs when the demand for water exceeds the available amount during part of the year or when water quality is insufficient to meet demands. More 70% of total water abstractions in Greece and Spain are for irrigation purposes (IEEP, 2000). The extraction of water from very limited groundwater supplies is an issue of concern as it can occur at unsustainable levels, may lower the water table and can affect wetlands adversely. This applies particularly to the coastal strips of southern Europe, subject to intensive cropping as well as several other parts of the EU where water for irrigation purposes is primarily drawn from groundwater reserves. Salinisation of water supplies is an important regional issue in several areas, where salts are either drawn in from nearby seawaters, or dissolved from newly exposed soil strata.

There are no rules at the EU level to restrict the quantity or quality of water used in irrigation, but a variety of policies are in place in different Member States, generally reflecting the degree to which water shortage is an issue and irrigation is common. In many Member States, water is viewed as a public resource and permits are required to abstract water for agricultural use; however there are some regions where private abstraction is unregulated. Most state authorities can place restrictions on abstraction during critical periods of shortage. Charges are generally limited to the cost of water administration (e.g. issuing permits, maintaining infrastructure). Few Member States' charges reflect the opportunity cost of water for irrigation, though France and some regions of Spain are now starting to apply limited charges of this nature. Cost-sharing grant programmes and extension services exist in many regions to encourage more efficient use (e.g. adoption of drip systems, water metering). In one area of Spain, farmers are receiving payments for reducing their consumption of irrigation water under an agri-environmental scheme designed to protect an internationally important wetland (IEEP, 2000).

A few countries or regions have general prohibitions on all activities capable of destroying, altering or modifying wetland, marshes, springs and watercourses but it is unlikely that these impose significant operational restrictions on irrigating farms outside a few very sensitive areas.

Soil quality and soil erosion

Salinisation

Salt intrusion is an issue of concern in a relatively limited number of regions in the EU, mainly the south, especially in Spain along parts of the Mediterranean coast. Salinisation is particularly associated with irrigation, e.g. along the Mediterranean coastline, but also occurs in some drained areas in northern Member States, which use seasonal groundwater irrigation for the production of arable and horticultural crops. It is not a major concern at the EU level. Voluntary measures have been taken by the agricultural sector to address these problems.

Soil contamination and acidification

Contamination of agricultural soils by heavy metals and pesticides is an issue in the European Union, as is the diminishing organic content of soils. Soil contamination arises from different sources, including sewage sludge, pesticides and agricultural wastes. The use of sewage sludge has led to some significant incidences of soil contamination in particular regions, particularly in the former East Germany. There are limit values set for concentrations of a series of heavy metals. These apply to the soil itself, to sludge used in agriculture and to the amounts of sludge which may be applied annually to farmland, based on a ten-year average.

EU legislation requires that sewage sludge should be treated before use on farmland. Legal provisions set limits for heavy metals such as copper, nickel and lead, as well as the more toxic elements like cadmium and mercury. Harvesting should not take place less than three weeks after sludge has been spread and there are much tighter restrictions applying to its use on soils used for fruit and vegetable crops. Where limits are breached, sludge may not be applied and a system must be in place to ensure that heavy metal accumulation does not exceed the maximum loads.

Several Member States have adopted Codes of Practice for the use of sewage sludge on agricultural land. Most of the recommendations in Codes of Practice are aimed at those who undertake to apply the sludge to farmland and in the majority of cases this will be a water company or its contractors, rather than the farmer. This may be expected to affect the price at which sewage is supplied to the farm (in that it will include the impact of these extra costs), but generally, sludge is regarded as a low cost option among the choice of fertilisers available to agriculture. Usually, it is free.

Acidification is a concern in certain regions with vulnerable soils including parts of Scandinavia, but most of acid deposition is industrial or urban in origin. Emissions of all gases contributing to acidification are projected to go down and there is a major programme of legislation to reduce SO_2 and NO_x emissions. This should lead to improvements of soil quality. The relative contribution of agriculture to these emissions may increase, however, because of the difficulty to reduce the emissions of ammonia, relative to the emissions from industrial and urban sources.

Soil erosion

In contrast to other major agricultural regions of the world, soil erosion is not a significant problem for the EU as a whole, partly because so much of its farmland is under pasture, and partly because of its stable soils, generally temperate climate and long history of agricultural use. However, soil erosion may be increasing across the EU and the problem is greatest in the Mediterranean regions, in several of which it is a severe environmental concern. The most severe cases of water erosion in recent years have been confined to restricted areas, including parts of Italy and Spain, but have led to water pollution, flooding and slides, and damage to houses and infrastructure. Particularly in hilly drier Mediterranean zones, erosion is commonly associated with the abandonment of land and vegetation management caused by agricultural decline. Soil erosion is linked in part to crop rotations which do not include green cover crops during the winter period, substitution of traditional arable fodder crops by maize for silage and farming of uncultivated land.

Wind erosion is not as significant or widespread a problem in Europe as in dryer parts of the world, but it can cause major damage in small areas. The hazard is greatest in the lowlands of northwestern Europe, in places such as Lower Saxony in Germany (about 2 million ha of land), the Netherlands 97,000 ha, western Denmark (about 1 million ha), southern Sweden (170,000 ha) and south-eastern and eastern England (260,000 ha) (Riksen and De Graaff, 2001). The on-site effects are damage to crops, loss of fertile topsoil and loss of soil structure resulting in yield reduction and the need for additional inputs, and the long-term degradation of the soil. The off-site costs are mainly the result of dust penetration into residential areas and into machinery.

There are no soil erosion standards set at a European level and few standards exist at Member State level, despite the fact that this is an issue of significance to some, particularly southern Member States. There is some debate about the introduction of an EU level soil policy. No specific farm-level rules on sediments in water apply in the EU, although buffer zones and habitat protection measures may also reduce sedimentation in many cases. Soil conservation practices may be incorporated in voluntary farm assurance standards imposed by food processors and retailers for certain crops, and the issue is covered in Codes of Good Agricultural Practice which may be advisory in some Member States (e.g. UK) and quasi-regulatory in others (e.g. Germany).

Air quality

Odour, ammonia and noise

Odour and noise are issues of major concern as a source of nuisance at the local level, mainly in regions with intensive livestock production units. The policy response is also local and varies considerably within the EU. Ammonia is an issue of significant concern mainly in regions with a high concentration of intensive livestock production units, such as the Netherlands. Well over 90% of total EU emissions of ammonia and the subsequent acid deposition originate from agricultural activities. Soils, water and vegetation can all be affected adversely by acid deposition.

Preventive measures need to be taken against pollution by large intensive livestock production units, in particular through the application of BAT, and measures which are adapted to local circumstances and take into account contributions to transboundary pollution. Both voluntary and mandatory approaches to reducing emissions of ammonia may involve building standards, e.g. for pig housing. In the low countries - the Netherlands and Flanders as well as Denmark - northern Germany and some areas of north-western France and northern Italy, the use of shallow and deep injectors, trailing shoe spreaders and band spreaders is increasingly common. Such equipment is more expensive relative to more traditional approaches to spreading manure, but allows its immediate incorporation into soils, and so contributes to lower emission of ammonia and odour nuisance.

Odour is mainly dealt with by a variety of national and local development and land use planning laws, at Member State level. Permits are required for new installations of a certain size. Rules vary between Member States but most are tightened over time. Permits are generally given subject to conditions designed to minimise damage to air quality, including ammonia. Conditions may include the specification of slurry storage capacity, requirements for on-farm treatment facilities. These may give rise to significant on-farm costs.

Crop burning

Burning of crop residues such as straw remains an issue in parts of the EU, but is no longer practised in several Member States such as Germany and the UK, due to national legislation. However, elsewhere it can be a widespread practice reflecting farm specialisation and it can still cause significant nuisance to nearby settlements and road users. It is not a significant issue on the EU scale.

Several countries apply restrictions at a local level to ensure safe burning of crop residues. A total ban on the burning of cereal straw and stubble, as well as the residues of oil seed rape and field beans and peas in arable fields, applies in Germany, and there are similar in the UK and several other Member States.

Nature conservation, biodiversity and landscape

The loss of biodiversity and of cultural landscapes are both issues of major significance in the EU, and a high priority for policy. A significant proportion of flora and fauna depend upon semi-natural habitats and mosaics of farmed and forested land cover. More than a third of bird species are in decline, mainly due to habitat degeneration and land use changes caused by agricultural intensification. In Europe, many of the most valuable areas for wildlife and landscapes are those which have been settled and farmed for many centuries, in which species have co-evolved with traditional agricultural management and where landscapes are dependent on regular management for their variety and interest. Losses of biodiversity and landscape quality have also been widespread because of the removal of landscape features, loss of permanent grassland and destruction of other semi-natural habitats, as well as agricultural intensification on productive farmland. The policy response is divided between the EU, national and more local levels. Marginalisation and abandonment of traditional farming and inappropriate afforestation are also problems in some regions.

Most Member States have legislation designed to protect remaining examples of non-farmed, terrestrial valuable natural habitat, including the core areas of those national parks which meet the IUCN category 1 and 2 definitions (this does not apply to all EU Member States' national parks), wetlands of international importance (e.g. RAMSAR sites) and other key sites. The majority of these areas will be included in the priority habitats for consultation listed in Annexes to the Habitats and Birds Directives (92/43 and 79/409 respectively), which Member States are required to take action to protect. Controls will generally include requirements for environmental impact assessment of any significant proposed development and a presumption against both agricultural and non-agricultural development of these areas, except in cases of overriding national interest. However in terms of their area, these sites are less important in many Member States than those sites occurring on farmed land including large areas of grazed vegetation. Sweden and Finland, which still have a predominance of forested rather than farmed habitat, are notable exceptions to this rule.

Endangered species and alien species

Under the EC's Habitats and Birds Directives Member States face obligations to take steps to protect all plant and animal species of Community interest which are listed in annexes to the Directives, as well as the habitats upon which they depend for feeding and breeding. These species include both terrestrial and marine, indigenous species and important migrants, identified because they are rare, threatened or vulnerable at a European level. Protection measures include direct prohibitions on capture/collection, killing, disturbance and the taking of eggs or live specimens from the wild, and measures to protect the habitats of these species through the establishment of Special Protection Areas (in the case of birds) and Special Areas of Conservation (in the case of other species) under the two Directives. Once designated, these areas and the wider countryside must be subject to measures determined by the Member States that will ensure the maintenance of 'favourable conservation status' for all listed species. Such measures are enacted at Member State and at regional and more local levels. They commonly include a mix of regulations to prevent deliberate and irreversible damage to habitats and control direct damage to species, incentives to encourage appropriate management of habitats and species in the form of public-funded compensation schemes to farmers and other land managers, and advice from a variety of public and private sources. The costs of species protection on farmland will generally be shared between the public purse and landholders but the uncompensated cost to farmers is unlikely to be very high in the majority of cases, since many Member States allow for compensation wherever such costs are significant.

Member States are also required to consult the Commission on any proposals to introduce alien species into the EU, under these Directives.

Habitat conservation

In general terms, policies for nature conservation, landscape and biodiversity have increasingly come to have an impact at farm level through constraints imposed upon farming practices, farm development or land use management.

At the European scale, it is probably fair to say that the majority of broad but explicit constraints on land management for nature conservation or landscape purposes are generally compensated for. This is commonly through a direct management agreement with the farmer which involves payment either as compensation for lost income due to the imposition of the constraint, or as payment in return for providing an environmental management service, which covers the cost of provision (in terms of labour, materials etc). In the most extreme cases, compensation may involve a compulsory purchase of protected land by a public authority, but in many others payment is made annually and the farmer retains ownership and the ability to use the land for a level of agricultural production which is compatible with conservation requirements.

The main exception to this rule is the application of land use and territorial planning procedures. These vary greatly between localities, regions and Member States. In some regions, farmers are subject to constraints on the conversion of farmland into woodland and vice versa. On the ploughing of grassland, removal of certain landscape features, on the alteration of watercourses, construction of ponds, etc and on the construction of buildings and roads. In some cases, farmers face absolute prohibitions on certain changes in land use or management, in others they are subject to consent or authorisation procedures. Local authorities may exercise considerable discretion in deciding what forms of development are permitted. These controls often have a landscape or biodiversity element but are usually concerned with a wider range of environmental issues. Typically, constraints are not compensated. More generally, the balance of constraints and compensation arrangements varies between Member States.

At the European level, several mechanisms can potentially be used in order to compensate farmers for the cost of following environmentally-sensitive management practices in certain areas. Most notable of these to date have been agri-environmental schemes, which were first used in a limited number of areas in the late 1980s but which were made a compulsory element of the CAP following its reform in 1992. Agri-environmental aids can be offered to farmers through 5-year management agreements designed to achieve a variety of environmental benefits including reduced fertiliser and pesticide use, the maintenance or reintroduction of extensive farming methods, upkeep of semi-natural habitats and particularly valuable landscapes and the promotion of more sustainable practices including organic farming. The payments are designed to cover the costs of such management, which should go beyond the basic requirements of good agricultural practice. These payments have been used to enhance or maintain management on a significant proportion of EU farmland (now over 20%) including many zones and sites of particular value for nature conservation and landscape.

In relation to specific constraints for the protection of endangered species, the costs to farms will vary greatly according to the species concerned and the level of damage or risk that their protection might bring to farming operations. For example, a requirement not to deliberately damage the nest of a hedgerow bird is likely to cause few costs, whereas a requirement not to damage or disturb the large, labyrinthine burrows of the European badger which may extend some distance into a cultivated field and present a risk to heavy farm machinery, could prove quite costly.

Visual aesthetic

There is no legislation forming part of the Community's environmental policy exclusively concerned with landscape or countryside protection, although Directive 75/268 and Regulations 2328/91 and 2078/92 which form part of agricultural policy include countryside protection among their objectives. However landscape protection is a common element in most Member States' development control policies, and it has a particular link to protected area policies in the EU, many of which are based around the dual concept of landscape protection and the conservation of biodiversity and other environmental resources.

Typical measures which apply through development controls would include restrictions on the siting, design and size of a range of new buildings, including most agricultural ones, in order to minimise their adverse effects upon landscapes. In addition, in some Member States there are planning controls over new afforestation of farmland, and on the deliberate removal of certain landscape features including hedgerows and particularly large or valuable trees. In the Netherlands there is a general prohibition on the ploughing of all semi-natural areas of land cover (permanent pastures, heathlands, etc). In France, landscape protection measures apply to the process of land consolidation programmes, whereby fragmented landholdings are reorganised at Commune (Parish) level via a process of exchanges allowing farmers to rationalise their holdings. In the past this had led to widespread landscape damage as field boundaries and green lanes were removed once holdings were regrouped. Since 1984 a national law has introduced compulsory environmental impact assessment to the process and steps are now frequently taken to avoid these problems through prior agreement between farmers and the authorities.

Policies to protect habitats and landscapes within the farmed area

The level of on-farm measures applied to EU agriculture for reasons of biodiversity and landscape protection is particularly apparent. Many EU policies seek directly to modify farming practices in order to conserve biodiversity, protect species and maintain landscapes. As noted above, the EU's Habitats and Birds Directives place an obligation on all Member States to identify and maintain at 'favourable conservation status' a large number of important habitats and species, many of which are found on farmland. To date, most Member States are still in the process of devising measures to address this obligation fully, and it seems likely that compensation will be required in most cases to achieve the goals of the Directives.

Many of the EU's agri-environmental programmes, funded through the CAP, support the maintenance of traditional, extensive forms of farmland management in order to maintain semi-natural habitats such as heathland, permanent unimproved grassland, grazing marshes and extensive dryland arable systems. The prevailing economic trends towards farm enlargement, specialisation and intensification on the best land, and abandonment at the margins, has threatened many such habitats and public payment is used to maintain their economic viability and ensure their sensitive management. In an effort to reverse the significant decline in the extent of these habitats in recent decades, many agri-environment schemes also offer payments for habitat restoration and creation on farmland.

In addition, some Member States have imposed obligations upon farmers without compensation. For example, in the Netherlands farmers must not plough valuable meadowlands, which are managed for agriculture. In the UK, the most important hedgerows on farms are now protected from deliberate removal through legislation. Legislation is also in place to protect margins alongside watercourses in Denmark and several other Member States, and in some Regions there are specific local regulations to preserve landscape character, such as protection measures for trees, prohibitions on ploughing grassland and landscape conditions controlling the erection or style of new farm buildings. In Germany, some regions have lists of key 'biotopes', which must not be damaged by agricultural intensification, and these include managed as well as 'wild' habitats.

Genetically modified organisms (GMOs)

GMOs are defined in the EU as organisms in which the genetic material has been altered in a way that does not occur naturally by mating and/or natural recombination. There are currently widespread public concerns in many countries in Europe about the potential risks to human health and to the environment that may be posed by the production and consumption of genetically modified food products. Public concerns on GMOs have gained importance recently, following widespread publicity, because of food safety, environmental sustainability and ethical aspects related to the production methods applied (Blandford and Fulponi, 1999). Consumers, for example, have complained that genetically modified products have been forced upon them and that the labelling is not transparent. In the case of human-health, recent papers by scientists have offered conflicting assessments about the potential risk in relation to the environment. It is a fast moving major debate, with key measures adopted at the EU level, but important variations between countries.

As a result of all these concerns, adoption of the technology is currently minimal and likely to remain so for several years. Some 2,000 ha of *Bt* maize was under production in France in the late 1990s. By the end of October 1998, in total some 30 GM crops were approved or awaiting approval for commercial release in the EC (House of Lords, 1998). The development of GM crops for commercial use in the EU accounts for less than 0.1% of total land used to grow these crops, and the approval of new GM varieties was stalled by a decision by EU environment Ministers to revise the procedures. Some Member States have now signalled their domestic concern to restrict any further use of GMOs by refusing to authorise products already approved at EU level. In other Member States, testing is underway to determine the environmental impacts of crops for which the manufacturers have sought approval.

EC legislation requires that all GMOs and GMMOs (micro-organisms) in the EU must be approved for release before they can be marketed or used. The uptake remains very small. There is a system of controls over authorised use of GMOs at EU level. Several actions are required in the EU before the commercial release of a genetically modified organism on the market is allowed:

- Prior consent is required from the competent authority in the country where the product is intended to be used. The notification should contain a technical dossier of information, including an environmental risk assessment, appropriate safety and emergency response, and in the case of products, precise instructions and conditions for use, and proposed labelling and packaging. Consent can only be given by the competent authority once it is satisfied that the release will be safe for human health and the environment.
- Member States must undertake all appropriate measures 'to avoid adverse effects on human health and the environment' which might arise from the release or placing on the market of the product (Article 4 of Directive 90/220).
- Once these steps have been taken, Member States act as rapporteurs and can recommend approval of a GMO product to the European Commission. A qualified majority of Member States takes decisions on market approval for any recommended GMOs, which should then be authorised throughout the EU.

The relevant Directive, 90/220, was extensively amended in 2001 and new and generally more stringent procedures appear likely to be accepted. One example is labelling. A Recommendation was adopted by the European Commission in 1997 on the compulsory labelling of foods containing GMOs. This has been amended more recently and it is now proposed that all GMO contents in foodstuffs should be labelled wherever they constitute more than one per cent of the ingredients of that product. This legislation is likely to affect costs to food processors and retailers, particularly where they commonly use a significant proportion of imported ingredients from countries like the USA, where GMOs are commonly grown.

In response to consumer concerns, many major EU retailers and processors have begun to make pledges not to use any GM ingredients in their products and have taken steps to replace GM sources with GM-free sources. These are rational market responses which could have cost implications for farmers in that they could significantly reduce the on-farm constraints which otherwise arise from restricted access to GM varieties.

Animal welfare

As consumer concerns about human health risks from intensive livestock production have increased, so has awareness of the potential animal welfare issues surrounding modern production systems. Animal welfare is determined by the capacity of an animal to avoid suffering and sustain fitness. It can be observed through the observation of conditions of animals, their physiological performance and behavioural preferences. On the whole, animal welfare concerns tend to be strongest in the north of Europe, and particularly in those countries with higher levels of income per head, but they are increasing throughout the EU. Public concerns on animal welfare are expressed most powerfully in Sweden and in the UK, exemplified by wide protests against live export of animals. Also, there is an increasing demand for free-range meat in several EU countries. Animal welfare concerns now affect many aspects of modern livestock production, some of the most prominent include:

- Animal stress and increased disease incidence related to confinement and high-density stocking and rapid growth rates of intensive livestock production (poultry, pigs and veal).
- Practices which involve particular 'cruel' practices such as foie gras production from force-fed geese.
- Stress that is caused to animals through long-distance transport to slaughterhouses, and the treatment of animals at slaughter.

The main systems for housing laying hens, which are applied in European production systems, include:

- The battery cage system; more than 90% of all laying hens are kept in battery cages in the EU. This system consists of rows of metal and wire cages that can be arranged up to 6 tiers high. The legal space requirement is at least 450 cm^2 per bird, measured in a horizontal plane and which may be used without restriction. More space per bird will be required in future under legislation now discussed.
- The barn (perchery) system, where marketing rules allows for a maximum stocking density of 25 hens per m^2 of floor space available to the hens (or

400 cm^2 per hen). This system also requires that the housing systems have at least 15 cm of perch space to be provided for each bird.

- The deep litter system, which allows a maximum stocking density of seven hens per m^2 of floor space available to the hens (or $1,425 \text{ cm}^2$ per hen). Housing systems used under this system also need to have at least a third of the available floor area covered by litter material such as straw, wood shavings, sand or turf.
- The semi-intensive systems, which apply to poultry enterprises in which hens have continuous daytime access to open-air runs. The ground to which hens have access is mainly covered with vegetation, and the maximum stocking density is not greater than one hen per 2.5 m^2. The interior of the building must also satisfy the conditions applying to the deep litter and perchery enterprises.
- The free range system applies to hens that are kept for free range egg production with continuous access to the outdoors during the day. Housing conditions, which are laid down for barn or deep litter systems need to be satisfied for free range systems in meeting egg marketing rules. In addition, current marketing rules on the free range system require that hens must have continuous access during the day to an open air run, which must have a stocking density that does not exceed one hen per 10 m^2. In the EU, some 3% of the laying hens are kept in such system, and their shares of production is highest (between 10 and 20% of the total) in Denmark, the United Kingdom and Ireland. The share also is above average (6% of the national flock of kept laying hens) in Austria.

There are also measures setting standards for other animals. For example, there are measures setting down minimum standards for pigs, Directive 91/630. This too is likely to be amended soon but at present it sets out the following principal requirements:

- minimum surface requirements for different categories of pigs;
- a prohibition of the construction or conversion of installations in which sows and gilts are tethered;
- castration of male pigs aged over 4 weeks may be carried out only under anaesthetic by a veterinarian or a qualified person.
- a prohibition of routine tail docking and tooth clipping; if tooth clipping appears necessary, it must be carried out within 7 days of birth;
- piglets must not be weaned from the sow at less than 3 weeks of age.

The welfare of animals was the subject of a political agreement by the EU Member States at the Amsterdam Summit in 1997, resulting in a new, albeit very short, declaration being annexed to the Treaty. A Protocol was added to the Amsterdam Treaty on Improved Protection and Respect to the Welfare of Animals which states that:

> ... the European Community and the Member States shall in formulating and implementing the Community's agriculture, transport, internal market

and research policies, pay full regard to the welfare requirements of animals, while respecting the legislative or administrative provisions and customs of Member States relating in particular to religious rites, cultural traditions and regional heritage.

Transport of farm animals

The transport of farm animals is subject to EU legislation, notably an animal transport Directive. The following requirements apply to the transport of cattle, pigs and poultry:

- Journey times for animals shall not exceed 8 hours.
- The maximum journey time may be extended where the transporting vehicle meets the following requirements:
    - sufficient bedding on the floor;
    - sufficient feeding for the animals transported for the journey time;
    - there is direct access to the animals;
    - adequate ventilation is possible, which may be adjusted depending on the temperature;
    - there are movable panels for creating separate compartments;
    - vehicles are equipped for connection to a water supply during stops; and
    - sufficient water is carried for watering during the journey if pigs are being transported.

Pigs may be transported for a maximum period of 24 hours, on condition that, during the journey, they must have continuous access to water. Other animals need to be given a rest period of at least one hour after 14 hours of travel, sufficient for them to be fed and to be given liquid. They may again be transported for a further 14 hours.

General provisions that apply to the transport of farm animals include:

- Only healthy animals may be transported.
- It is prohibited to transport an animal in conditions liable to cause it unnecessary suffering.
- If emergency slaughter is necessary, this should be undertaken in a way, which does not cause animals any possible unnecessary suffering.

There is considerable dissatisfaction with these standards and amendments to the Directive, particularly for vehicles travelling more than 8 hours, are likely to be proposed in 2001.

Slaughter

The slaughter of farm animals is an important animal welfare issue. The general aim of policy is that animals must be spared any avoidable pain or suffering during slaughter. In the EU, this legislation requires that the construction, facilities and equipment of slaughterhouses, and their operations, shall be such as to spare animals any avoidable excitement, pain or suffering. Rules are applicable to solipeds, ruminants, pigs and poultry brought into the buildings for slaughter. Commission experts, who make on-the-spot checks intended to ensure relatively even application of the requirements, enforce legislation. A representative sample of establishments may be checked to ensure that the requirements are fulfilled. Special provisions apply to slaughter according to certain religious rites with arrangements giving religious authorities in the Member States a particular role in the application and monitoring of the special rules.

Human-health

Hormones and animal feed ingredients

In addition to the environmental concerns associated with farming systems, there has been an increasing range of public concerns regarding the risks to human health arising from modern farming methods. Generally speaking, European consumers have a relatively high level of concern about health issues related to residues in foodstuffs. Leaving aside GMOs there are significant issues of concern relating to:

- the potential residual effects of hormones which may be used to stimulate animal growth or milk outputs;
- antibiotics for veterinary purposes or for use as growth promoters in livestock feed;
- pesticide residues in food; and
- other veterinary residues.

Of increasing concern have been the risks of disease contamination in meat, eggs and dairy products. Salmonella in eggs and poultry, bacterial contamination of meat and tuberculosis in cattle have all been subject of numerous consumer campaigns. The use of bovine somatotrophin (BST) is considered unacceptable in the EU, and there is widespread antipathy to the use of growth-promoting hormones. There is a considerable body of EU legislation in this sphere, including bans on the use of BST and certain growth promoting hormones. The EU has banned imported beef produced with artificial growth hormones since the late 1980s because of potential dangers to human-health.

Both the placing on the market and the administration of BST is prohibited in the EU, as is the use of growth-promoting hormones in beef production is banned in the EU. The ban is applicable to the use of six hormones, including 17ß-oestradiol, oestradiol, progesterone, tesosterone, zeranol and trenbolone. The import of beef from the USA and Canada was subject to a virtual ban, because of the potential risks from the permitted use of 5 hormones in beef production.

Four antibiotics were forbidden as feed additives in the EU from the beginning of 1999, including bacitracin zinc, spiramycin, virginiamycin and tylosin phosphate. Circulation was permitted until the middle of 1999. Three EU Member States have implemented bans of their own separately from the EU legislation. Sweden has a total ban on the use of anti-microbial feed additives. The Danish farmers' union established voluntary programmes in 1997 to ban use of anti-microbial feed additives for all poultry, cattle and fattening pigs. Finland had banned two products, i.e. spiramycin and tylocin phosphate prior to the EU legislation (Agra Europe, 1998). Clearly these restrictions will affect the growth rate of livestock and their feed conversion efficiently, potentially reducing farm competitiveness.

Hygiene rules for dairy farming

Hygiene rules apply to a wide range of farm production methods and sectors, particularly cattle, pigs, poultry and other livestock. Hygiene rules applying to pigmeat, however are largely off-farm and apply to the processing stage. In the dairy sector, the relationship with the final product is often closer. The provision of milk produced in accordance with a high level of hygiene and food safety is of significant concern in the EU as elsewhere. Authorities are obliged to carry out controls to ensure that good hygiene practices are applied on dairy farms.

Member States may amend or introduce national hygiene provisions that are more specific than those laid down at EU level provided that these provisions are no less stringent, and do not constitute a restriction, hindrance or barrier to trade in foodstuffs.

Requirements apply to all enterprises involved in the production, processing and marketing of milk and milk products derived from cows, sheep and goats' milk up to the point of retail. They are applicable, inter alia, to cooling tanks, water use and herd standing.

- Animal health requirements for raw milk (e.g. raw milk from cows should be free of tuberculosis and should not show symptoms of infectious diseases communicable to human beings).
- Hygiene on the holding (raw milk must come from holdings which undergo regular veterinary inspections).
- Hygiene relating to milking, the collection of raw milk and its transport from the production holding to the collection or standardisation centre or to the treatment or processing establishment - these rules stipulate the materials used and standards required in milking parlour construction, milk storage facilities and cleaning requirements for buildings, equipment and animals.
- Hygiene of staff, specifically these milking and handling raw milk, who should wear suitable clean milking clothes.
- Standards to be met for the collection of raw milk from the holding or for acceptance at the treatment or processing establishment.

Veterinary requirements and conditions to control animal diseases

Veterinary requirements and conditions to control the occurrence of animal diseases are of significant concern in the European Union. Concerns involve the risks of disease contamination in meat, eggs and dairy products. Salmonella in eggs and poultry, bacterial contamination of meat and tuberculosis in cattle have all been the subject of numerous consumer campaigns. The growing use of a precautionary approach in human health issues pertaining to agriculture stems partly from the serious and lasting effects upon consumer confidence and agricultural markets of the problem of BSE. This is a prime example of an issue where previously lax policies encouraged an undue confidence in livestock production up to a point where the disease was able to become a serious epidemic in a few Member States and consumer confidence was badly affected across the whole EU. Control programmes for salmonella exist in Member States and they are aimed at reducing the extent of salmonella, and their related human health problems.

The EU has several programmes to control the occurrence of diseases in live animals and animal products:

- Salmonella is controlled either on a voluntary basis (e.g. in the Netherlands) or compulsorily (e.g. Sweden).
- *E. coli* is controlled in some countries. Farm sampling is required where *E. coli* is observed on the holding; with government covering the costs of sampling.
- Control of the New Castle disease, either through compulsory or voluntary vaccination policies. EU financial aid is provided for the operation of the Community Reference Laboratory for this disease.

The poultry sector in the Netherlands has developed a national programme to control salmonella in animals, and which is applicable to chicken meat and eggs with the main objective is to reduce the infection of animals from salmonella. The programme aim is to have less than 10% of poultry population infected (Verstegen, 1998).

The overall goal of the Swedish salmonella control programme is to provide consumers with food free from salmonella, and subsequently to prevent salmonella infections in humans. This overall goal is based on the consideration that no livestock products and animals sent for slaughter should be infected or contaminated with salmonella (NVI/SBA/NFA, 1995). A control programme is operational following the introduction of legislation in 1961 and subsequent adaptations. The following rules apply to the control programme:

- Sampling of fattening pigs and cattle slaughtered in Sweden is on a sufficient scale to detect a prevalence of salmonella of 0.1% at a 95% confidence level. Approximately 3,000 samples are taken annually.
- The salmonella control programme in poultry meat is designed to detect a prevalence of salmonella of 5% at a 95% confidence level. Sampling is carried out once per day in plants with a production capacity of at least 100 tonnes per week, but twice per year in plants with a production capacity of less than 5 tonnes per week.

The programme for control of salmonella in poultry (e.g. laying hens, broilers, and turkeys) includes prophylactic measures to avoid the introduction of salmonella in poultry including:

- rules for feed production and transport (heat treatment and hygiene control);
- hygienic rules to protect birds from salmonella infections from the surroundings;
- salmonella free chickens to be delivered from hatcheries;
- hygienic rules to stop spread of salmonella from an infected flock;
- all in - all out principle is applied in all categories of poultry production.

Swedish farmers get compensation for salmonella outbreaks of up to 70% for infected cattle and pigs. The remaining costs they need to meet themselves and they have the option of taking out insurance. Normally, farmers get compensation of 50% for poultry, but this rises to 70% if they apply to the voluntary programme. Participation in such programmes for chicken meat is more than 90% and for chickens themselves covers around 25% of the laying hens in Sweden.

ECONOMIC IMPLICATIONS OF COMPLIANCE WITH CONSTRAINTS

In general, provisions on the control of nitrate pollution in water appear to have the greatest impact in the most intensively farmed regions of the EC. Such areas generally have larger than average farm sizes and relatively low labour inputs per ha, which suggests a low socio-economic impact of environmental measures. However, the sectors likely to be most affected are arable, horticulture, pigs, poultry, dairy and beef. Of these, only cattle farming is characterised by relatively small family farms in some NVZ areas. Therefore, in these areas, the impact of the Nitrates Directive and related measures may be significant for a number of farmers.

Figures on the economic implications of compliance with constraints in NVZs have been assessed for England and Wales by the government (Table 4.2). Restrictions apply to the amount and timing of applications of fertilisers and manure. Total compliance costs are based on the requirements of implementing the Nitrates Directive according to the measures in force in England and Wales.

We will examine the ratio of the relevant total farm output values and net farm incomes derived from figures published by the Ministry of Agriculture, Fisheries and Food (MAFF) in the UK in their annual Farm Business Survey. This was done for each of the years 1994/1995, 1995/1996 and 1996/1997, and the average taken. The resulting average ratio was then applied to the relevant affected sectoral output figures (Table 4.3). Livestock farms, particularly intensive indoor units, look likely to face the biggest cost increases on this evidence. The most onerous obligations on producers will tend to be those that relate to manure use and handling, which is related to the provision of storage facilities, building requirements and other capital expenditure items on livestock farms.

Table 4.2   The cost of compliance in the UK under the Action Programme for
NVZ (England and Wales) in 1998 (in £)

Sector and measure	Proportion of farms affected (%)	Annual costs per farm affected
Dairy		
Storage	30	5,500
Transport	10	1,400
Beef		
Storage	10	1,000
Pigs		
Storage	7	9,750
Transport	10	2,100
Poultry		
Transport	100	2,100
All farms	100	150

Source: Ministry of Agriculture, Fisheries and Food.

Although the on-farm constraints related to irrigation, sediments and salinisation undoubtedly fall more heavily on crop farms than livestock farms, these constraints are generally less widespread and less onerous than the other kinds of constraints identified in relation to water issues.

Siting rules, building and environmental rules put significant constraints on new livestock installations in the EU, mainly in the form of conditions on planning consents. In relation to air pollution issues, these measures impact predominantly upon livestock production and particularly indoor units for pigs, poultry and cattle.

Table 4.3   Estimated impact of legislation of the Nitrates Directive in the UK

	Poultry	Pigs	Beef	Dairy
Total sectoral output, 1996 (£ million)	1934	1316	1962	3514
Estimated sectoral impact (£ x 1,000)	360	1686	134	1910
Number of affected holdings, 1996	235	555	422	310
% of total holdings	0.78	2.9	0.6	0.8
Value of affected output (£ million)	135	132	24	95
% of total sectoral output	7	10	1	1
Estimated affected net farm income (£million)	15	15	3	20
Impact as % of total sectoral output	0.02	0.13	0.01	0.05
Impact as % of total affected output	0.3	1.3	0.6	2.0
Impact as % of estimated net farm income	2.4	11.3	4.5	9.6
Average impact per holding (£)	1532	3038	317	6161

Source: Wilkinson (1998).

Animal welfare legislation clearly affect only animal producers but appears to be an area where significant new costs can arise as a result of complying with rising standards. Rules on cage systems for laying hens is one area where there will be cost implications as shown in data from the UK (Table 4.4). The cost implications of new housing systems for pigs may be significant as well. Typical recurring costs of confirming to the provision relating to the construction and size of battery cages are around 1.5% of annual turnover. Typical recurring costs of confirming to the provisions relating to stalls and tethering and space requirements for weaners and rearing pigs range between 3 and 11% of annual turnover.

Table 4.4   Estimated impact of the UK welfare of livestock regulations

	Laying hens	Pigs
Total sectoral output 1996 (£ million)	437	1316
Estimated sectoral impact (£ million)	16.7	7.9
Number of affected holdings, 1996	750	4,000
% of total holdings	2.7	41
Value of affected output (£ million)	387	1,250
% of total sectoral output	91	95
Estimated affected net farm income (£ million)	43.5	140
Impact as % of total sectoral output	3.8	0.6
Impact as % of affected output	4.2	0.63
Impact as % of estimated net farm income	38.4	5.6
Average impact per holding (£)	22,267	1,975

Source: Wilkinson (1998).

CONCLUSIONS

Several conclusions emerge from the identification of environmental and human-health related standards applying to agriculture in the EU.

The most important environmental and health-related issues, which are of major concern in the EU and with high priority in policy are:

* nutrient enrichment by nitrates and phosphates;
* pesticides (including pesticide drift and applicator safety);
* odour and nuisance;
* biodiversity and landscape;
* GMOs; and
* animal welfare

Regulation plays an important role in policies for pollution control (pesticides, water pollution, hygiene, and food quality). Legislation and command-and-control measures are relatively widely used in the EU to protect the physical environment, as well as to advance human-health and animal welfare objectives. In the EU a sizeable proportion of production originates from intensive farming systems and the average farm size is small. Most policies have been reactive in the sense that measures were introduced in response to perceptions of emerging or newly documented problems.

The EU seems to have made relatively widespread use of public funds to support agri-environmental programmes, which offer some financial compensation to producers in return for generating public policies.

The costs of compliance with nutrient regulation and measures to control odour and nuisance from intensive livestock production units are increasing in the EU. It is primarily a question of finding a location for the facility that reaps the available size economies and at the same time is far enough away from adjacent land uses. The compliance costs of producing pigs and poultry have increased during the past 10 years, and this may have a significant effect on the location of production in the future.

REFERENCES

Agra Europe (1998) Commission proposes new antibiotic feed ban. Agra Europe, London, 13 November, Volume 1824.

Bignal, E.M. and McCracken, D.I. (1996) Low-intensity farming systems in the conservation of the countryside. *Journal of Applied Ecology*, 33, 413-424.

Blandford, D. and Fulponi, L. (1999) Emerging public concerns in agriculture: domestic policies and international trade commitments. *European Review of Agricultural Economics*, 26 (3), 409-424.

CEC (1996) *Communication from the Commission to the Council and the European Parliament on Environmental Agreements.* Commission of the European Communities, Brussels, COM(96) 561, final.

CEC (1997) *Report from the Commission to the Council and the European Parliament on the application of Council Regulation (EEC) 2078/92 on agricultural production methods compatible with the requirements of the protection of the environment and the maintenance of the countryside.* Commission of the European Communities, Brussels, COM(97) 620 final.

CEC (1998a) *Guide to the approximation of European Union environmental legislation.* Commission of the European Communities, Brussels. Revised and updated version of SEC (97) 1608 of 25.08.1997.

CEC (1998b) *A partnership for integration: A strategy for integrating environment into EU policies.* Commission of the European Communities, Brussels, COM (98) 333.

CEC (1999) *Communication from the Commission to the Council, the European Parliament, the Economic and Social Committee and the Committee of the Regions: Directions towards sustainable agriculture.* Commission of the European Communities, Brussels, COM (1999) 22 final.

EEA (1999) *Environment in the European Union at the turn of the century: Summary.* European Environment Agency, Copenhagen.

EUREP (1999) EUREP GAP verification 2000. EHI-EuroHandelsInstitut, Köln.

Eurostat (2001) Organic farming. Office for Official Publications of the European Communities, Luxembourg, Statistics in Focus, Theme 8-5, 2001.

Heinz, I. (2002) Co-operative agreements to improve efficiency and effectiveness of policy targets. In: Brouwer, F.M. and Van der Straaten, J. (eds) *Nature and Agriculture in the European Union: New Perspectives on Policies that Shape the European Countryside*. Edward Elgar Publishing, Cheltenham (forthcoming).

House of Lords (1998) EC Regulation of genetic modification in agriculture. Select Committee on the European Communities. Session 1998-99, 2nd Report. London, Stationery Office, HL Paper 11-I.

IEEP (2000) The environmental impact of irrigation in the EU. A report to the Environment Directorate-General of the European Commission. Published on DG Environment's website at: http://europa.eu.int/comm/environment/agriculture/studies. htm.

NVI/SBA/NFA (1995) Swedish salmonella control programmes for live animals, eggs and meat. National Veterinary Institute, Swedish Board of Agriculture and the National Food Administration, Report 1995-01-16.

Riksen, M.J.P.M. and Graaff, J. de (2001) On-site and off-site effects of wind erosion on European light soils. *Land Degradation & Development*, 12 (1), 1-11.

Stanners, D. and Bourdeau, P. (Eds) (1995) Europe's Environment - The Dobríš Assessment. A report of the European Environment Agency, Copenhagen, Denmark.

Verstegen, J.A.A.M. (1998) Voorstudie Technology Assessment koersbepaling veehouderij. Agricultural Economics Research Institute, The Hague, Publicatie 3.170.

Wilkinson, D. (1998) The commercial effect of environmental and animal welfare regulations on UK agriculture. Environmental regulations.

USA

Chantal Line Carpentier and David E. Ervin

INTRODUCTION

Agriculture is a large sector in the US economy, estimated to deliver $220 billion in final output value in 2000, which accounts for 2.5% of Gross Domestic Product (GDP) (USDA, 2000)[1]. Agriculture is also one of the few net export sectors in an otherwise trade-deficit economy. Maize, wheat, soybeans and beef are among the export leaders of the agricultural industry. Hence, there is concern that the industry not be unduly hampered in competing abroad by excessive domestic environmental regulation.

Agricultural production is spread across a diverse natural resource base, from dryland farming in the Corn Belt to irrigated land in California's Central Valley. Not only are agricultural enterprises highly variable; they use very different production practices, depending on farm size, soil moisture and fertility, topography, cultural preferences, and a host of other factors. It is the joint distribution of production practices with natural resources, which vary also within and across regions, which defines the size and shape of agriculture's environmental footprint (Antle and Just, 1991). Different areas (ecosystems) respond quite differently to the possible stresses of agricultural production, depending on the land's resiliency (quality), water quantity and quality, plant and animal species diversity, and the character of the landscape.

The increased productivity of farming since 1950 has been achieved primarily through intensive management, including more application of fertilisers, pesticides, and irrigation. Intensive livestock rearing and feeding facilities have created waste management challenges that rival or exceed those facing large industrial plants and municipalities. Land intensification has been exacerbated by annual federal programmes that removed cropland from production for supply control purposes. This intensification is the most pervasive environmental

---

[1] This chapter uses information gathered in 1998 and was not updated. The basic patterns and trends, however, continue.

pressure from increased agricultural production, which became apparent during the last century. More and more output has been coaxed from each hectare of soil. From 1949 to (record-breaking) 1994, the average yield of maize per harvested ha rose from 2,388 kg to 8,663 kg, and the average wheat yield climbed from 979 kg per ha to 2,538 (USDA, 1994). Total agricultural land in crops has remained largely stable since World War II, although the regional composition of arable land has shifted somewhat. Converting some lands (e.g., wetlands) to production has also diminished natural resource functioning locally. Examples include wetlands drainage in the Prairie Pothole region and farmland conversion around rapidly expanding urban areas. The intensification and limited extensification processes have left a large environmental footprint across the countryside.

Soil erosion, declining water quality and quantity, air pollution, and reduced diversity of plant and animal species, in addition to food quality and safety have surfaced as serious issues in one or more regions. The diffuse sources of these problems are spread over nearly two million farms and ranches. Agriculture has been identified as a key source of some of the remaining problems by a growing body of science. Serious environmental problems caused by agriculture are real and prevalent, but not universal. Significant problems occur in most regions, but their scope and severity are uneven. They tend to concentrate where production pressure is intense and natural resources are vulnerable to damage. In response, several major agricultural programmes have been implemented. However, despite much progress in reducing point source pollution and reversing declines in many endangered species, serious and prevalent environmental problems persist.

The traditional approach has been for the federal government to institute broad (umbrella) programmes that are implemented in each state, but a trend to individual state and local programmes has emerged. For the most part, regulatory approaches with input or performance standards have not been applied to agriculture. Hence, US policies for environmental and health management in agriculture have imposed neither large costs on farmers nor burdens on export competitiveness. The technical difficulties and costs of regulating so many diverse operations and a strong agricultural lobby have combined to limit such standards. Instead, dating back to the Great Depression, a series of voluntary-payment programmes has been used to entice farmers to change their behaviour such that more environmental benefits - and fewer environmental evils - are produced. Most progress on environmental problems, such as erosion reduction and wildlife population increases, has been accomplished with measures such as the federal conservation and wetland reserve programmes. These and other programmes offset part or all of the costs of shifting management practices and land uses, through practice cost-sharing and land retirement payments. Human-health and animal welfare problems do not receive as much policy attention as environmental issues, with the exception of pesticide residues on food.

In four situations, when human or environmental health is perceived as seriously threatened, the public has supported the regulation of agriculture. These situations include:

- water pollution from large confined animal feeding operations (CAFOs);

- pesticide applications;
- wetland alterations; and
- endangered species habitat.

The evidence shows that the impacts of most of these programmes have not been widespread, with the exception of pesticide controls and other controls in certain geographic areas with intensive production in fragile environmental settings. The regulation of genetically modified organisms (GMOs) and animal welfare issues is limited. The composite picture that emerges from this review of issues and programmes is a collage of disparate efforts. Responding to periodic crises in production and environmental management, agriculture has been mostly enticed with voluntary payments, but also controlled by some significant regulations. Clear environmental performance targets do not guide most programmes. Rather, they have focused on controlling the quantity and quality of production inputs, exemplified by land retirement. The most severe constraints appear to apply to water-quality issues, especially for animal agriculture, and to pesticide use.

The trend to more regulation and environmental standards for agriculture is growing, however, along with the potential to impose on-farm operational constraints, especially at the state level. To illustrate, a major study of state animal-waste programs showed that 17 of 32 states reviewed had new legislation proposed in 1998 and 12 had passed new legislation during the prior 3 years (Animal Confinement Policy National Task Force, 1998). States are also increasingly suing or fining livestock operations not in compliance with the federal Clean Water Act (CWA). In addition, Congressional funding of the US Department of Agriculture's (USDA) flagship Environmental Quality Incentives Program (EQIP) declined in 1998, and the US Environmental Protection Agency (EPA) recently announced its intention to pursue more vigorously farms and ranches causing water pollution. Hence, concern is rising that US agriculture's enviable global competitiveness record may suffer.

ISSUES AND ON-FARM CONSTRAINTS

Agriculture is associated with soil erosion, land conversion, drainage, fertiliser use, pesticide use, animal wastes, and water used for irrigation, all of which affect water quality, sometimes with dramatic impacts on plants, fish, and water resources. Most scientific assessments of environmental problems associated with agriculture place water quality at the top of priorities for remedial and preventative actions. Therefore, water quality receives the most attention in this chapter.

Water

Overall, the US has not made significant progress in reducing agricultural pollution to surface waters since the 1972 CWA established goals of swimmable and fishable waters. Approximately 64% of the nation's surveyed rivers, 61% of lakes, and 62% of estuaries met the 'swimmable and fishable' quality goals in 1996 (US EPA, 1998a). Of the sampled waters not meeting the goals, farming and ranching were sources of water-quality impairment in 70% of the river miles, 49% of lakes, and 27% of estuaries. Federal, state, and local governments have invested billions of dollars in municipal wastewater treatment during the last 20 years, but very little - by comparison - to control pollution from about 2 million farms and ranches (Ervin, 1995).

Findings from the US Geological Survey's (USGS, 1998) National Water Quality Assessment studies affirm the conclusion from the weaker EPA survey data that agriculture plays a significant role in surface-water quality. USGS scientists have estimated that 71% of total cropland lies in watersheds where at least one agricultural pollutant violates criteria for recreation or ecological health (Smith et al., 1994). Another assessment identified regions where these problems were most severe (US Congress, 1995). Impairment of surface waters was considered particularly significant in the Corn Belt, where fertiliser and pesticide residues and sediment are concentrated in many streams, rivers, and lakes.

Contamination of groundwater sources of drinking water is also an issue of concern, with nitrates from fertilisers and livestock manure causing serious problems. For instance, the issue that draws most public attention at the national level is the periodic spills and water pollution from concentrated livestock operations.

Irrigated agriculture remains the primary consumptive use of freshwater (USDA, 1997a). The withdrawal of water from surface water or groundwater sources can reduce stream flows that degrade aquatic habitat or cause saline intrusion. The runoff and leaching of irrigation water can cause environmental problems, such as sediment, and nutrient and pesticide residues in return flows. Most of the environmental effects from irrigated agriculture occur in the western part of the USA, Florida, and Arkansas.

Water quality (nutrients)

Nutrients enter surface water and groundwaters via three routes:

- runoff from cropland, pasture, or animal feedlots;
- run-in or direct transport of chemicals into groundwater through sinkholes or fractures; and
- leaching of pollutants through the soil profile caused by rainfall or irrigation.

Nitrogen and phosphorus, primarily occurring as nitrate and phosphate, are the most important nutrient sources of water-quality problems in the USA (USDA, 1997a).

Nutrient enrichment can be found in most states. The problems are both regional and local in character. In two high-profile cases, it is clearly a regional concern. The federal government and states surrounding the Chesapeake Bay have undertaken a massive programme (albeit mostly voluntary to date) to reduce nutrient pollution in the Bay, about one third of which is attributed to agricultural sources, cropland runoff and animal manure. Although there is as yet no formal initiative, a similar approach could easily be adopted to address the Corn Belt region (US Congress, 1995) from which nutrients flow into the huge Mississippi River watershed. USGS (1999) found that hundreds of thousands of metric tonnes of agricultural contaminants end up in Louisiana's Gulf Coast estuaries, contributing to an off-shore 'dead zone'. USGS scientists concluded that 70% of the total nitrogen delivered to the Gulf originates above the confluence of the Ohio and Mississippi Rivers, transported over distances of 1,500 km or more (Alexander et al., 1997; Rabelais et al., 1997). About 90% of these nutrients were estimated to come from non-point sources, primarily agricultural runoff and atmospheric deposition. These nutrients encourage algal growth in the Gulf, leading to an oxygen deficit that kills shellfish and other aquatic organisms if they remain in that area.

The same scientists, using 10-year water-quality records, have estimated the extent to which agricultural pollutants (such as nitrates, phosphorus, or herbicides, including atrazine) found in rivers of each state originate in other states (Smith et al., 1996). Over 16 states receive more than half of their concentrations of atrazine via watersheds in other states. The prevalence of transboundary linkages reveals that many individual states are unable to control water quality within their boundaries, thus necessitating interstate co-operation or federal initiatives. In other cases, nutrient pollution is largely a local concern with excessive concentrations in certain lakes, rivers, and wells.

In addition to degrading surface water quality, nutrients from agriculture can affect groundwaters through leaching and run-in. Groundwater supplies half of the population with drinking water and is the sole source for most rural communities. For this reason, contamination of groundwater often tops the list of public environmental concerns. The extent of groundwater pollution from agricultural nutrients is less well documented than for surface waters. The most serious problem appears to be nitrates from inorganic fertiliser and animal wastes. The levels of nutrients applied to farmland via commercial fertilisers are five times as large as those from animal manure (NRC, 1993).

Early findings from a national water-quality assessment showed that fully 12% of domestic wells in agricultural areas exceed the maximum contaminant level (MCL) for nutrients (Mueller and Helsel, 1996). This rate of contamination was twice as high as for domestic wells in other land uses. Later analysis of groundwater nitrate sampling data concluded that areas showing the highest levels have high nitrogen input (such as from fertiliser and animal wastes), well-drained soils, and less-extensive forested areas relative to cropland (Nolan et al., 1998). Analyses of nitrates in surface waters point also to a heavy role for agricultural practices (Puckett, 1994).

Animal operations were found responsible for at least 50,000 km of impaired waters in 22 states that categorised impacts by type of agriculture (USDA, 1998b). Nationally, feedlots are considered to be the principal source of 16% of the waters impaired by agricultural practices, the third leading agricultural source of water pollution after non-irrigated crops and irrigated crop production (USDA, 1998b).

Water quality (sediments)

Land cultivation, irrigation, and grazing can cause excessive soil erosion and deliver sediment to surface waters. In surveys of the state water-pollution problems, siltation is reported as one of the leading problems in rivers and streams, and among the top four problems in lakes and estuaries (US EPA, 1995).

The impairments from sediment damages due to erosion have been estimated to cost from $2 to $8 billion per year across the country (Ribaudo, 1989). The impairments are diverse, including damages and costs to navigation, reservoirs, recreational fishing, water treatment, water conveyance systems, and municipal and industrial water use. An Office of Technology Assessment (OTA) study identified ten areas most susceptible to erosion damages, most of them in the Great Plains, Corn Belt, and Mississippi River Valley regions (US Congress, 1995).

Cropland erosion is one of the principal causes of sediment and siltation. Soil erosion from agricultural lands has fallen significantly over the last decade. Hence, it is not surprising that the USGS found that suspended sediment levels trended downward over the 1980s (Smith *et al.*, 1993). This decline was most apparent in the Ohio-Tennessee and upper and lower Mississippi River Valley regions. The total erosion from agricultural lands is estimated to be approximately 1.8 billion metric tons, down from 2.7 billion in 1987 (USDA, 1997a). Nonetheless, some serious cropland and pasture erosion problems remain in certain areas and have a high priority in policy. There is also concern that the decline has stopped or has been reversed. For instance, conservation tillage use has levelled off and 2 million ha of croplands temporarily retired in the Conservation Reserve Program (CRP) had been returned to production in the 1990s.

Water quality (pesticides)

Much public attention has focused on pesticide residues in water resources. Overall, monitoring data do not reveal widespread occurrences of these residues in surface water or groundwater resources above MCL. However, the science needed to fully understand the extent of pesticide pollution of waters and long-term implications for human and environmental health is still evolving. Pesticide residues in food and their potential human heath consequences are also high priority issues. Potential harmful effects on water quality, as related to pesticide drift, and safety measures related to pesticide usage are also issues of concern.

The evidence of pesticides in groundwaters is sparse but growing. A 1988 survey by the EPA showed that less than 1% of the wells sampled contained pesticides that exceeded the applicable MCLs (US EPA, 1990). Following a comprehensive review of past research, Barbash and Resek (1996) concluded that pesticide contamination from agricultural sources is likely to be a localised problem. Therefore, national surveys may not detect the full pattern of contamination unless sufficient samples are analysed to detect local patterns of pollution.

The latest findings from USGS analyses convey a different picture (Gilliom, 1997). Samples from 48 agricultural areas reveal pesticides in 59% of shallow wells. The concentration of individual pesticides is usually below drinking water standards, but the co-occurrence of multiple pesticides is common and little is known about the effect of these joint occurrences. Atrazine, widely applied on maize, is the most frequently detected compound. USGS (1998) reports that in agricultural areas, 38% of stream samples and 5% of shallow groundwaters contained a mixture of atrazine, deethylatrazine (DEA), and metolachlor. Atrazine and DEA were also often found in the same waters as simazine and prometron. Fifty percent of all streams sampled contained more than five pesticides and 25% of all groundwaters sampled contained two pesticides.

This 'pesticide soup' is more pronounced in streams, as the data above indicate. About half of the compounds found in these mixtures, including significant numbers of degradation products or metabolites that often exhibit toxic effects, have not yet been assigned MCLs. There is insufficient science to assess whether multiple pesticides in low levels pose substantial or insignificant risks to human or environmental health. Although such pesticide contamination may ultimately prove serious, nitrates are - for now - the most widespread, documented chemical pollution problem in agriculture. National economic estimates of the damages from nutrient or pesticide contamination of surface water or groundwaters are not available.

Water quantity (irrigation)

The quantity of water used by agriculture is also an issue of growing significance. The effects range from diminished instream flows for threatened or endangered species to high rates of withdrawal from groundwater aquifers. Irrigation water mainly originates from surface water and is supplemented with more expensive groundwater during periods of drought. Competition and problems relating to overuse of water for agriculture are particularly serious in a few regions, e.g., California, the southern Plains' Ogallala Aquifer area, and Florida.

In the case of groundwater, a high-profile case is the huge Ogallala Aquifer stretching across Kansas, Nebraska, and Colorado. Over time, intensive irrigation has drawn down the aquifer, which is composed largely of ancient water left behind by glaciers, and thus posing the potential for significant future shortages and reduced productivity (Opie, 1993; White and Kromm, 1995). Further west, US Bureau of Reclamation projects have dammed river systems, such as the Colorado and Columbia, and diverted billions of m^3 of water to agricultural irrigation, at prices averaging one-tenth those charged to non-agricultural users (Frederick, 1990). These water transfers, many implicitly or explicitly subsidised by the federal government, raise doubt that the true scarcity value of America's water resources has been recognised.

Irrigation-water costs vary across the nation but are invariably subsidised. Anderson and Snyder (1997) report Bureau of Reclamation Irrigation Project's subsidies between 57% (Imperial Valley, California) and 93% (Truckee-Carson, Nevada). Farmers pay between $0.10 per hectare-cm (100 m^3) in the Grand

Valley and $1.28 per 100 m^3 in Westlands, California. Until recently, the in-stream flow of water was not recognised as a beneficial use, and thus water could not be maintained instream to protect species living in these rivers and streams. This is changing, as reflected in the evolution of the Bureau of Reclamation's mission from one of building and managing large irrigation projects to one of managing a scarce resource. For instance, the Central Valley Project (California) enacted legislation in 1992 to allocate around one billion m^3 of water annually to fish, habitat, and wildlife protection. In addition, a surcharge is imposed to cover other environmental programmes, along with a tiered pricing system to encourage water conservation (NRC, 1996).

Public pressure to reverse the subsidisation of irrigation water for farming is rising as the demand for other water uses by growing non-farm populations rises. An increasing number of states are restricting irrigation-water use to serve environmental purposes, such as recreation, wildlife habitat, and endangered species (USDA, 1997a). One of the most notable examples is the 'Bay-Delta' agreement by California, in collaboration with the federal government, to reserve in-stream flows for endangered species protection, and decrease allowable withdrawals by agriculture during certain low-flow periods. Salinity problems are mostly local in nature.

Air

Odour and air quality problems from agriculture are mostly localised. Examples include the burning of crop residues, noxious odours, nitrogen emitted into the air from CAFOs and deposited, particulate dust from cropland tillage, and wildfires in dry areas (USDA, 1997a). Total estimated ammonia emissions from agriculture in the North Carolina Coastal Plain are 97.7 million kg of nitrogen. Hog operations were responsible for 63% of these emissions (Rudek, 1997). Exposure to hog farm fumes, for example, has caused human-health problems, and is rising in importance at the local and state level.

The USA has phased out much crop burning where it was a problem, for example, for grass straw in Oregon and rice straw in California. Thus, few problems stemming from crop burning remain in the USA.

Biodiversity

The loss of biodiversity is an issue of significant concern in the USA, especially in grassland areas. As the rising global population steadily increases the pressure on food production, biodiversity is threatened by agricultural extensification and intensification. Several factors contribute to the loss of biodiversity. The conversion of many grasslands and wetlands to crop production, increased field sizes, reduced crop diversity, elimination of woodlands and field edges, reduced crop rotations, and increased fertiliser and pesticide use have had dramatic impacts on animal and plant populations, even among species well-adapted to agricultural land uses, such as cottontail rabbits, quail, and pheasants (US Congress, 1995). Farmers do not, in general, have incentives to conserve critical plant and animal species. Species dependent on grasslands have suffered the most dramatic declines (Knopf, 1994). Samson and Knopf (1994) report that in the Great Plains, 99.9% of native tall-grass prairie and 30% of short-grass prairie has been converted to intensive crop production. As a direct result, at least 55 grassland wildlife species are now listed as threatened or endangered and 728 are candidates for listing.

The CRP has shown how converting fields from row crops to grassland and woodland affects overall species abundance. As of 1993, the regions with the highest levels of CRP enrolments, the Midwest and Great Plains, showed the most dramatic improvements in a variety of nesting species (Allen, 1993; Berner, 1994). The Midcontinent Ecological Science Center of the National Biological Service, in Fort Collins, Colorado, began in 1988 a CRP-monitoring exercise extending over 30 states, and found significant improvements in the level and suitability of habitat for a variety of game and non-game species. Separately, Johnson and Schwartz (1993) studied the response of grassland birds and found that several species that had declined dramatically from 1966 to 1990 were now common in CRP fields.

A non-admirer of songbirds or cottontails might well ask whether their survival is necessary or even important. The answer is that we do not know, although their consumption of a wide range of insects and plants harmful to commercial crops is an important contribution to agricultural productivity. A more utilitarian argument for wildlife diversity can be made on behalf of hunters and other conservationists, who support measures to preserve and protect uplands and wetlands.

In the case of crop diversity, we have more compelling historical experience. Historical examination of the systematic elimination and propagation of crop varieties, at first through selective cultivation and then by plant breeding, has shown that a narrowed genetic base can have catastrophic consequences by creating susceptibility to a variety of plant diseases (Duvick, 1996). In 1970, southern maize leaf blight attacked the US maize crop, and was turned back through the use of crop varieties held in storage by seed companies. Modern awareness of the need to conserve a diverse store of germplasm, not only in natural environments (*in situ*) but also in 'banks' (*ex situ*), grew correspondingly (Tripp and Heide, 1996). However, the capacity of market forces alone to preserve this diversity is questionable. Donald Duvick, formerly chief plant breeder for Pioneer Hi-Bred International, the dominant world maize seed company, argues that a strong public role is necessary for plant germplasm conservation (Duvick, 1996).

Of the 663 animal and plant species listed as endangered, 380 are listed, at least in part, due to agricultural activities (USDA, 1997a). Exposures to fertilisers and pesticides have contributed to the listing of 115 of these species. In those cases, pesticide application rates may need to be lower than those set by the Federal Insecticide, Fungicide, and Rodenticide Act (FIFRA) in any county identified as critical habitat. The conversion of land to agriculture has reduced the number and size of habitats for many listed species. Though few listings have affected agriculture to date, restrictions on private property to preserve endangered species habitat has the potential to affect states in the Midwest and Pacific regions considerably.

An issue of widespread current policy attention related to landscape and nature conservation involves farmland conversion. As urban areas have expanded into productive farmlands, one set of pressures on landscape, habitats, and biodiversity has given way to another - leading to wholesale conversion of productive farmland and potentially valuable wetlands and riparian areas to urban uses. Not only has this reduced the available area of farmland, it has destroyed valuable natural areas and their biodiversity as well. In the Central Valley of California, which is the leading agricultural producing area by value in the nation, low-density urban sprawl is expected to consume more than 0.4 million ha of farmland by 2040, which could be reduced by more than half if policies favouring more compact growth were to be adopted (Thompson *et al.*, 1994).

Wetlands

Heimlich *et al.* (1998) estimate that wetland area in the 48 contiguous states has decreased from 88.4 million ha in 1780 to 49.6 million ha in 1992. Eighty percent of these wetlands were converted to agricultural production. Wetlands are now found mainly in the Southeast, Midwest, and Delta and Gulf Regions (Heimlich *et al.*, 1998). Until the late 1970s, cost-share money and direct federal assistance were available to drain wetlands. These subsidies, plus federal income tax incentives and potential profits from crop production, made agriculture the number one converter of wetlands prior to 1980. In 1985, the Food Security Act Swampbuster provision, which denied agricultural programme payments to farmers who inappropriately converted wetlands to production, was approved. In 1986, the federal tax law was reformed to eliminate some tax incentives favouring conversion. The scope and effect of Swampbuster was reduced in 1990 and further reduced in 1996 following pressure from farmers and ranchers.

Transgenic crops

In 1999, the USA devoted approximately 30% of its total' cropland planted to major crops to transgenic crops. This is more than any other country (USDA, 2000). Transgenic plants used commercially in the US include herbicide-resistant crops, insect-resistant crops, and transgenic plant products (Krimsky and Wrubel, 1996). Herbicide-resistant crops include traits for glyphosate tolerance (e.g., maize, canola, cotton, soybean, sugar beet), bromoxynil tolerance (e.g., cotton), glufosinate tolerance (e.g., maize, canola, sugar beet, soybean), and sulfonylurea tolerance (e.g., cotton, flax). All commercially available transgenic insect-resistant plants contain gene(s) transferred from *Bacillus thuringiensis* (Bt). In data presented to the EPA by Monsanto for approval of a Bt crop, Bt was described as selective and having an impact only on the primary pest, and not harming secondary pests or beneficial insects. Evidence of harmful effects on beneficial insect predators, including lacewings (Hilbeck *et al.*, 1998) and ladybird beetles (Birch *et al.*, 1997), monarch butterfly larvae (Losey *et al.*,

1999), and soil biota (Watrud and Seidler, 1998) is now emerging. Although initial developments were oriented to enhancing production efficiency, such as reducing pesticide costs, new developments are aimed at the consumer, such as improving nutritional quality.

Genetically modified crops have been adopted more rapidly than any previous agricultural technology. Part of the reason for their rapid introduction and spread in the US is that about one-half the time and one-fifth to one-seventh the cost are required to approve a new biotech product, compared to approval of new chemical pesticide compounds (Ollinger and Fernandez-Cornejo, 1995). The most important transgenic crop in terms of percentage of production is cotton, with 61% of ha planted, followed by soybeans at 54%, and maize at 25%. The use of other transgenic crops is still quite limited.

Transgenic crops can have many possible effects on the environment (Ervin et al., 2000). Potential environmental benefits include:

- the use of fewer, less toxic, or less persistent pesticides;
- increased crop yields (which may reduce the need to convert pasture or other lands to agricultural production);
- decreased water use; and
- reduced soil tillage.

Potential risks include:

- uncontrolled flows of genes to wild relatives;
- development of herbicide, pest, and virus resistance in wild relatives;
- reduced in situ crop genetic diversity; and
- adverse effects on organisms that are not pests, such as beneficial insects.

A relatively small but expanding body of evidence exists to assess these effects. Early estimates suggest that transgenic crops will confer environmental benefits in some areas and/or in some years - for example, some crops appear to induce reduced use of toxic pesticides and slightly increased yields on average. The early experimental findings also show that using transgenic crops will, in certain circumstances, increase some environmental risks, such as gene flow and harm to species that are not pests.

Most studies of the environmental effects of transgenic crops have been confined to laboratories or small fields. The lack of detailed environmental impact data required for commercial approval and release in the USA has hindered risk and benefit assessment efforts. Monitoring is not being conducted on the potential environmental impacts or on the interactions of multiple transgenic plants in ecosystems. Farmers will likely monitor the impacts on their farms, but usually will not extend their oversight beyond farm boundaries. Remedying the lack of environmental science will require more than simple increases in funding for current public research efforts (Ervin et al., 2000).

Animal welfare

Animal welfare generally is less of a public policy issue in the USA than in EU and other countries. Cases of mistreatment of farm animals periodically surface in the media and engender public reaction, but sustained public interest in altering common animal husbandry practice remains mostly a local issue in selected areas.

Human-health

Food safety issues often arouse concern by consumers. While an extraordinary food network makes US supermarkets the envy of the rest of the world, long marketing channels create numerous handlers, increasing the need for preservation and chances for contamination. Occasionally, outbreaks of foodborne illness, such as *E. coli* and salmonella contaminations, alarm the public and lead to calls for better oversight of food safety. The concerns have, to date, surfaced mostly at the food-processing level. Controls at the farm level to reduce risks posed by microbial pathogens are not common (Roberts, 1999).

Many health concerns related to the food supply focus on chemicals applied at the farm level and the residues that remain within food when it is consumed.

Classified broadly as pesticides, they include insecticides, herbicides, and fungicides. The growth in reliance of American agriculture on pesticides since 1960 has been dramatic. In 1995, 146 million kg of active ingredients (a.i.) of herbicides, 32 million kg a.i. of insecticides, 20 million kg a.i. of fungicides, and 57 million kg a.i. of other pesticides were applied on major crops (USDA, 1997a). The total was 255 million kg (a.i.) compared to 97 million kg in 1964. Herbicide use has jumped almost seven-fold since the early 1960s. For example, between 1964 and 1991, herbicide use on maize grew from 12 million kg (a.i.) to 95 million kg, and on soybeans from 2 to 32 million kg. Analysts attribute this growth to substantial increases in planted hectares of these crops (from 26 to 30 million ha of maize and from 13 to 28 million ha of soybeans), but also to substantial increases in the proportion of fields treated (now about 95%), and rates of application (from 1.37 to 3.27 kg a.i. for maize, and from 1.14 to 1.37 kg a.i. for soybeans) (Whittaker *et al.*, 1995). Insecticide use has declined from its 1976 peak of 59 million kg a.i., while use of fungicides and other compounds has risen without decline over the past 40 years.

Livestock and poultry producers in the USA commonly use antibiotics and hormones such as rBST (recombinant bovine somatotrophin) in dairy production. These compounds are regulated by the US Food and Drug Administration (FDA) to be safe for human consumption. The genetically modified version of a naturally occurring protein hormone in cows that induces milk production, rBST is injected into dairy cows to boost production. It was approved by the FDA in 1993 and is used on 30% of dairy herds in the USA.

The use of antibiotics in animal feed has recently become a high-profile public-policy issue in the USA. Fred Angulo, an official of the federal Centers for Disease Control and Prevention (CDC), has stated publicly that there is a relationship between widespread antibiotic use in animal agriculture and the remarkable increase in resistant foodborne pathogens. He went on to say that the EU's recent ban of four animal antibiotics is scientifically justified. A revision of FDA policy on antibiotics in animal agriculture has not yet occurred.

On-farm constraints

We have reviewed on-farm agricultural constraints derived from environmental policies at the federal, state, and local levels. It is clear that water-quality policies are the most constraining, especially for CAFOs. Water quantity and air pollution restrictions both have the potential for large impacts in the future, but currently impose narrow or insignificant constraints on agriculture. Similarly, the protection of endangered species has had limited impacts on agriculture in the past, but has the potential for greater impact in the future, both in terms of limiting agricultural land and limiting water that can be extracted for irrigation.

Few of the standards applied to crop production 'have teeth', and when they do, it is usually because the problem is locally so serious that everyone, including farmers, agrees that something needs to be done. In most cases, problems and policies are highly localised and practices are highly subsidised. In some cases, farmers' incomes increase after the investment in new practices. Conservation tillage and integrated pest management are prominent examples. This economic improvement following changes in management practices to remedy an environmental problem suggest farmers were operating at sub-optimum before the change. Whether farmers operated at sub-optimum because they lacked required knowledge, are risk averse, responded to distorting policies, or simply preferred the status quo, the results are often the same - a win-win situation in the long run (although profits may fall in the short run) for farmers and the environment.

Federal pesticide regulation has been implemented rather uniformly across all states. Studies show that pesticide regulation has not imposed large costs on most commodity sectors, except for minor crops. These regulations may not be more costly than in other trading countries, and thus are not expected to have a tremendous impact on trade. In the short run, allowed technologies such as transgenic crops and rBST may give American farmers an advantage over farmers in trading countries that cannot use them, but the long-term impact may be different once we understand the human and environmental costs of these technologies. For example, studies have shown that transgenic Bt crops can be detrimental to non-targeted pests, and the limited amount of refugia established may lead to increased resistance, and thus reduce their future efficiency. Animal welfare issues do not play large roles in the USA and are not expected to lead to serious on-farm constraints in the near future. Indeed, the trend is to reduce the reach of states' anti-cruelty laws.

A comprehensive review of federal and state environmental programmes that affect farming and ranching portrays the broad potential for imposing costs. Federal legislation covers air, surface water, groundwater, coastal-zone, pesticide, wetland, and endangered species programmes. This chapter finds that the impacts of these programmes have been confined to small parts of the industry or to specific (localised) geographic areas. For example, the regulation of animal waste discharge into surface waters pertains only to CAFOs with more than 1,000 animal units (AU). CAFOs are a small proportion of the nation's animal feeding operations, varying from 0.6% of the beef and dairy industries to 4% of the broiler industry, but produce a disproportionately large share of industry output. In addition, the latest data from the EPA indicate that only 30% of CAFOs operated under a permit in 1992. Most beef and all dairy CAFOs had permits by 1992. Yet large poultry and hog operations were virtually unregulated with less than 10% of broiler, layer, and turkey CAFOs and 13% of hog CAFOs with permits. In all cases, these permitted CAFOs represented less than 1% of total operations. The permitting of CAFOs likely has increased since 1992, especially for swine CAFOs, but data are not available to capture the extent of change.

The Endangered Species Act (ESA) is the piece of federal legislation that holds the most potential to constrain on-farm production activities. Currently, farmers are not compensated for income losses they face, but ways to provide compensation are being discussed. The protection of farmed landscapes is only a significant issue with respect to the control of urban development.

ENVIRONMENT AND HEALTH LEGISLATION

Most environmental and health policies that apply to US agriculture began around 1970. The primary legislation includes the CWA, Federal Water Pollution Control Act amendments, Coastal Zone Management Act (CZMA), ESA, FIFRA, Safe Drinking Water Act (SDWA), National Wetland Policy Act, and Clean Air Act (CAA). Recent Congressional and administrative actions promise to strengthen federal controls for animal wastes and pesticides. The Clinton administration had developed a Unified Animal Waste Management Strategy that would implement controls on animal wastes from operations with more than 1,000 AU more uniformly across states. These efforts respond to spills and perceived threats from the increasing concentration of CAFOs. The new unified strategy, under review and comment during 1999-2000, has not been implemented, and therefore, is not analysed here. Similarly, the 1996 Food Quality Protection Act (FQPA) that reforms pesticide registration and use is still under development. Although the infamous 'Delaney Clause' that prohibited the use of cancer-causing compounds was eliminated by FQPA, the law has the potential to ban whole classes of compounds, such as organophosphates, which play large roles in agriculture. These potential outcomes are driven by new FQPA requirements to impose a 10-fold safety factor for children and protect against risks of endocrine disruption, which have not been regulated previously.

In many cases, these pieces of federal legislation set minimum environmental and health standards or guidelines, such as maximum contaminant levels of pesticide residues in drinking waters, and delegate responsibility to the states to design and implement programmes that enforce them. State actions often are stricter than the federal standards (Lester, 1994). The states, in turn, may delegate responsibility to local governments for controlling some environmental problems, such as zoning for land use conflicts. The chain of devolution of responsibilities means that, ideally, federal, state, and local government actions should be inventoried to identify the full range of on-farm operational constraints relating to farm production.

A public information base that describes the full set of on-farm environmental and health requirements applied in the 50 states and 3,068 counties does not exist. The approach adopted here is to selectively review the federal and state programs that appear to hold the largest potential to impose costs and exert trade effects. Even this subjective sampling may be impossible for most states and counties, given the lack of data. The analytical task for the national and state governments is made somewhat manageable in that binding environmental and health standards apply less often as a rule to agriculture than to other industries. The chapter's discussion covers on-farm constraints related to water, soil, air, nature conservation, animal welfare and human health.

Water quality

Most operational constraints at the farm level result from federal or state water-quality standards. These constraints are derived from standards to protect waters from nutrients, sediments, or pesticides, or aimed at restricting the allocation of scarce water.

Nutrients

On-farm constraints to protect surface waters from nutrients are binding for livestock and crop production, while constraints to protect groundwater mainly affect crop production. Discharge permits are required only for large animal units.

• *Livestock production*
Operational standards affecting livestock production vary greatly by state and by county within these states. To keep this discussion manageable, we limit our review to the top five states in terms of cash receipts from each livestock category. This review included 18 states that account for more than 57% of livestock receipts (with the exception of sheep) and more than 73% of hog production, most of which is concentrated in a few states (see Carpentier and Ervin, 1999, for details).

The review of standards affecting livestock production and waste management is based on three recently published sources, including NASDA (1998)[2], the Animal Confinement Policy National Task Force (1998)[3], and Marks

---

[2] NASDA surveyed the State Departments of Agriculture in 1998 to collect information about CAFO regulations. The survey covers all states except Arkansas (which does not have a State Department of Agriculture). Responses to some questions are missing.
[3] The Animal Confinement Policy National Task Force (ACPNT) survey conducted in late 1998 covered the states of Alabama, Alaska, Arizona, Arkansas, Colorado, Connecticut, Georgia, Hawaii, Idaho, Illinois, Iowa, Kansas, Kentucky, Maine, Maryland, Michigan, Minnesota, Mississippi, Missouri, Nebraska, New Hampshire, New York, Ohio, Oklahoma, Oregon, Pennsylvania, South Carolina, South Dakota, Tennessee, Vermont, Virginia, and Washington.

and Knuffke (1998)[4]. The information to characterise these and other new state environmental programmes for farming had not been available before. This study collected, summarised, and analysed information from myriad sources to describe the coverage and stringency of these states' environmental programs, and their on-farm operational constraints.

The state and local laws usually require some or all of a series of requirements, such as:

- a construction permit that may include some siting requirements;
- an operation permit that may include design standards on the structure and/or size, carcass disposal requirements, and manure management plans, which usually includes a waste management plan (treatment, collection, storage, allowable seepage from waste lagoons, transport);
- separation distances (to both water and properties);
- manure utilisation, usually through a nutrient management plan (for land application); and
- land area requirement.

Very few have odour or air pollution requirements or groundwater monitoring or testing requirements.

The entries in Table 5.1 are state regulations for the states having the most animal production. Each entry includes the year the legislation was passed, regulations that apply, and to which operations, enforcement measures, and type and level of public assistance to comply with the legislation. The number of full-time equivalent (FTE) staff available to enforce CAFO regulation is reported when available (in the enforcement column) to give an indication of the commitment and potential to apply the law. Arkansas does not impose fines despite violations. In California, actions are taken mainly once a farm has been detected as polluting, and enforcement is targeted toward dairies. In Texas, no significant monitoring is required for farms that do not discharge 'directly' into the waters. The states of Arkansas, Illinois (for lagoon), Iowa, Kansas, Minnesota, Nebraska (inspection and application) and Ohio assess fees during the approval of manure management process or facility construction.

National Pollutant Discharge Elimination System (NPDES) permits must be obtained by CAFOs to operate unless deemed non-discharging, in which case no permit is required. In Table 5.2, 11 states require a permit to operate a CAFO[5].

Most states start with the NPDES permit and add additional requirements. Arkansas (for hogs), California, Florida, Indiana, Minnesota, Mississippi, Montana, North Carolina, North Dakota, Texas, and Washington apply their regulations to AFOs with less than 1,000 AU and thus are stricter than the

---

[4] The Clean Water Network and Natural Resources Defense Council contacted state activists in 30 states to conduct their survey. Data for Arkansas, Arizona, Connecticut, Delaware, Florida, Georgia, Hawaii, Idaho, Louisiana, Maine, Massachusetts, Michigan, Nevada, New Hampshire, New Jersey, New Mexico, New York, Oregon, Rhode Island, South Carolina, and Utah were not obtained.
[5] Not in the Table, Florida also requires a permit for large dairies (and all new dairies in the Lake Okeechobee area) and Washington requires permits for all dairies.

Table 5.1 CAFO's laws, enforcement and assistance for the 18 most important livestock states

State	Target	Enforcement	Assistance
Alabama	>1,000 AU	Complaints; routine check	
Arkansas	>1,000 AU others	1 of 3 years; self-monitoring	S, L[d] cost-share
Regulation Five	<1000 AU hogs	no fines[b]; 15 FTE[c]	tax credit
California Porter-Cologne Water Quality Control Act	Dairy construction >1,000 AU	Complaints	S, F cost-share; incentive fund
Colorado	New, hogs >1000 AU	Complaints	NRCS[e] cost-share
Georgia	New/E[f]	2 FTE	None
Illinois Livestock management Facilities Act	Discharging >1,000 AU: NMP[g] >7,000 AU: NMP	Complaints permitted lagoon 1/yr; no fines; 5-7 FTE	EQIP
Title 35 Section 505	>50 AU: buffer		
Indiana Animal Waste 1	>600 swine >300 cattle >30,000 poultry	Complaints; routine; 9 FTE; $375 for fish kill[h]	EQIP
Iowa Senate File 473 Groundwater Act 1987; Animal Factory Operation; Act 95; New Manure rule 1997	F >1,000 AU	Complaints; was understaffed; now 18 FTE, fine up to $19,500	EQIP; Organic Nutrient Management Fund for NM[i]
Kansas Statute annotated 65-171.d and HB2950	>1,000 AU	Complaints; routine monitoring; hog more often; 19 FTE	S, F cost-share; free assistance; NRCS
Minnesota Rule Chapter 7020	All	Complaints	S, F cost-share; low interest loans
Mississippi *Moratorium on hogs until 2000*	>1,000 AU	Complaints; Routine 1/yr; 2 FTE; fines	None
Nebraska Livestock waste management Act; Local moratoria	Based on initial inspection	Complaints; routine, when staffing allows; 12.5 FTE	EQIP

North Carolina Senate Bill 1217 Moratorium on hogs until autumn 1999 (still in effect)	>250 hogs, >100 cattle, >30,000 birds (liquid waste)	2 visits/yr if >250 hogs; $10,000; visit all CAFO once; visit CAFO once a year	75% of costs, max $75,000 per year; technical assistance; total 1996 $8.6 million
New York	nr[a]	Complaints; $20,000	Technical assistance; cost-share; BMPs; S incentive
Ohio Adm. Code Chapter 3745	>1,000 AU	Complaints; routine; 32 FTE Spot check; fine up to $113,000	SWCD; low-interest loans
Pennsylvania	>1,000 AU	Complaints; inspections	Cost-share
Texas TAC Sub B	> 250 dairy or AFOs; >1,000 AU other	Complaints; routine, new polluters	Some EQIP; S cost-share
Wisconsin	>1,000 AU	Complaints; inspections	S, F cost-share

a) nr = not reported.
b) No fines indicate that spills have occurred but no fines were charged.
c) FTE is full-time equivalent employees working on enforcing the CAFO rule.
d) L = local, S = state, F = federal.
e) NRCS = Natural Resource and Conservation Service
f) New/E = new and expanding farms.
g) NMP = Nutrient Management Plan
h) Maximum fine ever imposed on CAFOs.
i) NM = Nutrient Management
Source: Animal Confinement Policy National Task Force (1998); Marks and Knuffke (1998); NASDA (1998).

national standard dictated by the CWA (Table 5.1). North Carolina, for example, requires a permit for all hog operations with more than 250 hogs (100 AU).

Kansas is the only state having a permit requirement for poultry (America's Clean Water Foundation, undated). Poultry operations were to start being regulated in 1999 in Alabama and Oklahoma. Illinois requires a permit for poultry raised using lagoons and Arkansas requires a permit from poultry operations that manage manure in liquid form (America's Clean Water Foundation, undated). Maryland enacted a new law in 1998 that requires all operations with sales exceeding $2,500 to have (by 2001) and implement (by 2002) a nitrogen- and phosphorus-based nutrient management plan. The small geographic coverage of regulation suggests that most poultry operations around the country are not subject to any standards, although most of them have more than 1,000 AU.

In addition to sound manure-storage facility construction regulated through design standards, land application of manure is the most important activity that can create pollution problems. Although most states require a nutrient- or waste-management plan (NMP or WMP) to acquire their permit, few states require and monitor the implementation of the NMP.

Of the 18 states in Table 5.2, 17 require a nutrient- or waste-management plan and 12 have mandatory design standards. Eleven states require setbacks from properties, 9 require setbacks from water resources, and 10 states require setbacks from wells. The remaining five states leave it up to the local zoning ordinance to decide whether to have setbacks, and, if so, what dimensions. Depending on local environmental problems, states require various application standards. For example, in the flood plains of Iowa manure must be injected; a soil test for manure application is required, and manure must be applied within 30 days of nutrient uptake, in Kansas; and manure cannot be applied within 3 days of forecasted rain in Alabama. In Georgia, the application of manure is limited by the level of nitrate in the groundwater, not to exceed 10 mg L^{-1} as nitrogen (the National Primary Drinking Water Standard).

The states of North Dakota, South Dakota, Nebraska, Kansas, Mississippi, Minnesota, Iowa, and Wisconsin prohibit large publicly traded corporate investment in agriculture. Dairy and swine appear to have the most stringent regulations. Detailed descriptions of selected states with strong requirements can be found in Carpentier and Ervin (1999). Other important state environmental programs include those for nutrient leaching, erosion, irrigation withdrawals, wetlands protection, and pesticide registration. Again, most appear to exert localised cost effects. The unfolding redistribution of irrigation water from agriculture to urban and environmental uses in the West may be an exception.

- *Crop production*[6]

There are no federal permits required by crop producers for nutrient discharges, and there are no mandatory federal standards. Permits and mandatory requirements can be found at the state level only. Regulations are mainly used as backup authority to other (usually voluntary) programmes to control pollution, or to deal with extreme cases where no other programme exists (US EPA, 1997). Most enforceable state laws and/or statutes are reviewed below. This list is not comprehensive, but includes standards and programmes that are most likely to have impacts on farmers.

In Arizona, general permits requiring best management practices (BMPs) are needed for application of nitrogen fertiliser or for CAFOs, as designated by the Aquifer Protection Act (A.R.S. 49-241). The permit requires the use of practices including timing, precise application, irrigation water management, testing, and tillage practices that maximise nitrogen uptake. If farmers fail to follow BMPs, an individual permit that requires record keeping and reporting replaces their general permits. Administrative orders and criminal fines are imposed when farmers fail

---

[6] This section draws from Environmental Law Institute (1997) and (1998).

Table 5.2   CAFO standard description by state

	Permit Construction	Permit Operation	WMP[a]	Registration	Buffers	Others
Alabama	x		x	CAFO		Register CAFOs, broiler composters for dead birds
Arkansas	x	field application	N[b]		stream, prop.	Test site for lagoon, rec. keeping limit of permits for hog operations
California		x	N		water, prop., well	
Colorado		plan	N		150 ft to well	
Georgia	x		x		few	Monitor groundwater. Appl. of nitrate < 10 mg L^{-1}
Illinois	x		N	new lagoon	if > 50 au	Soil boring for lagoon
Indiana		NPDES[c]	N		water, prop., well	Inject in flood plain, 100-years flood plain protection
Iowa	approval		N		water, prop., well	Soil test or N<250 kg ac^{-1} year^{-1} inject in flood plain
Kansas	x		N or P		water, prop., well	Land loading
Minnesota	x		N or P		water, prop., well	Record keeping
Mississippi		NPDES	N or P		well	
Nebraska	initial inspection	NPDES	N		well	Minimum storage, permeability test
North Carolina		application	N		well	

	Permit		WMP[1]	Registration	Buffers	Others
	Construction	Operation				
New York		NPDES PS				
Ohio		NPDES	N			Lagoon plan
Pennsylvania		application	N		prop.	WMP if >4.94 AU ha^{-1}
Texas		NPDES	N or P		water, prop., well	Record Keeping, BMPs
Wisconsin		NPDES	N		water, prop., well	
					few	

a) WMP = waste or nutrient management plan. N or P preceding NMP or WMP indicates N-based or P-based plan, respectively.
b) N = nitrogen, p = phosphorus
c) NPDES = National Pollutant Discharge Elimination System

to follow individual permits.

In Colorado, a permit is required to add chemicals to irrigation water (chemigation) and the permit can be suspended if surface water or groundwaters can be shown to have been contaminated. BMPs to protect groundwater from salts and nitrates can be made mandatory if water-quality standards are not met. North Dakota also regulates chemigation to minimise groundwater and surface water pollution. The rule includes application rates, and installation and maintenance of irrigation equipment. South Carolina requires anti-siphon devices when irrigation systems are used to chemigate.

In Florida, a watershed approach using BMPs makes farmers in the Everglades responsible for reducing phosphorus loadings. A scheduled, phase-in tax increases over time (until 2013) unless the loading reductions are met. A higher tax is automatically imposed on operations where prescribed results are not achieved (US EPA, 1997). The state relies on voluntary BMPs, although it recommends using accepted BMPs to minimise risk of liability for damage compensation if waters are shown to have been polluted by their activities. Florida has collected a $0.62 tax on each tonne of nitrogen fertiliser sold since the 1995 Commercial Fertilizer Law.

Kentucky passed the Kentucky Agriculture Water Quality Act (KAWQ) in 1994. Under this law, which must be implemented by October 2001, farmers and silviculturalists with more than 4 ha are informed of all state water-quality requirements and must choose the appropriate BMPs for their operation. Farmers failing to develop and implement an agricultural water-quality plan can be fined a civil penalty of $1,000, plus an additional civil penalty of up to $1,000 per day, if found in violation. Once a water-quality problem is traced back to a property, the landowner must design and implement a new plan, with help from the conservation district and the Division of Water of the Kentucky Department of Natural Resources and Environmental Protection. Failure to comply may result in the farmer being fined up to $25,000 (NASDA, 1997). Water priority protection areas can be created to revert local water-quality problems. Once the area is designated as a priority, financial assistance is made available to help farmers to adopt their plan. Although the programme is voluntary and cost-shared, if the desired water quality has not been achieved within 5 years, farmers are reviewed and offered help again; if new plans are not adopted, farmers may be subject to a 'bad actor' law.

In Maine, pollution control is voluntary except for local problems or individual farm problems, in which case mandatory, site-specific BMPs must be adopted after a complaint has been confirmed. Adopting the practices protects farmers under the Right-to-Farm Law, while failure to adopt the practices subjects them to enforcement actions. This law applies to nutrients, pesticides, and animal waste problems.

All Maryland farms with land within 39 m of the Chesapeake Bay or its tributaries, or farms benefiting from a public drainage, must have an individual soil conservation and water quality plan, including BMPs for nutrients, pesticides, and sediments. Enforcement actions can also be taken if it is shown that nutrients or pesticides are improperly applied.

BMPs are mandatory in Nebraska areas where nitrate levels in surface water or groundwaters are higher than 10 mg L^{-1}, the national standard. These BMPs include irrigation scheduling, timing of fertiliser application, and pesticide application and management. The Central Platte Natural Resources District (one of the districts with power to mandate BMPs) developed a four-tier programme whereby stricter practices must be adopted as the level of nitrate in groundwater rises.

- If the nitrate level is less than 12.5 ppm, commercial fertiliser cannot be applied on sandy soil before 1 March, and autumn and winter applications are prohibited.
- When nitrate reaches between 12.6 and 20 ppm, additional restrictions apply. Restrictions include that (1) commercial fertiliser cannot be applied after 1 November, unless an inhibitor is used (this restriction does not apply to heavy soils); (2) fertiliser application must be made by a certified applicator; (3) fertilisation recommendation must take into account nitrate in the irrigation water (obtained by testing the water annually); and (4) the fertiliser plan and yield record must be submitted.
- When nitrate concentration exceeds 20 ppm, deep soil analysis must be done annually, and fertiliser must either be applied in a split, pre-plant, and sidedress application, or a nitrogen inhibitor must be used in the spring.

Nebraska, Michigan, and Montana can take legal action also when nitrates are found in groundwater. Prescribed actions may include prescriptions on fertiliser application rates, location of applications, and other practices. BMPs may be mandatory in West Virginia if a significant groundwater problem is identified.

North Carolina has started to apply water quality-based controls (when technology-based control has failed, as mandated in the CWA). A 30% nitrogen reduction was set for the Neuse River Basin, which must be met by (non-)point sources. Point sources are regulated through 5-year renewable NPDES permits, while non-point source control is voluntary and cost-shared. Livestock production is regulated as defined in the previous section; landowners must have a minimum buffer strip of 19.5 m, with the 15.6 m closest to the water planted to trees (conservation easements and financial assistance are available). A 3.9 m buffer is also required of farmers in critical areas within a water supply watershed. The Tar-Pamlico River estuary, the Chowan River, and the New River are all watersheds designated as nutrient-sensitive waters for which reduction must be established and a plan to achieve it put in place.

Oregon uses a targeted approach whereby water-quality management areas can be designed to correct water-quality problems using total maximum daily loads of nutrients (TMDL) under the CWA. As part of the restoration plan, agricultural operations within the area must implement BMPs, although practices can be prohibited only if it can be proven that they have caused water-quality problems. Fines can be applied if non-compliance continues.

Vermont requires that all fields bordering permanent waters have a vegetative buffer strip, and all farmers must adopt Accepted Agricultural Practices (AAPs) to protect water from sediments, animal waste, fertilisers, and pesticides. When it is determined that AAPs (non-structural practices) do not suffice, stricter BMPs might be imposed. Initially, BMPs are voluntary. Sufficient financial assistance (up to 85%) must be available before BMPs are made mandatory.

Virginia has had a complaint-based programme since 1996 by which the conservation districts may rule, after investigation, that a farm causes pollution, in which case the farmer must develop and implement a conservation plan. Farms in counties draining into the Chesapeake Bay must have either a 39 m buffer or a 9.75 m buffer and adequate BMPs, in addition to a water conservation plan.

- *Conservation-compliance*

The federal conservation-compliance programme is perhaps the federal programme with the largest reach of on-farm operational constraints. In 1997, a total of 30.4 million ha of highly erodible land had to be placed under specified management to be eligible for other government financial assistance. On 96% of the land, an approved plan was actively applied to minimise erosion. A combination of conservation cropping and crop residue use is by far the largest system used (27.5%), followed by conservation cropping and conservation tillage (10.8%) and conservation cropping alone (7.8%). These practices are expected to reduce erosion from an average of 33.8 kg ha^{-1} year^{-1} to 13.5 kg ha^{-1} year^{-1}. The combined wetland and cross-compliance programmes are expected to cost US farmers $3.1 million in the long run, with half of this cost borne by Corn Belt farmers (USDA, 1997b).

Pesticides

A range of chemicals is used in the production, processing, handling, and distribution of agricultural products, and is subject to numerous layers of regulation, inspection, and oversight. In the 1960s, the public perceived serious human and environmental risks from inappropriate use of pesticides. Through an elaborate registration process, compounds were reviewed to ensure that they met minimum environmental and human health criteria. The primary framework for this oversight is FIFRA and the Federal Food, Drug, and Cosmetic Act (FFDCA). The agencies responsible for enforcing these laws are the EPA and FDA. In addition, the USDA maintains food inspection and safety services responsible for monitoring and inspecting both domestic and imported foods, including fruits, vegetables, grains, and meats.

FIFRA regulates the process of pesticide approval and use through labeling of individual pesticides. All pesticide products must, by law, be registered with the EPA and used according to the label. For registration, a manufacturer must prove that, when used as directed, the product will be effective; will not injure humans, crops, livestock, or wildlife, or damage the environment; and will not produce illegal residues in food or feed (Pedigo, 1989). After the FQPA criteria and standards are determined, current pesticides in use will undergo re-registration.

Pesticides are classified either as general use or restricted use. Restricted pesticides can be applied only by a certified applicator in the state where the application occurs. In all cases, applications must follow the label instructions and applicators must keep records for 2 years of the kind of pesticide, amount, date, and place of application. In addition to the trade name, manufacturer's name and address, EPA registration number, and EPA formulation manufacturer establishment number, the pesticide label includes: ingredients (kg L^{-1}); precautionary statements; whether the pesticide is restricted-use; how soon fields can be reentered; directions for use, storage, and disposal; crops for which the pesticide is registered; and how the pesticides can be used (protective clothing, buffers to water or structure, equipment, and drift control recommendations). Residue monitoring is performed to ensure that residue tolerances are not exceeded. A violation rate of about 15% on average is found for agricultural pesticide use (US EPA, 1998b), although the nature of the violations is not explained.

The special review of high-risk pesticides leads to pesticide use reduction (for carbofuran, iprodione, vinclozolin), voluntary cancellation for some or all uses (carbofuran, methamidophos, vinclozolin), pre-harvest application restrictions added (ipodione), special packaging/engineering controls (carbofuran, methamidophos, methyl parathion), stronger use directions (copper and zinc naphthenate), spray drift labeling (copper and zinc naphthenate), and environmental safeguards (carbofuran). All of the triazines, aldicard, 2,4-D and its derivatives, 1,3-dichloropropene, dichlorvos, oxydemetron-methyl, TPTH, and tributyltins are in a special review process. Some re-registrations will result in various actions. Alachlor's review is completed and its use will be subjected to State Management Plan rule. Cadmium, captafol, and carbon tetrachloride were already voluntarily cancelled years ago. Captan was cancelled for certain food uses, and carbofuran can be used on rice only under special local need registration. Chlorobenzilate was cancelled for all uses except citrus. Cyanazine was cancelled for all uses starting December 1999; existing stocks can be used until 2002 (US EPA, 1998b).

California and Arizona each maintain a list of pesticides with the potential to leak to groundwater. In these states, a pesticide is subject to regulation within pesticide management zones if it is found below the root zone or in groundwater. The pesticide label can be modified to change the use or application rate, or registration can be cancelled. Pesticide rules such as these likely have little economic impact on farmers since a pesticide registration can be removed only if there are alternative products or practices available that avoid severe economic hardship for farmers and are relevant to areas with chronic problems.

In 1987, Iowa passed the Groundwater Protection Act, which established a two-tiered system to monitor and protect groundwater quality. The first tier relies on voluntary BMPs, such as on-farm assessment, revised education for certified applicators, and transmission of information to pesticide handlers and users about vulnerable areas and container recycling. If desired goals are not achieved with the voluntary approach, mandatory measures such as setback areas; pesticide restrictions based on soil type; changes in application rate, method, or timing, including moratoria or local ban of the chemical, are authorized. Since 1990, for example, the rate of application of atrazine is limited to 3.4 kg ha^{-1} in pesticide management zones (half the label recommendation for other areas in the state).

Kansas is also starting to designate areas as pesticide management zones. Within the zone, atrazine cannot be used within 195 m of public water supplies, within 39 m of public drinking water supply wells, and within 19.5 m of all wells. Mandatory actions can be taken if water-quality goals are not achieved. Legal actions can be taken in Georgia if negligence in handling and applying pesticides can be proven, and in Iowa if voluntary measures fail to achieve the desired outcome.

Wisconsin has a Groundwater Law (1983) that becomes stricter when threshold levels are exceeded for regulated substances such as pesticides. For instance, the standard for atrazine is 3.5 ppb; thus the first threshold is 0.35 ppb (10%). An area with a concentration of 0.35 ppb is designated as an Atrazine Management Area, while an area reading 3.5 ppb is designated an Atrazine Prohibition Area (where no atrazine can be applied). The high occurrence of atrazine above the standard level also prompted a state-wide atrazine restriction on application rates and time based on soil types, and prohibition on applying atrazine through irrigation systems. Additional restrictions apply in Atrazine Management Areas (Nowak *et al.*, 1993).

If contamination has occurred in Michigan, Montana, or Nebraska, pesticide use may be regulated. In Montana, for instance, a groundwater management plan specific to the chemical and geographic area must be developed when: (1) the level of an agricultural chemical reaches 50% of the standard; (2) a clear increase in the presence of the chemical in groundwater is observed; or (3) chemicals have been shown to migrate from a point of application to groundwater. The same is true for commercial fertilisers if residues are found in groundwater. The plans include BMPs such as buffers, lower application rates, and alternative nutrient sources and pest control.

In 1996, the EPA proposed a regulation that would require states to develop groundwater management plans as a condition to continue the use of five compounds: (1) alachlor, (2) atrazine, (3) cyanazine, (4) simazine, and (5) metalachlor. Since then, however, fear of further regulations has spurred some changes in the industry. DuPont voluntarily agreed to phase out production of cyanazine for use in the USA by the end of 1999, at which time acetachlor, a substitute for atrazine and metolachlor, was introduced (Ribaudo, 1999). In 1993, Ciba-Geigy, the producer of atrazine, voluntarily withdrew atrazine as an industrial weed control and made label changes that reduced application rates, eliminated autumn application, required buffer zones to water, and designated atrazine as a restricted-use pesticide (US EPA, undated).

Water quantity

Irrigation return flows are excluded from the definition of point sources under the CWA (US EPA, 1997), and thus are not regulated at the federal level. Although states may require permits for water extraction, farmers mostly have not been affected by water restrictions. Only Nevada regulates irrigation return flows if negative impacts are demonstrated. Texas requires reasonable practices to protect water supplies when irrigation systems are installed.

Generally, the situation is the same across the country - the demand for water is increasing because of growing populations in irrigated areas and in-stream uses for nature and species conservation, while water supplies are decreasing because of high withdrawal rates, and few new dam and canal projects are being implemented. These changes push the costs of water withdrawals and water prices up, which in turn drive incentives for more efficient irrigation systems, the fallowing of marginal irrigated land, and the switch to high-value crops in irrigated areas. However, few states or areas have imposed water-quantity restrictions, and adoption of more efficient irrigation systems is voluntary and usually cost-shared. For example, in Washington State, federal, state, and local financial assistance is available to improve irrigation system efficiency. Low interest rate loans are offered in the Ogallala Aquifer to encourage investments in more efficient irrigation systems (NRC, 1996). Texas also provides low-interest loans to invest in more efficient water systems.

Some areas have started adjusting to higher costs and new demands. In California's Central Valley, high irrigation costs have begun to cause the substitution of high-value fruits and vegetables for field crops (CAST, 1996). Drip-irrigation is spreading more quickly in Florida and California than in the Mississippi Delta, where water is cheap and land continues to be irrigated with traditional gravity-powered (high water volume) systems (NRC, 1996).

Water-quality standards for irrigation return flows in the San Joaquin Valley were put in place to mitigate the impact of high selenium concentration and resulting birth defects and deaths of migratory birds (NRC, 1996). The water thus saved becomes the property of the state. Otherwise, given the current 'use-it-or-lose-it' and seniority laws, little private incentive for conservation exists. California has also forced the Imperial Irrigation District to increase its efficiency. The impact on farmers is limited at this time.

Oregon has proposed setting maximum amounts of water use, and Arizona has gradually reduced withdrawals from groundwater in active management areas for use on crops (NRC, 1996). Many irrigation water authorities now require water withdrawal measurement. Requirements for the protection of endangered species are also emerging as a driving force in future water management in the West (NRC, 1996). The demand for water to protect endangered species, especially endangered salmon species, may have greater impact on water availability in the Pacific Northwest. Recovery actions call for limited future water withdrawals, flow augmentation, water acquisition, new storage assessment, and uncontracted storage space, which affect (or will affect) mostly irrigators in Oregon, Idaho, and Washington.

In the Great Plains, irrigators almost exclusively depend on the large, 455,000 km^2 Ogallala Aquifer. Included are irrigators in Colorado (4%), Kansas (10%), Nebraska (65%), New Mexico, Oklahoma (3.5%), South Dakota, Texas (12%), and Wyoming. Water extraction exceeds recharge and water levels are declining in the region. Texas, Oklahoma, and Kansas management districts control the issuance of drilling permits, controlling well spacing, developing workable recharge, installing water meters, and preventing water waste (NRC, 1996). Some are regulatory; for example, all three states require spacing for new wells, Kansas requires meters on wells, and towns in Kansas and Nebraska use fresh water and deliver wastewater to agriculture. Overall, however, farmers' practices have been affected only by higher water extraction and use costs, rather than regulations.

Water management districts (WMD) in Florida face pressures similar to their California counterparts, from population growth and environmental demands. Under the Florida Water Resources Act of 1972, five water districts regulate water quality and quantity. Under the same law, producers who implement BMPs for nitrate reduction are given a waiver of liability from future remediation and are assumed to be compliant with state groundwater standards (NASDA, 1997).

The only Florida areas experiencing water-quantity problems are the South (Lake Okeechobee) and Southwest Districts. No new permits can be issued in Southwest Florida WMD, and water-conserving technologies are required from new and existing irrigators (NRC, 1996). Farmers may suffer economic losses due to lack of water only when a drought is worse than one in ten (the standard on which the permit is based). District farmers are also pressured (but not obligated) to adopt water-conserving irrigation systems. The major difference between Florida and the irrigation situation in the western US is that, in Florida, the water belongs to the state, permits must be obtained to use water, and permits must be renewed every 5 to 10 years. Water-quality problems associated with irrigation, however, are different than in the West. Stormwater runoff mostly causes pollution in Florida, instead of pollution from return flows in the West.

Wetlands

The only data available concerning state-level wetland programmes are from Kusler et al. (1996), cited in Heimlich et al. (1998). The survey highlighted 44 states that have wetland statutes or laws, including 18 that regulate both coastal and freshwater wetlands, seven that regulate only coastal wetlands, and four that regulate coastal and part of their freshwater wetlands (Heimlich et al., 1998). However, only 33 states track and enforce wetland permits, and only 26 penalise violators of the wetland laws (Heimlich et al., 1998). According to Kusler's 1996 survey, the states of Massachusetts, Maine, and South Carolina have the most comprehensive wetland regulations. Massachusetts and South Carolina, however, exempt agricultural activities from their regulations. States with regulations require permits to modify wetlands. Some states have included wetlands in their definition of waters of the commonwealth, and thus a wetland's water quality is protected under the state no-discharge laws. A group of states, including Massachusetts, has mitigation and no net loss laws in place, and even net gain goals for the future.

The only on-farm constraints are some setbacks, and prohibitions against filling in wetlands for other uses.

Soil

Sewage sludge

Concern regarding contamination of agricultural soils mainly relates to the application of sewage sludge. An issue of some concern relates to chemicals and metals in the sewage, and the cumulative chemical loading applied on the land surface. The application of sewage sludge to agricultural land is subject to EPA restrictions on the concentrations of various chemicals in the sewage, and the cumulative chemical loading applied to a hectare over a year. Sewage sludge must be applied at a nitrogen-based agronomic rate and may not be applied: (1) where it is likely to affect endangered species or their habitat; (2) to flooded areas, or frozen or snow-covered ground, or where it can enter a wetland; or (3) within 10 m of any waters. Sewage sludge must be labeled for its content and records of application kept for 5 years. Sites where sewage sludge is applied must also be monitored at a frequency that varies with the amount of sludge applied.

Excessive accumulation of salt

Roughly 30% of land in the contiguous USA, mostly in the West, has at least moderate potential for salinity problems (NRC, 1996). Areas such as the upper Colorado River Basin, northern Great Plains, and San Joaquin Valley suffer from salinity or drainage problems. In the Colorado River Basin, technical assistance and grants are available from the USDA's Natural Resources Conservation Service (NRCS) for the seven states draining into the basin (Arizona, California, Colorado, Nevada, New Mexico, Utah, and Wyoming). In California, 1.8 million ha of irrigated cropland are affected by saline soils or by saline irrigation water. A large proportion of this land is found in the San Joaquin Valley, a highly productive valley with groundwater depletion (CAST, 1996).

The former Colorado River Basin Salinity Control Program (now part of EQIP) was funded at $14 million a year during the early 1990s, then dropped to $4.5 million before it started increasing again to $5 million in 1999 (under EQIP).

Cost-share money is provided under this project to farmers in Arizona, California, Colorado, Nevada, New Mexico, Utah, and Wyoming, to improve water delivery systems and the efficiency of on-farm irrigation systems to replace the old practice of flooding fields. The project is part of a larger international agreement to deliver to Mexico a fixed amount of water (via the Colorado River) per day and with a specified quality. The quantity of water extracted is not affected by this project because it is the jurisdiction of the states to determine water-quantity limits (Mason, 1999). Under EQIP, a farmer can receive 75% cost-share, for a maximum of $50,000 per year, to install recommended practices.

Erosion

Erosion control has received more assistance than any other US agricultural conservation practice. Farmers in California, Florida, Georgia, Kentucky,

Massachusetts, New York, North Carolina, Pennsylvania, Rhode Island, Vermont and Washington must have a plan to prevent sedimentation or damage to off-site properties. The plans are enforced at the local level. A limitation of some of the plans, including New York's, for example, is that only the preparation of the plan is mandatory. Implementation is not required.

A number of states have instituted laws to control erosion and sediment loading, which do not exempt agriculture and forestry (US EPA, 1997). Most states with laws that reach into the agricultural sector have less stringent requirements for agriculture than for other industries, such as encouraging voluntary use of BMPs. Most legislation includes permitting requirements, buffer zones, and enforceable obligations that apply at the watershed level to protect an impaired or valuable water resource. Through a sediment and erosion control law, or through power granted by the state as part of a planning and zoning code, most regulations are implemented at the local level (US EPA, 1997).

Some states take enforceable actions after pollution has occurred, such as with the 'bad actor' laws in Maine, New Hampshire, Texas and Virginia. Other states make BMPs more than voluntary by requiring that farmers be held responsible for damages if they are proven to cause pollution (e.g., Georgia), by prohibiting states to take action against a farm that has implemented BMPs even if the farm is causing pollution (Delaware, Florida, Idaho, Illinois, Iowa, Kentucky, Maine, Montana, Pennsylvania, Texas, and Vermont), or by making BMPs a protection against nuisance actions (Maine, Michigan, Ohio, and Vermont) (US EPA, 1997). One third of the states (including Florida, Idaho, Maine, Maryland, Mississippi, Montana, Ohio, Pennsylvania, South Dakota, Utah and Wisconsin) have a 'no more stringent' than the federal law clause, which requires that state laws be less stringent than federal laws, or if they are more stringent, requires detailed justifications for such actions.

Ohio has an erosion law whereby farmers must install and use BMPs to limit soil erosion to the T-level (where T-level is the rate at which long-term soil productivity may be depleted). Enforcement is complaint driven and no individual plans are necessary if practices used on the farms are in the local NRCS technical guide. Practices must be adopted only if 75% cost-share assistance is available.

Nebraska has an Erosion and Sediment Control Act that sets the maximum soil-loss rate by soil type, to be achieved through conservation plans. However, farmers must adopt the plan only if 90% of the cost to install permanent structural practices is available as cost-share. 'Bad actor' laws can also be used in Maine, New Hampshire, Texas, and Virginia to mandate management plans if it can be proved that an agricultural activity is creating or will create pollution. In Wisconsin, the bad actor law can be invoked only in priority watersheds, and farmers must be given a year to comply. Most of these states integrate local Soil and Water Conservation Districts (SWCD) into the implementation and enforcement of the plans. Indiana has a voluntary programme to reduce soil erosion on cropland, which may become mandatory if the voluntary programme does not achieve desired outcomes. Since 1994, Maryland has had an Agricultural Sediment Control Law that is complaint driven. Farmers reported to cause pollution problems must adopt recommended BMPs or face enforcement actions.

Air

Odour and noise

Exposure to swine odour has been reported to cause migraine headaches, nausea, diarrhea, and dizziness (Meador, 1998). In addition to odour, swine operations release ammonia (NH_3) and hydrogen sulphide (H_2S). These unpleasant experiences have given rise to many nuisance suits in the USA. Most swine operations, however, are protected by the right-to-farm laws present in each state, although an Iowa farmer was charged $45,000 because he caused a temporary nuisance (Meador, 1998). CAFOs are not regulated by the CAA, but some states are taking action. In North Carolina, if economically feasible odour control technology is made available, it must be incorporated into the regulation by spring 1999 (Meador, 1998). The Minnesota Pollution Control Agency (MPCA) monitors and identifies livestock facilities that exceed the ambient air quality standard for H_2S. Livestock producers in Illinois must use odour control methods when handling manure, and new or expanding facilities must adopt odour control methods and technology to prevent air pollution (Meador, 1998). New, expanding, and renewing CAFOs in Kansas must prepare an odour control plan to obtain a permit. In Kentucky, odour is included among air contaminants, and thus is covered by air pollution regulation. Both Oklahoma and South Carolina CAFOs must have an odour abatement plan. Many bills to control odour and air pollution at the state level are being proposed and may be adopted in the near future. Generally, they seek to increase setback distance and subject livestock operations to air and odour restrictions (e.g., Arkansas, Colorado, Illinois, Indiana, Iowa, Kansas, Maine, Maryland, Mississippi, Nebraska, and Oklahoma).

Burning

The 1991 California Burning Reduction Act called for the ban of rice straw burning in the Sacramento Valley Basin by the year 2000, with a gradual phase out of burning (10% reduction per year) by 2000 (Wrysinski *et al.*, 1999). Regulation AB-318 was enacted to reduce air pollution in the Valley. Rice growers are allowed to burn up to 36,000 ha in the fall and 44,000 ha in the spring, to control for disease. After 2000, a safe harbour of 25% of the area is permitted to burn to control for diseases that cause yield losses. Additional conditional burning permits can be obtained for a fee (that covers the county's cost) if no other economically and technically feasible alternatives are available to control diseases. The permit cannot exceed 25% of the farmers' rice fields, or a total of 50,000 ha in the Valley. Also, growers with less than 160 ha may burn their entire field once every 4 years if they are in compliance with the remaining clause of the law. Rights to burn are not transferable. The rice area has continued to increase since 1992, expanding from 157,600 ha to 200,000 ha in 1996 (California Rice Industry Association and California Rice Promotion Board, 1998). Today, however, approximately 90% of the rice straw is incorporated or rolled with the soil and flooded, instead of burned. This change in technology costs farmers between $100 and 200 per ha, depending on the incorporation practice and the cost of water to flood fields (Garr, 1999). Most of the costs are borne by farmers. Because water is left on the field for the whole winter, these flooded areas provide large ecological benefits via habitat for migratory waterfowl and endangered species.

The 1997 California SB-318 authorised regulations that required 62% of the rice fields not to be burned, but few off-farm alternatives for the resulting rice straw were found. A fund was created to find alternatives to burning the rice straw, which was burned by approximately 99% of rice growers at that time (Garr, 1999). The fund is cost-shared by the state rice board, rice growers, and other public funds. Rice straw tax credits are also given to off-farm users of rice straw to develop incentives for its use.

Nature conservation

Endangered species

Most endangered species depend heavily on private lands (Bean, 1999). Texas is the first state to have a programme that pays landowners up to $10,000 to conserve rare species and their habitats. Otherwise private landowners are not compensated for their efforts and may try to hide evidence that an endangered species is on their property in order to avoid the need to protect its habitat. Given that 701 plant species and 474 animal species are listed in the USA, and 120 species have designated critical habitat, the potential cost to farmers and ranchers of species conservation could be high. As of September 1998, 243 incidental take permits or 'Safe Harbors' have been issued for habitat conservation plans and another 200 are at various stages of development. The Safe Harbor clause, first initiated in 1995, limits a farmer's liability to an environmental baseline established when he/she is enrolled. Under the Safe Harbor clause, the farmer's disincentive to attract species or protect habitat beyond initial conditions is removed (US FWS, 1998).

Transgenic crops

Within the USA, transgenic crops are subject to a tripartite regulatory system. The Animal and Plant Health Inspection Service (APHIS) of the USDA, the EPA, and the FDA each play a part in this system, which focuses on determining the relative safety of biotechnology's end products and uses, and not on the process by which they are created (NRC, 1989). APHIS' mandate is to ensure that a transgenic product is as safe to grow as are conventional varieties. The EPA's primary responsibility is to ensure the environmental safety of new plant-pesticidal substances (a mandate that includes human-health effects). The FDA's regulatory process focuses on how foods made from transgenic crops might affect human-health. The tripartite regulatory process initially conducted product-by-product reviews of transgenic crops. Now, the process may assess the possible risks from new transgenic products based on the experience gained in reviewing earlier products. The implication is that some crops might be approved, or disapproved, without actual testing in fields. Critics worry that the APHIS regulatory process for release of transgenic crops is simplified for crops with which the agency is familiar. However, some actions may be required by the agencies overseeing the reviews before release is approved. For example, the EPA allows commercialisation of Bt crops only if an insect resistance-management plan is in place.

The National Research Council (NRC) recently concluded that the overarching regulatory framework for transgenic plants resistant to pests must be completed as soon as possible. In the NRC's view, the diversity of bio-engineered crops gaining commercial status, as well as the rates at which they are being adopted, argue strongly for action now (NRC, 2000). Ervin *et al.* (2000) recommend four improvements:

- increased roles for environmental scientists;
- designation of a lead environmental agency;
- more transparency and public access to regulatory information; and
- inclusion of social values in setting risk decision levels.

Animal welfare

The welfare of animals is affected at the rearing, transportation, and slaughtering stages. On-farm constraints implied by federal and state laws, or by industry at each stage of production, are reviewed below.

Rearing of farm animals

The welfare of farm animals in the USA relies mainly on industry standards rather than government-imposed on-farm constraints. The underlying hypothesis is that by increasing animal welfare, farmers increase productivity, and thus it is in their best interest to treat animals 'right' (Reynnells and Eastwood, 1997). Battery cages are the dominant method to maintain laying hens. Recommended guidelines for cage size is 312 cm^2 per bird. There are no commonly accepted definitions of free-range chickens in the USA. Debeaking layers and broilers, forced molting of laying hens, lack of access by layers to nest boxes or housing that allows a full-body range of motions, and disposal of male chicks are common practices for fowl.

Pigs are castrated and tails are removed without anaesthetics, and gestating and farrowing sows are housed in stalls where they are unable to turn around. Calves are transported on their first day of life; they are housed in stalls where they cannot turn around, and they may be fed a liquid diet (without forage). Dairy cows are dehorned without anaesthetic, and some dragging of downed cows for selling or slaughter is permitted. Cattle are also dehorned without anaesthetic and are branded with a hot iron. Geese are force-fed via a pump down their throats to produce 'foie gras'. A common practice across species is genetic selection for production traits that may render animals more violent (e.g., pigs). Because these are common farming practices they cannot be prevented by law. Washington and Puerto Rico are the only states with requirements for adequate light, and Wisconsin is the only state requiring clean living conditions. No requirements for adequate exercise, space, or ventilation are found in state anti-cruelty laws.

As for rearing, most federal and state laws that pertain to animal welfare exempt farm practices or only applies minimally. Comprehensive data about on-farm constraints for slaughtering animals are not available, but the restrictions are not thought to exert widespread effects.

Human-health

Hormones and animal feed requirements

Currently the following antibiotics are in use in USA feed for livestock and poultry: chlortetracycline, procaine penicillin, oxytetracycline, tylosin, bacitracin, neomycin sulphate, streptomycin, erythromycin, linomycin, oleandomycin, virginamycin, and bambermycins. Also, arsenical, nico-furan, and sulfa compounds are often used in animal feed. Antibiotics are regularly used at a rate of 2-50 grams per tonne to improve performance in animals. The levels are increased to 50-200 grams per tonne when specific diseases are targeted. As indicated in the section on transgenic crops above, rBST is used widely in US dairies.

IMPACT OF ENVIRONMENTAL AND HUMAN-HEALTH LEGISLATION

Although some theoretical models predict large trade impacts (including relocation of industries) resulting from high environmental standards (Pethig 1976; Porter 1991), the empirical evidence is not convincing (Ervin, 1999; Grossman and Krueger, 1992; Jaffe *et al.*, 1995; Lindsey and Bohman, 1997; Palmer *et al.*, 1995). Many reasons have been given for this relative resiliency of trade to environmental standard costs. The first and most important is that compliance costs are usually small relative to total costs (Grossman and Krueger, 1992). Second, the difference between air and water pollution regulations among the major industrialised countries is slight (Palmer *et al.*, 1995).

The reasons compliance costs are small in agriculture are manifold.

- First, the agricultural sector benefits from having voluntary-payment programs instead of mandatory standards that apply to other industries (Ervin, 1997). The review of environmental standards affecting agriculture presented earlier supports this argument.
- Second, most mandatory programs are targeted to limited regions, areas, or watersheds with acute environmental problems and are reactive instead of preventive. For example, cropland practices are targeted to highly erodible lands and to areas with high water tables susceptible to leaching, and

pesticide banning is targeted to pesticide management zones where pesticides have been found in groundwater.

- Third, mandatory programmes are also targeted to industries with the most pollution potential, such as those handling livestock manure in liquid forms, including dairy and swine operations, and not poultry firms that mostly handle manure in solid form.
- Fourth, the large-scale livestock operations, more likely to be regulated, can capture economies of scale in their production and compliance costs, thus even further reducing the impacts of regulation (Leatham *et al.*, 1992; for dairy, see Wright, 1998).

No systematic collection of compliance costs for the different agricultural sectors exists; therefore, collecting this information for all agricultural sectors, although a noteworthy exercise, was beyond the scope of this chapter. Because, *ceteris paribus,* higher compliance costs will be more likely to exert effects on trade, agricultural activities in states judged to have some of the highest environmental compliance costs were selected for detailed analyses. This exercise is not a formal econometric analysis of the costs of regulations, but rather an interpretation of existing estimates in the literature; it suggests hypotheses to be tested instead of trying to test these hypotheses.

Table 5.3 presents a summary of estimated production and environmental costs by commodity. The ranges presented include data from a national review and 'upper bound' for the states believed to have the largest compliance costs, along with the government subsidies (referred to as Producer Support Estimate, or PSE) (see Carpentier and Ervin, 1999, for details). The states and sectors most likely to have large compliance costs are: (1) dairy, grape, tomato, and orange in California; (2) hog in Iowa and North Carolina; and (3) maize and soybean in Nebraska.

California, Iowa, and North Carolina have witnessed major manure spills over the last few years, have water-quality problems related to animal waste discharge, and have the second, sixth, and third largest livestock cash sales in the USA, respectively. These state situations capture the high end of regulation and compliance costs that are most likely to induce competitiveness effects.

California has the largest dairies in the USA and contained 25% (1,800 out of the 7,250) of the dairies with more than 200 head in 1997-1998. The costs for the California dairies are only potential, because attempts to regulate the dairy industry in California have not been successful. However, the crackdown on non-compliant dairies started in 1998 by increasing enforcement capacity, and giving jail sentences, as well as fines, to dairy operators who polluted water, in addition to the added resources and number of inspectors. Compliance costs for dairies depend on initial practices; whether only some BMPs must be used, or if the whole manure management plan must be redesigned; and which policy is used to regulate (limited number of cows per ha, design standard, etc.).

Less than 5% of Iowa's 8.8 million ha of row-crop land is covered by the 580 manure management plans written before 1998 (NASDA, 1997). Many CAFOs are located near agricultural drainage wells (ADWs), the most vulnerable areas

Table 5.3   Commodity production and environmental costs and subsidies, 1997

	Production costs ($/unit)		Social	Subsidies		Coverage	
	Private	Social	(% of private)	Production (% PSE)	Environmental (%)	Animals (M head or ha)	Farms
Dairy (head)	2,815	20-691	0.7-24.5	47	0-75	4	2,257[a]
Poultry (bird)		0-0.13		3		326	3,890
Cattle (bredcow)	757.94	1.8-30	0.2-4	4	0-75	26	6,024
Swine (head)	145.28	0.3-9.2	0.2-6.3	5	0-75	37	5,540
Sheep[b] (lamb)	84.00	--	0-5.4	6	0-75	2	189
Maize (hectares)	876.2	0-52.50	0-0.05	17	<75	30.2	na
Wheat (ha)	451.2	0-?	0 -	32	<75	23.5	na
Apples (ha)[c]	6,343	87.50	0-1	6.5	small	0.24	na
Grapes (ha)[d]	6,403	33-4,310	7.5-67	5.5	small	0.4	na
Oranges (ha)[e]	8,918	30	0-0.003	8.6	small	0.4	na
Tomatoes (ha)[f]	4,435	50	0-0.01	4.2	small	0.16	na

a) Includes farms and dairy cows on farms with more than 500 head (700 head needed for 1,000 AU)
b) 1992 feeder lamb budget, high concentrate diet, Pennsylvania
c) PSE for deciduous trees in California 1994-1996 from Sumner and Hart (1997).
d) PSE for grapes in California 1994-1996 from Sumner and Hart (1997).
e) PSE for citrus and olive in California 1994-1996 from Sumner and Hart (1997).
f) PSE for tomatoes in California 1994-1996 from Sumner and Hart (1997).
Source: Carpentier and Ervin (1999).

because they provide a direct link to the aquifers, rivers, and lakes supplying drinking water to thousands of people. Stricter standards on feeding operations near ADWs may be costly to the livestock industry, given the large number of wells and the limited locations left for new sites. Iowa contains 1,200 swine operations with more than 2,000 head of swine, and 600 operations with more than 1,000 head of cattle (USDA, 1998b). If these operations overlap with an area having environmentally sensitive ADW, they could incur relatively high costs of compliance. However, experts who are studying the state livestock waste problems do not perceive that the regulations are imposing excessive costs; indeed, the swine operations are 'over-complying' with existing standards (Babcock, 1999). This seemingly irrational economic behaviour may be quite rational in that their actions to add extra safety for spills may avert more stringent future regulations and minimise legal suits from the general public.

North Carolina's moratorium on new, large confined animal feedlots is the well-publicised case that has shaken the hog industry. Strict standards now exist for hog operations of more than 250 hogs (100 AU). These operations are inspected twice a year and fines of up to $10,000 can be imposed. North Carolina is home to 1,300 swine operations with more than 2,000 head (23% of all large swine operations).

Several observations emerge from Table 5.3. First the variation in regulation across commodities is large, with the hog and dairy industries most affected, followed by cattle, maize, and wheat, and then by the poultry sector, with fruit and vegetable production having the least costs. Given the levels of pesticides used in fruit and vegetable production, they may become more regulated as pesticide residue standards get stricter. In the livestock industry, the dairy sector has the highest absolute and most variable compliance costs. Cattle follow with a variation of $2 to $30 a head, and swine generally have compliance costs of $3 to $9 per hog. As a percentage of production costs, the upper bounds for dairy and grape compliance costs lead with up to 25 and 67%, respectively. Swine and maize follow with compliance costs representing up to 6 and 5% of total costs, respectively. Compliance cost information for sheep and poultry are not readily available because they have not been widely regulated. However, we know that close to 6,000 poultry farms and 326 million birds, and 189 sheep farms and 2 million sheep are eligible to be regulated under the CAFO rule.

For the same commodity, compliance costs can range from near zero to thousands of dollars per unit, such as for dairies. This variation is due to (1) site-specific biophysical characteristics, (2) the media being protected, (3) geographical/cultural variation in regulation levels, (4) the methods by which the media are being protected, (5) the size of the farm being regulated, and (6) the amount of R&D allocated to agri-environmental technological innovation. Although the range in Table 5.3 starts from zero, studies have shown that after nutrient and/or pesticide management is implemented, long-term profits sometimes go up and the costs could be negative (see Carpentier and Ervin, 1999, for details).

Table 5.3 does not reveal that some farms, especially smaller farms, have negative economic returns. These farms have very small 'profit' margins and the smallest increase in cost may put them out of business. Government subsidies are highest for the dairy industry, followed by wheat and maize. All other supports are less than 9%. Support to fruits varies from a low of 5.5% for grapes to a high of 8.6 (for disaster relief) for citrus. Tomatoes, poultry, and cattle receive the lowest support, varying from 3 to 4.2%, while swine and sheep receive 5 and 6%, respectively.

For these reasons, it is unclear whether agricultural environmental regulations have displacement effects as in the pollution haven hypothesis. For instance, claims have been made that corporate swine farms have started to move out of North Carolina because of stricter environmental regulations, and that dairies have moved out of California's Central Valley, but no empirical evidence of the causal effect has been provided. It is clear that some displacement and/or status quo will occur in the USA given the recent actions by three local areas in Idaho, Minnesota, and Nebraska; by three counties in Colorado, Georgia, and Washington; and by three states (Arkansas, Mississippi, and North Carolina) to pass moratoria on confined hog production (Animal Confinement Policy National Task Force, 1998). However, trade impacts may be small since three of the four largest pork-exporting countries, USA, Canada, and Denmark, have strict legislation, and Taiwan is fighting an outbreak of foot-and-mouth disease. In Denmark, the problem arises because of a shortage of land and resulting high level of manure application, and in Canada and the USA because of the concentration of the pork production in a few counties (USDA, 1998a). Uncertainty and litigation costs are other costs brought about by the upsurge of environmental restrictions. Mississippi, for example, just imposed a $14 million fine on a poultry operation for discharging manure into state waters. In North Carolina, Smithfield Foods Inc., the largest hog producer and integrator, entered into an agreement with the state of North Carolina to avoid being sued. Such restrictions are likely to set a trend, which some argue, will push the swine industry to countries with milder environmental standards.

In summary, the compliance costs imposed on agriculture by current environmental standards appear limited in scope, as a percentage of total costs, and relatively small in magnitude for the commodities under examination. As with other industries in the USA, compliance costs vary widely as a percentage of private production costs, and some particular industry segments may be subject to extreme cost stress that causes their migration to other domestic or foreign areas. This may be an 'optimal' adjustment if the appropriate environmental charges are applied in both settings. The apparently limited current competitiveness effects may not persist. More and more stringent standards are being imposed on farms and ranches in the USA, especially by states. The fundamental forces driving that shift, including structural shifts in the industry, rising per capita incomes, immigration to rural areas by non-farm residents, and increasing recreation in rural areas, appear robust. Maintaining negligible trade competitiveness effects for farming will require more environmental subsidies, and/or more cost-effective approaches, including more research and technology development, to satisfy the standards in low-cost ways than currently exist (Ervin et al., 1998).

CONCLUSIONS

In the short run, concerns that the agriculture industry might be unduly hampered by excessive domestic environmental regulation are unfounded. The policy approach so far has been mostly one of voluntary actions through payments such as the conservation and wetland reserve programmes. However, agriculture has been identified as a key source of some of the nation's remaining pollution problems by a growing body of science. Serious environmental problems tend to concentrate where production pressure is intense and natural resources are vulnerable to damage. The resource most affected is water quality, and nitrogen and phosphorus, primarily occurring as nitrate and phosphate, are the most important nutrient sources of contamination.

We have reviewed on-farm agricultural constraints derived from environmental policies at the federal, state, and local levels. It is clear that water-quality policies are the most constraining, especially for confined animal feeding operations. Regulation of confined animal operations is spreading at the state level and being revised at the federal level, while pesticide use regulations are being reviewed at the federal level through reregistration of all pesticides. Nationally, CAFOs are considered to be the principal source of 16% of the water impaired by agricultural practices, the third leading agricultural source of water pollution after non-irrigated crops and irrigated crop production. Studies show that pesticide regulation has not imposed large costs, except for minor crops, and does not affect trade because pesticide regulations may not be more costly than in other trading countries.

The reasons compliance costs are small in agriculture include:

- voluntary-payment programmes exist instead of mandatory standards that are applied to other industries;
- most mandatory programmes are targeted to limited regions, areas, or watersheds with acute environmental problems and are reactive instead of preventive;
- mandatory programmes are also targeted to industries with the most pollution potential; and
- the large livestock operations, more likely to be regulated, can capture economies of scale in their production and compliance costs, thus even further reducing the impacts of regulation.

Few of the standards applied to crop production 'have teeth', and when they do, it is usually because the problem is so serious in the local area that everyone, including farmers, agrees something must be done.

Water quantity and air pollution restrictions both have the potential for large impacts in the future, but currently impose narrow or insignificant constraints on agriculture. Similarly, the protection of endangered species has had limited impacts on agriculture in the past, but has the potential for greater impact in the future, both in terms of limiting agricultural land and limiting water that can be extracted for irrigation.

Acknowledgement

The authors acknowledge and appreciate the excellent editorial assistance of Suzanne DeMuth.

REFERENCES

Alexander, R.B., Smith, R.A. and Schwartz, G.E. (1997) The regional transport of point and nonpoint-source nitrogen to the Gulf of Mexico. In: *Proceedings of the First Gulf of Mexico Hypoxia Management Conference, 5-6 December 1995, Kenner, Louisiana.* EPA-55-R-97-001, US Environmental Protection Agency, Washington, D.C. Web site http://pelican.gmpo.gov/nutrient/front.html (viewed April 1999).

Allen, A.W. (1993) *Regional and State Perspectives on Conservation Reserve Program (CRP) Contributions to Wildlife Habitat.* US Fish and Wildlife Service Federal Aid Report, National Ecology Research Center, Fort Collins, Colorado.

America's Clean Water Foundation (undated) *State Programs for Poultry Production.* America's Clean Water Foundation, Washington, D.C.

Anderson, T.L. and Snyder, P.S. (1997) *Priming the Invisible Pump.* Policy Series No. 9, Political Economy Research Center, Bozeman, Montana. Web site http://www.perc.org/ps9.htm (viewed December 1998).

Animal Confinement Policy National Task Force (1998) *National Survey of Animal Confinement Policies.* http://cherokee.agecon.clemson.edu/confine.htm (viewed April 1999).

Antle, J.M. and Just, R.E. (1991) Effects of commodity program structure on resources and the environment. In: Just, R.E. and Bockstael, N. (eds) *Commodity and Resource Policies in Agricultural Systems.* Springer-Verlag, New York, pp. 97-127.

Babcock, B. (1999) Personal communication, 4 February. Iowa State University, Ames, Iowa.

Barbash, J.E. and Resek, E.R. (1996) *Pesticides in Groundwater: Distribution, Trends, and Governing Factors.* Ann Arbor Press, Ann Arbor, Michigan.

Bean, M.J. (1999) Endangered species, endangered act? *Environment,* 41(1), 12-18, 34-38.

Berner, A.H. (1994) Wildlife and federal cropland retirement programs. In: *When Conservation Reserve Program Contracts Expire: The Policy Options: Conference Proceedings, 10-11 February 1994, Arlington, Virginia.* Soil and Water Conservation Society, Ankeny, Iowa, pp. 70-75.

Birch, A.N.E., Geoghegan, I.E., Majerus, M.E.N., Hackett, C. and Allen, J. (1997) Interactions between plant resistance genes, pest aphid populations and beneficial aphid predators. *Scottish Crop Research Institute Annual Report 1996/97.* Scottish Crop Research Institute, Dundee, Scotland, pp. 68-72.

California Rice Industry Association and California Rice Promotion Board (1998) *Rice Straw: Facts, Opportunities, Perspectives.* Sacramento, California.

Carpentier, C.L. and Ervin, D.E. (1999) *Environmental and Health Standards: Compliance Costs for US Agriculture.* Policy Studies Working Paper, Henry A. Wallace Institute for Alternative Agriculture, Greenbelt, Maryland.

CAST (1996) *Future of Irrigated Agriculture.* Task Force Report No. 127, Council for Agricultural Science and Technology, Ames, Iowa.

Duvick, D.N. (1996) *Biology, Society, and Food Production: New Concepts, Old Realities.* Unpublished manuscript (March), Iowa State University, Ames.

Environmental Law Institute (1997) *Enforceable State Mechanisms for the Control of Nonpoint Source Water Pollution.* Environmental Law Institute, Washington, D.C.

Environmental Law Institute (1998) *An Almanac of Enforceable State Laws to Control Nonpoint Source Water Pollution.* Environmental Law Institute, Washington, D.C.

Ervin, D.E. (1995) A new era of water quality management in agriculture: From best management practices to watershed-based whole farm approaches? *Water Resources Update,* 101, 18-28.

Ervin, D.E. (1997) *Agriculture, Trade and the Environment: Anticipating the Policy Challenges.* OECD/GD(97)171, Organisation for Economic Co-operation and Development, Paris.

Ervin, D.E. (1999) Toward GATT-proofing environmental programs for agriculture. *Journal of World Trade,* 33, 63-82.

Ervin, D.E., Batie, S.S., Welsh, R., Carpentier, C.L., Fern, J.I, Richman, N.J. and Schulz, M.A. (2000) *Transgenic Crops: An Environmental Assessment.* Policy Studies Report No. 15, Henry A. Wallace Center for Agricultural and Environmental Policy at Winrock International, Arlington, Virginia. Web site http://www.winrock.org/transgenic/pdf.

Ervin, D.E., Runge, C.F., Graffy, E.A., Anthony, W.E., Batie, S.S., Faeth, P., Penny, T. and Warman, T. (1998) Agriculture and the environment: A new strategic vision. *Environment,* 40(6), pp. 8-15, 35-40.

Frederick, K.D. (1990) Water resources. In: Sampson, R.N. and Hair, D. (eds) *Natural Resources for the 21st Century.* Island Press, Washington, D.C., pp. 143-174.

Garr, J.D. (1999) Personal communication. Ducks Unlimited, Colusa, Calif.

Gilliom, R.J. (1997) *Pesticides in the Nation's Water Resources: Initial NAWQA Results.* US Geological Survey, Reston, Virginia.

Grossman, G.M. and Krueger, A.B. (1992) *Environmental Impacts of North American Free Trade Agreement.* Discussion Paper, Princeton University, John M. Olin Program for the Study of Economic Organization and Public Policy, Princeton, N.J.

Heimlich, R.E., Wiebe, K.D., Claassen, R., Gadsby, D. and House, R.M. (1998) *Wetlands and Agriculture: Private Interests and Public Benefits.* Agricultural Economic Report No. 765, US Department of Agriculture, Economic Research Service, Washington, D.C.

Hilbeck, A., Baumgartner, M., Fried, P.M. and Bigler, F. (1998) Effects of transgenic *Bacillus thuringiensis* maize-fed prey on mortality and development of immature *Chrysoperla carnea* (Neuroptera: Chrysopidae). *Environmental Entomology,* 27, 1-8.

Jaffe, A.B., Peterson, S.R., Portney, P.R. and Stavins, R.N. (1995) Environmental regulations and competitiveness of U.S. manufacturing: What does the evidence tell us? *Journal of Economic Literature,* 33, 132-163.

Johnson, D.H. and Schwartz, M.D. (1993) The Conservation Reserve Program and grassland birds. *Conservation Biology,* 7(4), 934-937.

Knopf, F.L. (1994) Avian assemblages on altered grasslands. *Studies in Avian Biology,* 15, 247-257.

Krimsky, S. and Wrubel, R.P. (1996) *Agricultural Biotechnology and the Environment: Science, Policy, and Social Issues.* University of Illinois Press, Urbana.

Kusler, J.A., Ray, C., Klein, M. and Weaver, S. (1996) *State Wetland Regulation: Status of Programs and Emerging Trends.* Association of State Wetland Managers, Berne, N.Y.

Leatham, D.J., Schmucker, J.F., Lacewell, R D., Schwart, R.B., Lovell, A.C. and Allen, G. (1992) Impact of Texas water quality laws on dairy income and viability. *Journal of Dairy Science,* 75, 2846-2856.

Lester, J.P. (1994) A new Federalism? Environmental policy in the States. In: Vig, N.J. and Kraft, M.E. (eds) *Environmental Policy in the 1990s: Toward a New Agenda.* Second edition. CQ Press, Washington D.C., pp. 51-68.

Lindsey, P.J. and Bohman, M (1997) Environmental policy harmonization. In: Loyns, R.M.A., Knutson, R.D., Meilke, K. and Sumner, D.A. (eds) *Harmonization/Convergence/Compatibility in Agriculture and Agri-food Policy: Canada, United States and Mexico: Proceedings of the Third Agricultural and Food Policy Systems Information Workshop.* University of Manitoba, Winnipeg, Canada, pp. 93-111.

Losey, J.E., Raynor, L.S. and Carter, M.E. (1999) Transgenic pollen harms monarch larvae [scientific correspondence]. *Nature,* 399(20 May), 214.

Marks, R. and R. Knuffke (1998) *America's Animal Factories: How States Fail to Prevent Pollution from Livestock Waste.* Clean Water Network; Natural Resources Defense Council, Washington, D.C., p. 183.

Mason, D. (1999) Personal communication. US Department of Agriculture, Natural Resource and Conservation Service, Washington, D.C.

Meador, M.M. (1998) *Regulation of Air Emissions from Concentrated Swine Feeding Operations: Woo Pig Poooieeee,* 30 April. Research Paper in Partial Fulfillment of the Requirements for the Degree of Master of Laws, University of Arkansas School of Law, Fayetteville.

Mueller, D.K. and Helsel, D.R. (1996) *Nutrients in the Nation's Water: Too Much of a Good Thing?* Circular 1136, US Geological Survey, Washington, D.C.

NASDA (1997) *Innovative Approaches to Natural Resource Protection: A Summary of Successful State Comprehensive Resource Management Planning Initiatives.* National Association of State Departments of Agriculture Research Foundation, Washington, D.C.

NASDA (1998) *State Survey on Waste and Manure Management Regulations.* Draft (December), National Association of State Departments of Agriculture, Washington, D.C.

Nolan, B.T., Ruddy, B.C., Hitt, K.J. and Helsel, D.R. (1998) A national look at nitrate contamination of groundwater. *Water Conditioning and Purification,* 39(12), 76-79.

Nowak, P.J., Wolf, S., Hartley, H. and McCallister, R. (1993) *Assessment of 1992 Wisconsin Atrazine Rule (Ag 30): Final Report.* University of Wisconsin, College of Agricultural and Life Sciences, Madison.

NRC (National Research Council, Board on Biology) (1989) *Field Testing Genetically Modified Organisms: Framework for Decisions.* National Academy Press, Washington, D.C.

NRC (National Research Council, Board on Agriculture) (1993) *Soil and Water Quality: An Agenda for Agriculture.* National Academy Press, Washington, D.C.

NRC (National Research Council) (1996) *A New Era for Irrigation.* National Academy Press, Washington, D.C.

NRC (National Research Council, Board on Agriculture and Natural Resources) (2000) *Genetically Modified Pest-Protected Plants: Science and Regulation.* National Academy Press, Washington, D.C.

Ollinger, M. and Fernandez-Cornejo, J. (1995) Regulation, Innovation and Market Structure in the U.S. Pesticide Industry. AER-719. U.S. Dept. of Agriculture, Economic Research Service, Washington, DC.

Opie, J. (1993) *Ogallala: Water for a Dry Land.* University of Nebraska Press, Lincoln.

Palmer, K., Oates, W.E. and Portney, P.R. (1995) Tightening environmental standards: The benefit-cost or the no-cost paradigm? *Journal of Economics Perspectives,* 9, 119-132.

Pedigo, L.P. (1989) *Entomology and Pest Management*. Macmillan, New York, pp. 646.

Pethig, R. (1976) Pollution, welfare, and environmental policy in the theory of comparative advantage. *Journal of Environmental Economics and Management*, 2, 160-169.

Porter, M.E. (1991) America's green strategy. *Scientific American*, 264(April), 168.

Puckett, L.J. (1994) *Nonpoint and Point Sources of Nitrogen in Major Watersheds of the United States*. Water-Resources Investigations Report No. 94-4001, US Geological Survey, Reston, Virginia.

Rabelais, N.N., Turner, R.E. and Wiseman, W.J. Jr. (1997) Hypoxia in the northern Gulf of Mexico: Past, present, and future. In: *Proceedings of the First Gulf of Mexico Hypoxia Management Conference, 5-6 December 1995, Kenner, Louisiana*. EPA-55-R-97-001, US Environmental Protection Agency, Washington, D.C. Web site http://pelican.gmpo.gov/nutrient/front.html (viewed April 1999).

Reynnells, R.D. and Eastwood, B.R. (1997) *Animal Welfare Issues Compendium: A Collection of 14 Discussion Papers*. US Department of Agriculture, Cooperative State Research, Education and Extension Service, Plant and Animal Production, Protection and Processing. Web site http://www.nal.usda.gov/awic/pubs/97issues.htm#doctop (viewed April 1999).

Ribaudo, M.O. (1989) *Water Quality Benefits from the Conservation Reserve Program*. Agricultural Economic Report No. 606, US Department of Agriculture, Economic Research Service. Washington, D.C.

Ribaudo, M.O. (1999) Personal communication. US Department of Agriculture, Economic Research Service, Washington, D.C.

Roberts, T. (1999) Personal communication (14 January). US Department of Agriculture, Economic Research Service, Washington, D.C.

Rudek, J. (1997) Atmospheric nitrogen deposition and ecosystem health in North Carolina: A public perspective. In: *Proceedings: Workshop on Atmospheric Nitrogen Compounds; Emissions, Transport, Transformation, Deposition, and Assessment, 10-12 March 1997, Raleigh, North Carolina*. North Carolina State University, Department of Marine, Earth, and Atmospheric Sciences, Raleigh.

Samson, F. and Knopf, F. (1994) Prairie conservation in North America. *BioScience* 44, 418-421.

Smith, R.A., Alexander, R.B. and Landfear, K. (1993) Stream water quality in the United States: Status and trends of selected indicators during the 1980s. In: *National Water Summary 1990-91: Hydrologic Events and Stream Water Quality*. Water-Supply Paper No. 2400, US Geological Survey, Reston, Virginia, pp. 111-140.

Smith, R.A., Schwarz, G.E. and Alexander, R.B. (1994) *Regional Estimates of the Amount of U.S. Agricultural Land Located in Watersheds with Poor Water Quality*. Open-File Report No. 94-399, US Geological Survey, Reston, Virginia.

Smith, R.A., Alexander, R.B. and Schwarz, G.E. (1996) *Quantifying Fluvial Interstate Pollution Transfers*. Unpublished manuscript, US Geological Survey, Reston, Virginia.

Sumner, D.A. and Hart, D.S. (1997) Government policy and California agriculture. In: Siebert, J.B. (ed) *California Agriculture: Issues and Challenges*. University of California, Division of Agriculture and Natural Resources,

Thompson, E.A., Sorensen, A.A., Harlan, J. and Greene, R. (1994) *Farming on the Edge: A New Look at the Importance and Vulnerability of Agriculture Near American Cities*. American Farmland Trust, Washington, D.C.

Tripp, R. and Heide, W.V.D. (1996) *The Erosion of Crop Genetic Resources: Challenges, Strategies and Uncertainties*. Natural Resources Perspectives No. 7, Overseas Development Institute, London.

US Congress, Office of Technology Assessment (1995) *Targeting Environmental Priorities in Agriculture: Reforming Program Strategies.* OTA-ENV-640, US Government Printing Office, Washington, D.C.

US EPA (undated) *Notes on Atrazine.* US Environmental Protection Agency, Washington D.C. Web site http://www.epa.gov/owowwtr1/info/NewsNotes/issue25/nps25atr.html (viewed April 1999).

US EPA (1990) *National Pesticide Survey Phase I Report.* EPA-570/9-90-003, US Environmental Protection Agency, Washington D.C.

US EPA (1995) *National Water Quality Inventory: 1994 Report to Congress.* EPA-841-R-95-005, US Environmental Protection Agency, Washington D.C.

US EPA (1997) *Nonpoint Source Pollution Control Program.* US Environmental Protection Agency, Office of Water, Washington, D.C. Web site http://www.epa.gov/OWOW/NPS/elistudy/nonpoin3.html (viewed April 1999)

US EPA (1998a) *National Water Quality Inventory: 1997 Report to Congress.* EPA-841-R-95-005, US Environmental Protection Agency, Washington, D.C.

US EPA (1998b) *Status of Chemicals in Special Review.* EPA-738-R-98-001, US Environmental Protection Agency, Office of Pesticide Programs, Washington, D.C.

US FWS (1998) *Endangered Species General Statistics.* US Fish and Wildlife Service, Division of Endangered Species, Washington, D.C. Web site http://www.fws.gov/r9endspp/esastats.html (viewed January 1999).

USDA (1994) *Agricultural Resources and Environmental Indicators.* Agricultural Handbook No. 705, US Department of Agriculture, Economic Research Service, Washington, D.C.

USDA (1997a) *Agricultural Resources and Environmental Indicators, 1996-97.* Agricultural Handbook No. 712, US Department of Agriculture, Economic Research Service, Washington, D.C.

USDA (1997b) *Status Review of Conservation Compliance Data.* US Department of Agriculture, Natural Resources Conservation Service, Washington, D.C.

USDA (1998a) World hog production: Constrained by environmental concerns? *Agricultural Outlook,* 249, 15-19.

USDA (1998b) *Livestock, Dairy, and Poultry.* US Department of Agriculture, National Agricultural Statistics Service, Washington, D.C. Web site http://jan.mannlib.cornell.edu/reports/erssor/livestock/ldp-mbb/livestock_dairy_and_poultry_08.17.98_updated_08.19.98 (viewed April 1999).

USDA (2000) *Agricultural Outlook.* May. US Department of Agriculture, Economic Research Service, Washington, D.C.

USGS (1998) *Pesticides in Surface and Ground Water of the United States: Summary of Results of the National Water Quality Assessment Program (NAWQA).* US Department of the Interior, US Geological Survey, National Water Quality Assessment Program, Pesticides National Synthesis Project, Reston, Virginia. Web site http://water.wr.usgs.gov/pnsp/allsum/ (viewed December 1998).

USGS (1999) *The Quality of Our Nation's Waters: Nutrients and Pesticides.* Advance copy subject to revision, US Department of the Interior, US Geological Survey, National Water Quality Assessment Program, Reston, Virginia.

Watrud, L.S. and Seidler, R.J. (1998) Nontarget ecological effects of plant, microbial, and chemical introductions to terrestrial systems. *Soil Chemistry and Ecosystem Health.* Special Publication 52. Soil Science Society of America, Madison, Wisconsin, pp. 313-340.

White, S.E. and Kromm, D.E. (1995) Who should manage the High Plains aquifer? The irrigators' perspective. *Water Resources Bulletin,* 31, 715-727.

Whittaker, G., Lin, B.H., and Vasavada, U. (1995) Restricting pesticide use: The impact on profitability by farm size. *Journal of Agricultural and Applied Economics*, 27, 352-362.

Wright, P. (1998) Manure spreading costs. In: *Manure Management Proceedings*. Presented at Manure Management Conference, Managing Manure in Harmony with the Environment, 10-12 February 1998, Ames, Iowa. Web site http://www.ctic.purdue.edu/FRM/ManureMGMT/paper8.html (viewed April 1999).

Wrysinski, J.E., Garr, J.D., and Bias, M.A. (1999) Rice straw decomposition and development of seasonal waterbird habitat on rice fields. *Valley Habitats: A Technical Guidance Series for Private Land Managers in California's Central Valley*, 1, 1-8. Ducks Unlimited Western Regional Office, Sacramento, California.

Canada

Glenn Fox and Jennifer Kidon

INTRODUCTION

This chapter examines the relationship between agricultural production and the protection of environmental values in Canada. We begin with an overview of the main features of agricultural production, trade and of domestic agricultural policies. We then present an overview of the most important environmental issues that have been linked to agriculture in the country. This is followed by a description of the institutional framework which regulates environmental aspects of Canadian agriculture. The penultimate section of the chapter presents a synopsis of the available estimates of the environmental compliance costs faced by Canadian farmers. The main motivation for this book is to explore the implications of domestic environmental issues and policies in agriculture for the ongoing process of trade liberalisation in agriculture. The information presented in this chapter illustrates the need to understand the variability in intensity of environmental problems attributable to agriculture across the major agricultural trading partners. It also documents the complexity of domestic institutional responses to environmental problems. This complexity underscores the importance of thorough comparative research before conclusions about the effects of environmental regulations on competitiveness are reached.

Canadian agriculture is diversified. The leading commodities, in terms of the gross value of primary commodity sales in 1997, were beef cattle ($4.7 billion[1]), dairy products ($3.7 billion), wheat (exclusive of durum, $3.0 billion), hogs ($2.9 billion), canola ($2.0 billion) and poultry products ($1.3 billion). Exports are an important source of revenue for the agricultural products not regulated under supply management. This includes, again for 1997, exports of grains valued at $6.5 billion, of oilseeds valued at $2.3 billion, of live cattle and beef valued at $1.0 billion and of live hogs and pork valued at $1.1 billion.

---

[1] Prices, revenues and costs reported in this chapter are in $ Cdn, unless otherwise indictated in the text. The current exchange rate with respect to the USA dollar is about $0.65 Cdn per $1.00 USA.

The regional distribution of agricultural commodity production in Canada reflects a pattern of comparative advantage and regional specialisation influenced by variations in climate, soil conditions, farmland area, urban settlement patterns and historical government policy. A map of Canada in which provincial and territorial borders are shown is presented in Figure 6.1. Most agricultural production in Canada is located in southern regions of the country. The Atlantic provinces of Newfoundland, Nova Scotia, New Brunswick and Prince Edward Island and on the west coast, British Columbia, are home to a small share of the national farmland area and produce relatively limited volumes of agricultural commodities, however Nova Scotia is an important region for apple production and New Brunswick and Prince Edward Island are important potato growing regions. Wheat production, especially hard red spring wheat, is dominated by the prairie provinces of Manitoba, Saskatchewan and Alberta. These three provinces accounted for over 95% of Canadian wheat production in 1997. Producers in these regions, assisted by crop breeders at agricultural federal and provincial research institutes, have developed a combination of cultural practices and grain varieties that are appropriate to the typically low moisture and short growing conditions of the region. Ontario and Quebec produce an even higher share of the national grain maize crop (over 97% in 1998), primarily due to a limited number of hybrids that can be grown further north where growing seasons are shorter.

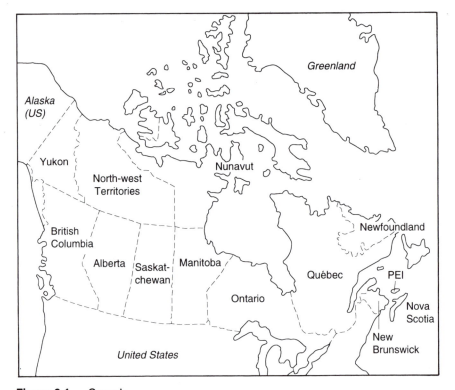

Figure 6.1   Canada

Turning to livestock products, Ontario and Quebec dominate production of chicken meat and milk, producing over 60% of national chicken meat output in 1996 and over 72% of national butterfat output in 1997. Beef and hog production are more evenly distributed between Ontario, Quebec, Manitoba, Saskatchewan and Alberta. Sheep production for either wool or meat has traditionally been a marginal industry in Canada.

Agricultural census data indicate considerable variation in farm size across Canada. The concept of a 'census farm' was defined in the 1996 census as any organisation that intends to sell any agricultural produce sometime in the next twelve months. This definition is overly inclusive and distorts average farm size calculations by including many small part time operations as farms. Nevertheless, similarly overly inclusive definitions have been employed in previous censuses. The average land area per farm in Quebec in the 1996 agricultural census was about 96 ha, and 83.4 ha in Ontario, while the average farm size in Saskatchewan in that year was about 466 ha. The actual difference in farm sizes for commercial full time farms is even greater. For example, beef cattle feedlots with 30,000 head of cattle on feed in one location are increasingly common in Alberta and feedlots with 10,000 head capacity are commonplace. Few cattle feedlots in Ontario have more than 2,000 animals on feed in one location at a given time. There are several reasons for these differences. First, land values are much lower in rural areas of the prairie provinces than they are in southern Ontario and Quebec. This is partly due to less demand for land for non-agricultural uses and partly due to lower productivity of land in grain production because of less advantageous climate. As a result, it is less costly to assemble the land base necessary to exploit the available economies of size in cattle feeding in western Canada. Furthermore, lower population density on the prairies makes it easier to satisfy separation distances between livestock production facilities and rural residences than it is in Ontario or Quebec.

Exports to the United States of America (USA) dominate Canadian agricultural foreign sales. Beef exports amounted to $500 million in 1996 and over 90% of these exports went to the USA. Pork exports exceeded $600 million in that year and over 60% went to the USA. Part of the trend of growing beef and pork trade between Canada and the USA can be attributed to the reduced trade tensions over red meats that has coincided with the implementation of the North American Free Trade Agreement (NAFTA). Reduced prospects of subsidy wars and countervailing duties have allowed producers to adjust production and trade to better accommodate continental transportation costs. An increasing volume of beef and pork is being exported from western Canada into the western USA while at the same time exports of beef and pork from the midwestern USA are serving markets in eastern Canada.

Exports of chicken meat, dairy products and eggs have typically been much less important than exports of beef and pork. The aim of supply management in the dairy and poultry industries has been to limit production to the expected level of domestic demand, and so it is not surprising that the export revenues are small for these industries, especially relative to the total value of domestic production. In contrast, wheat exports have ranged in value from $2.5 billion to $4 billion in recent years, making wheat Canada's largest source of agricultural export revenue.

GOVERNMENT POLICY

Government involvement in agriculture in Canada is bifurcated. One group of commodities, including dairy products, chicken and turkey meat, eggs and tobacco are regulated under a policy regime called supply management. The details of policies differ among commodities, but the basic structure involves production quotas, formula prices and import controls. Total national production is regulated by a national quota. This national quota is divided among the individual provinces. Within provinces, quota is distributed among individual producers. Initially, quotas were allocated among provinces and farms based loosely on their historical share of production prior the imposition of supply management. Commodity prices are set on the basis of an administrative formula, at the national level for milk destined for uses other than the fluid milk market and at the provincial level in other cases. In principle, the formula price is intended to reflect estimated production costs. Individual quotas can be bought and sold as separate assets by individual milk producers. In the cases of the other commodities, transfer of quotas among producers has been linked to the sale of other farm assets. Movement of quota among provinces as demand and supply conditions have changed since the inception of supply management in the 1970s has been a contentious issue. Only limited inter-provincial adjustments in quota allocations have been made. The third element of supply management is border controls. Imports priced below the formula price would obviously undermine the viability of the system. Under the most recent GATT agreement, national import quotas on supply managed commodities have been converted to tariffs. In some cases the tariff equivalent of these quotas is as high as 200%.

The remaining major agricultural commodities produced in Canada, beef, pork, food and feed grains, and fruits and vegetables are subject to competition from imports or compete for export markets. Canadian beef and pork producers compete with their counterparts in the USA in what is an increasingly integrated North American market. Aggregate national net trade data understate the level and nature of this integration. Increasingly, beef and pork produced in western Canada is exported into western US markets and the same commodities produced in the US midwest are exported into the eastern Canadian market, a situation that reflects underlying regional comparative advantage and transportation costs (see Runge and Fox, 1999). For the most part, Canadian producers of hogs and beef receive a price that reflects the price of these commodities in the North American market or in export markets in Asia. Maize and soybean producers operate either on an export or import basis with the US market depending on demand and supply conditions in Ontario and Quebec relative to the price of those feedgrains in the USA midwest.

Wheat production in Manitoba, Saskatchewan and Alberta has been subject to the policies of the Canadian Wheat Board and a series of transportation policies originating in the Crow's Nest Pass agreement of the late 19th century.

The Wheat Board is a statutory single desk monopoly grain marketing system that acquires all wheat sold from farms in the prairie provinces and sells that wheat to domestic and foreign buyers. The Board pools revenues from the grains that it sells and distributes those revenues to producers in the form of an initial and a final payment for each crop. With the exception of a few shortfalls when the initial payment turned out to be higher than the total overall revenues received by the Board when it sold grain the operations of the Board are self-financing. On those occasions when an unforseen drop in grain prices during the year revealed that the initial payment to producers was too high, the federal government subsidised the shortfall.

The Crow's Nest Pass agreement of 1897 imposed a ceiling on the nominal price that could be charged to transport grain from the prairie provinces to export points on Lake Superior and on the Pacific Coast by rail. By the middle of the 20th century, this ceiling began to fall short the railways costs of transporting this grain. The Government of Canada responded with a series of measures that subsidised grain transport and hence exports. These policies had a number of effects (see Klein *et al.*, 1991a, 1991b; Kerr *et al.*, 1991). Given that the farm price of grains in western Canada is largely determined by the price of grain in export markets less the costs of transporting that grain to those markets, the effect of subsidised transport was to make the farm price of grain in the prairies higher than it otherwise would be. This discouraged livestock feeding in the region since the price of feedgrains was increased. Some analysts have also argued that this policy encouraged grain production at the extensive and intensive margins so that more land was brought into grain production and inputs were applied more intensively on grain land than would have been the case in the absence of this subsidy. As of 1997, grain producers have been required to pay something closer to the actual cost of transporting grain to export markets. In anticipation of the demise of these subsidies, aided and abetted by Provincial government policies to offset the effects of the Crow subsidies on the costs of livestock farms, livestock production has grown rapidly in recent years in Alberta, Manitoba and Saskatchewan.

National unity issues have also played an important role in agricultural policy in Canada. The bulk of industrial milk production originates in the province of Quebec, making national dairy policy more than a purely economic or agricultural issue. Protection of dairy producers from foreign competition is in part a concession to secessionist factions.

ENVIRONMENTAL AND HUMAN-HEALTH ISSUES

The environmental impact of agriculture in Canada was examined in a national research workshop in 1992 (Milburn *et al.*, 1992) and more recently has been studied as part of Agriculture and Agri-food Canada's Agri-Environmental Indicator Project (McRae *et al.*, 2000). The 1992 workshop limited its attention to the effects of agriculture on water quality. The more recent Indicators project was more comprehensive in its scope. In addition to these national assessment efforts, individual studies of groundwater quality in rural areas have been conducted in

Nova Scotia (Moerman and Brigg, 1999; Government of Nova Scotia, 1990; Milburn *et al.*, 1995), Prince Edward Island (Somers, 1998; Milburn *et al.*, 1995), New Brunswick (Milburn *et al.*, 1995; Albert, 1996; Ecobichon *et al.*, 1996, 1999; Milburn, 1996), Quebec (Masse *et al.*, 1994; Lapp *et al.*, 1998), Ontario (Goss *et al.*, 1998), Manitoba (Racz, 1992; Pupp *et al.*, 1995a), Saskatchewan (Pupp *et al.*, 1995b; Harker *et al.*, 1997), Alberta (Harker *et al.*, 1997) and British Columbia (Zebarth, 1992; Zebarth *et al.*, 1998).

It is difficult to assess the relative importance of individual environmental issues linked with agricultural production. For a long time, public perception has seemed to indicate that pesticide residues in ground and surface water and in food products themselves were of the greatest concern to Canadian consumers. However, available evidence suggests that bacterial contamination of ground and surface water, excess nitrate in groundwater and displaced sediment and phosphorus in surface water pose more tangible and immediate risks to environmental quality and human-health. In addition, the emphasis that has been placed on developing separation distance policies for livestock operations suggests that air quality problems to do largely with odour are also important in a policy context. Separation distances specify how close a new livestock facility can be located to existing neighbours, roadways, property boundaries and watercourses.

Off-site water quality degradation from soil erosion from cropland has been an important problem regionally. Agricultural land in watersheds draining into Lake Erie has been identified as an important source of sediment and phosphorus in the Lake. Although the Sparrow Report (1984) argued that the productivity effects of erosion on cropland were ominous, subsequent research has not found this to be the case (Fox, 1992). But the offsite surface water quality effects of erosion are important, at least in some regions (Fox and Dickson, 1990; Fox *et al.*, 1995).

The available evidence seems to suggest that the leading cause of groundwater contamination in rural Canada is bacterial contamination, followed by nitrate contamination with pesticide contamination a distant third. Whether this bacterial contamination can be attributed to agriculture continues to be a controversial question. Rural residential septic systems are another potential source of this contamination. The potential for nitrate contamination varies across regions as a result of differences in intensity of manure and fertiliser application as well as because of precipitation differences and local variations in soils and topography. Harker *et al.* (1997) concluded that

> Within the context of the Canadian Water Quality Guidelines, there is
> no significant body of evidence to indicate the *wide-spread*
> contamination of surface and ground waters from agricultural activities
> on the prairies (Harker *et al.*, 1997, p. vii).

This is not to say that there are no local problems of contamination or that such problems cannot or will not occur in the future. Sediment loadings in major rivers in the region are identified as a seasonal problem. Few pesticide residues have been detected in surface or groundwaters, and concentrations have only rarely exceeded current guidelines. Nitrate contamination of groundwater was identified as one of the more common water quality problems associated with agricultural production in the prairies under intensively fertilised and irrigated croplands. Phosphorus in surface water is evident in the region, although attribution of phosphorus loadings among the various sources, including agriculture, has been problematical.

According to Harker *et al.* (1997), agricultural input intensity is low in Canada relative to much of the European Union (EU) and the USA. For example, the average application of pesticides per hectare of agricultural land in Canada, in terms of kg per ha, is about 40% of the corresponding value in the USA. The average application level on the prairie provinces is only about 25% of that of Ontario. The reasons behind the relatively less intensive use of pesticides in Canadian agriculture have not been fully investigated. We suspect that compared to crop producers in the USA and especially in the EU, lower product prices translate into a lower marginal value product of inputs generally and hence, for similar input prices, less use (Fox and Barber, 1991). Climate may be another contributing factor. One benefit of harsh winters is their adverse effect on pest populations.

McRae *et al.* (2000, p. 162) report estimates of average excess nitrogen levels for Canada that are about one third of those of the USA and as little as one tenth of the estimated values for Japan and Denmark. McRae *et al.* (2000, p. 167) concur with Harker's assessment that, for much of prairie agriculture, the risk of contamination of groundwater by excess nitrogen is low, however apparent trends toward higher excess nitrogen levels in other regions of Canada suggest that this issue will become more important in the future. Linking agricultural sources of nitrogen to excessive levels of nitrate in groundwater has proven to be problematic. Because natural background levels of nitrate nitrogen in groundwater in the prairie provinces are high, it is often difficult to identify the incremental contribution from agricultural sources (Harker *et al.*, 1997, p. 44). For example, high groundwater nitrate levels, from 100 to 500 ppm, have been detected in some shallow soils but these levels are attributable to geological not agricultural sources. With respect to groundwater quality in Alberta, Harker *et al.* (1997, p. 45) report findings by Henry (1995) that in one survey of 12,342 well water samples collected on farms in Alberta over a 6-year period prior to 1995, 4% of the wells sampled exceeded the 10 ppm limit for nitrate nitrogen concentration. Long term evidence summarised by Henry (1995) and Henry and Meneley (1993) indicates that the incidence concentrations of nitrate in groundwater on the prairies has remained roughly constant since the 1940s. Nevertheless, baseline data for individual aquifers is limited and further research is indicated, especially for areas of intensive land use, for locations where aquifers are shallow and for areas with high precipitation or where irrigation is used intensively (Harker *et al.*, 1997, p. 49). The combination of high levels of manure application and irrigation has been linked to nitrate levels of 500 ppm in groundwater in an experimental trial near Lethbridge, Alberta (Chang and Entz, 1996).

The Walkerton, Ontario groundwater contamination emergency in the spring of 2000, which is thought to have caused several deaths and led to the shutdown of the municipal water supply for several months, has raised public awareness of the risks from bacterial contamination of groundwater. A provincial inquiry is still underway as of this writing. New standards for management of livestock waste in the province of Ontario are pending.

REGULATION, CODES OF PRACTICE AND RIGHT TO FARM

Agriculture is subject to environmental policies and regulations at several levels in Canada (Milburn *et al.*, 1992). Jurisdiction is divided among the federal, provincial and municipal levels of government and certain provisions of common law have important implications for the effects of agriculture on the environment. Most provinces have their own legislation to protect water and air quality, public health and other environmental values that might be impaired by agricultural production. Most provinces also use legislation to authorise municipalities to regulate local agricultural environmental issues. Municipalities, in turn, use zoning, land use planning and permitting requirements to regulate the location and size of agricultural operations, especially livestock farms. Farming practices are also subject to scrutiny under trespass, nuisance and public nuisance provisions of common law as well as liability for accidents and damage caused by stray livestock.

Environmental regulation of agricultural production in Canada is undergoing dynamic change. The federal pesticide registration and regulation system was restructured in 1995. Several regional environmental projects and programmes directed toward primary agriculture were undertaken under the federal Green Plan in the early 1990s. The producer initiated Environmental Farm Planning process has been widely embraced in Ontario and is gathering momentum in the Atlantic provinces. Local and provincial guidelines and approval processes are evolving in light of recent structural changes in the swine and cattle industries. Most provinces have passed so-called right to farm legislation to clarify the standards for normal or reasonable farming practices to be applied in nuisance actions.

While it is difficult to draw generalisations in the midst of such rapid change, certain themes can be identified. First, environmental regulation of agriculture in Canada is undertaken as a combination of actions under federal and provincial legislation and regulation, municipal zoning and permitting processes and common law litigation and liability with respect to nuisance, public nuisance and riparian rights, as well as through the influence of peer pressure from other producers involved in voluntary stewardship initiatives. Multiple regulatory forces act in concert to address potential and existing problems. Canadian constitutional law grants primary jurisdiction to the provincial governments for environmental issues. The federal role has evolved to provide a national system for pesticide registration, to setting national water and air quality standards for selected contaminants and to providing financial assistance for regional agricultural environmental projects and programmes such as the Soil and Water Environmental Enhancement Program (SWEEP) and the Clean Up Rural Beaches (CURB) programme. In most of the provinces, with the noteable exception of

Quebec, enforcement of environmental agricultural standards is devolving to the municipal level. Provincial guidelines for agricultural codes of practice inform local by-laws, zoning policies and approval processes at the municipal level and criminal prosecution under provincial legislation acts as a backstop for local enforcement. Some regional programmes have been initiated based on cost sharing between governments and producers to address diffuse surface water quality problems from off-site erosion damage and excess nutrient levels from crop production or from runoff from manure from livestock operations. But most municipal standards are compulsory and reinforced by substantial fines for non-compliance. Furthermore, water and air quality degradation that occurs from producer actions that do not conform with applicable provincial Codes of Practice can be liable to criminal and civil penalties.

Federal policies and programmes

Federal jurisdiction

Agriculture in Canada is subject to regulation under the *Canadian Environmental Protection Act*, the *Fisheries Act* and the *Pest Control Products Act*. The federal Pest Management Regulatory Agency was established within the federal Department of Health in 1995 as part of a restructuring of the pesticide registration system. In addition, certain provisions of the *Canada Wildlife Act*, the *Migratory Birds Act*, the *Water Act* may also have implications for agricultural producers. But the primary responsibility for the environmental regulation of agriculture in Canada rests with the provincial and municipal levels of government. Federal standards for water quality have been established, however, for nutrients, bacteria and pesticides. These standards are summarised in Tables 6.1, 6.2 and 6.3. Standards for aquatic life are stricter than those for human drinking (Table 6.3). The explanation for this apparent paradox is that aquatic organisms are surrounded by water continuously, which results in a permanent and complete exposure to the contaminants present in that water, whereas ingestion of water is a temporary and limited exposure. Finally, producer groups have initiated the Environmental Farm Planning process, especially in Ontario and in the Atlantic provinces. This voluntary and confidential programme encourages environmentally responsible production practices.

Biodiversity

Although Canada was the first industrialised country to ratify the *Convention on Biological Diversity* developed at the 1992 Rio de Janeiro United Nations Conference on the Environment and Development, there is no federal endangered species legislation in Canada. A bill was introduced in 1996 but it was not passed into law before a national election was called in 1997. A second bill was introduced in 1999, but it too failed to pass before the federal election of 2000. A third bill is currently before Parliament.

Biotechnology

Agricultural biotechnology is regulated through a number of means at the federal level. Depending on the technology in question, the *Feeds Act*, the *Fertiliser Act*, the *Health of Animals Act,* the *Seeds Act*, the *Canadian Environmental Protection Act* or the *Food and Drugs Act* and related regulations may apply. A Plant with Novel Traits (PNT) is defined as a plant variety/genotype possessing characteristics that demonstrate neither familiarity nor substantial equivalence to those present in a distinct, stable population of a cultivated species of seed in Canada and that have been intentionally selected, created or introduced into a population of that species through a specific genetic change. If a biotechnology product is licensed by the USDA, the complete data provided to the USDA must be submitted to Agriculture and Agri-Food Canada (AAFC) authorities for the evaluation and licensing of the product in Canada. Similar protocols may be used by manufacturers of other countries. Additional field trials or other experiments may need to be done to prove safety of the product under Canadian environmental conditions.

Table 6.1   Canadian water quality guidelines for selected nutrients (maximum acceptable concentration - in parts per million)

Nutrient	Human drinking	Livestock drinking	Aquatic life	Irrigation
Total Nitrogen (TN)	N/G	N/G	1.00	N/G
Total Phosphorus (TP)	N/G	N/G	0.05	N/G
Nitrate + Nitrite ($NO_3 + NO_2$)	10	100	N/G	100
Total Ammonia (NH_3)	N/G	N/G	1.13-1.18[a]	N/G
Nitrate (NO_3)	10	100	N/G	N/G

[a] temperature and pH dependent; N/G = No guidelines
Note: There are no recreation guidelines for nutrients.
Source: Canada-Alberta Environmentally Sustainable Agriculture Water Quality Committee (1998).

Table 6.2   Canadian water quality guidelines for selected bacteria (maximum acceptable concentration - in counts per 100 ml)

Bacteria	Human drinking	Livestock drinking	Aquatic life	Irrigation	Recreation
Fecal Coliforms	0	N/G	N/G	100	400
Total Enterococci	N/G	N/G	N/G	N/G	70
E. coli	N/G	N/G	N/G	N/G	400

N/G = No guidelines
Source: Canada-Alberta Environmentally Sustainable Agriculture Water Quality Committee (1998).

Veterinary products

The majority of the veterinary biologics sold in Canada are imported from the USA. It is likely that for a Canadian licensed product, the biggest share of the export market will be in the USA. For the veterinary biologics to move freely, it is desirable that our guidelines for licensing of veterinary biologics be in harmony, especially with those, of the USA and other trading partners namely, the United Kingdom, other EU countries, Australia and New Zealand. Veterinary biologics produced by new techniques are now being evaluated on a case-by-case basis using the same standards for product safety, purity, potency and efficacy required for licensing of conventionally produced products. Depending on the complexity of the rDNA product, it may be necessary that the proposal be forwarded to a committee of experts. This committee would be chaired by a senior officer of the Animal and Plant Health Directorate, Food Products and Industry Branch and have members from the Veterinary Biologics and Biotechnology Section, the Health of Animals Laboratory Division, other areas of AAFC, other federal departments and interested groups as needed. The membership of the committee depends on the expertise required for the evaluation of a particular product. The final approval for field trials and licensing of the product is the responsibility of the Animal Health Division.

Table 6.3 Canadian water quality guidelines for selected pesticides (maximum acceptable concentration - in parts per billion)

Pesticide	Human drinking	Livestock drinking	Aquatic life	Irrigation
Atrazine (H)	60	60	2	10
Bromoxynil (H)	5	11	5	0.35
Dicamba (H)	120	122	10	0.006
2,4-D (H)	100	100	4	100
Diclofop-methyl (H)	9	9	6.1	0.18
MCPA (H)	UR	25	2.6	0.03
Lindane (I)	4	N/G	0.01	N/G
Triallate (H)	230	230	0.24	N/G
Trifluralin (H)	45	45	0.1	N/G
Ethalfluralin (H)	N/G	N/G	N/G	N/G
Imazamethabenz (H)	N/G	N/G	N/G	N/G
Picloram (H)	190	190	29	N/G
Fenoxaprop-ethyl (H)	N/G	N/G	N/G	N/G

UR = Under Review; N/G = No Guidelines; H= Herbicide; I= Insecticide
Note: There are no recreation guidelines for pesticides.
Source: Canada-Alberta Environmentally Sustainable Agriculture Water Quality Committee (1998).

The *Health of Animals Act* regulates those products used for the diagnosis, treatment or prevention of infectious diseases in animals, and applies to transgenic animals that are resistant to certain diseases. The objectives of the *Act* are to minimise the risk of introducing foreign animal disease, to prevent the spread of disease-producing micro-organisms in Canadian livestock as a result of contaminated products and to reduce the risk of exposing people to animal diseases that may be transmitted to humans.

Before they are registered for commercial sale, all biological products used for veterinary purposes (such as vaccines) must be shown to be safe, potent, pure and effective. To prove the safety of a product, the manufacturer must demonstrate that it will not endanger the environment, or animal- and human-health.

Veterinary biological products are broadly classified into two groups according to their characteristics and the safety concerns attached to them: Class I includes products prepared from inactivated organisms such as viruses and bacteria or their derivatives or toxoids that have been manufactured using genetic engineering techniques. Also in this group are monoclonal antibodies used to diagnose and treat diseases. Live products in this class come from organisms in which a single gene has been altered. These are very similar to modified vaccines already in commercial use. In summary, the inactive and live Class I substances pose little environmental risk or new safety concerns.

Class II includes products containing live micro-organisms that have been modified by introducing DNA with genetic material from different organisms or different strains. Substances in this group may also use a live delivery agent such as a virus to carry genetic material, infect the host animal and, in this way, immunize it. Government, industry and university researchers are now testing vaccines genetically engineered to prevent infectious diseases that afflict cows and other domestic animals. At the same time, many laboratories are working on diagnostic kits to detect various infectious diseases. Live products of Class II require special attention and data to prove that they will not harm human and animal health, and pose minimal threat to the environment. These products are assessed under the Environmental Assessment Review Process (EARP).

Import permits are required for all veterinary biologics, animal tissues, genetically engineered animals, infectious organisms and related materials. Research permits are granted for the field testing of genetically modified micro-organisms and naturally occurring micro-organisms following review by the Canadian Food Inspection Agency and its advisors in Health Canada, Environment Canada and the Canadian Forestry Service. Registration guidelines for naturally occurring micro-organisms also include requirements for genetically modified micro-organisms.

Voluntary animal welfare *Codes of Practice* have been developed at the federal level in conjunction with the Canadian Federation of Humane Societies. In addition, criminal sanctions against abusive treatment of animals can be invoked under the criminal code. *Codes of Practice* have been developed for all of the major livestock commodities. As of this writing, a revised *Code* for the transportation of animals is being prepared.

Food safety

The new Canadian Food Inspection Agency (CFIA), is the lead agency responsible for regulating agricultural products to assess whether new products are effective and safe to humans, animals and the environment. Before 1 April 1997 this responsibility fell to the Food Production and Inspection Branch of AAFC.

Agricultural biotechnology products are regulated under different acts. Plants with novel traits are regulated under the *Plant Protection Act* and the *Seeds Act*; biofertilisers are regulated under the *Fertilisers Act*; livestock feeds are regulated under the *Feeds Act*; and veterinary biologics are regulated under the *Health of Animals Act*. These products of biotechnology are regulated alongside similar products developed using traditional technologies.

Before these products may be used, they must be reviewed for human (occupational and food safety), animal and environmental safety, and for efficacy.

The safety-based approach to regulation works in the following way. First, a pre-regulatory review is undertaken to determine if a risk assessment is required. If a new product can be shown to be 'substantially equivalent' to a product already approved by the CFIA, then it will not require a risk assessment. Commodity-specific guidelines are in place to help determine if a novel product is substantially equivalent to those already approved in Canada.

The risk assessment is conducted by evaluators at CFIA using data and information provided by the product developer. The information required is detailed in guidelines that have been developed through multi-stakeholder consultations as well as national and international expert input. Potential risks are identified for each novel product on a case-by-case basis, and information and data addressing these are reviewed. If the novel product is found to pose no potential safety concerns to human, animal or environmental safety when compared to its traditionally developed counterparts in use in Canada, it will be considered acceptable.

Provincial policies

The Atlantic provinces

Provincial environmental statutes are a leading instrument of managing environmental issues in agriculture. In Newfoundland the relevant legislation includes the *Environment Act (1995)*, the *Waste Disposal Act (1995)*, the *Pesticides Control Act (1995)*, the *Environmental Assessment Act (1995)* and the *Well Drilling Act (1981)* (Khan, 1992; Eastern Canada Soil and Water Conservation Centre, 1997). In addition, provincial guidelines for agricultural development have been published. The *Environment Act* and its regulations prohibit the pollution of any body of water and regulate the discharge of any materials that might impair water quality. The *Department of Health Act (1990)* and the *Municipalities Act (1990)* also contain provisions to regulate well location and construction, septic system location, farm building location and separation distances as well as manure spreading standards.

In Prince Edward Island, the primary legislative instrument is the *Environmental Protection Act (1988)*, although the *Water and Sewerage Act (1988)* also contains provisions that relate to the environmental regulation of agriculture in that province (Eastern Canada Soil and Water Conservation Centre, 1998). *Inter alia*, the *Environmental Protection Act* regulates well construction, waste treatment and water supply systems and modifications to watercourses. While none of the provisions of the *Act* address agricultural practices specifically, it does not distinguish between types of environmental contaminants. The *Act* has been applied to water quality problems associated with agriculture, including groundwater contamination from inadequate manure storage facilities and surface water quality problems arising from cattle access to watercourses.

In Nova Scotia, agriculture is subject to environmental regulation under the *Agricultural Operations Protection Act, 1986*, the *Ditches and Watercourses Act*, the *Environment Act*, the *Fences and Detention of Stray Livestock Act*, the *Fences and Impounding of Animals Act*, the *Forest Act* and the *Municipal Act*, as well as under provincial legislation related to regional municipal governments (Moerman, 1997). Several aspects of agricultural water use are regulated under the *Environment Act*, including an approval process for extractions of groundwater or diversions of surface water in excess of 23 m^3 per day or for the construction of water storage structures larger than 25 m^3 in capacity, standards for the location and construction of wells, ditches, dams, and for the modification of watercourses. Additional regulations related to watercourses have been promulgated under the *Fisheries Act* and under the *Ditches and Watercourses Act*. Regulation of separation distances for manure storage facilities on livestock farms are imposed at the municipal level in the province. Setbacks from adjacent residences, wells, roads, watercourses and property lines vary substantially across municipalities. Setback requirements are implemented through the use of development or building permits. The provincial *Agricultural Operations Protection Act, 1986*, states that 'generally accepted farming practices' will not be actionable as nuisance.

In New Brunswick, agricultural practices are subject to environmental regulations under the *Clean Environment Act*, the *Clean Water Act*, the *Pesticides Control Act*, the *Health Act* and the *Agricultural Land Protection and Development Act*. Two regulations under the *Clean Water Act* affect agricultural operations, the Watercourse Setback Designation and the Watercourse Alteration Regulation (Vanderlaan, 1992). The aim of the Watercourse Setback Designation is the protection of water quality for water sources for municipalities. The aim of this policy is to prevent the contamination of surface water from nitrates, bacteria, pesticides, petroleum products and displaced sediment. Under this provision, a watershed that serves as a municipal water supply can be designated as a protected area. Land and water use activities in protected areas are subject to additional regulations above and beyond the normal provincial environmental and health measures. The provincial *Pesticides Control Act* prohibits the use of pesticides that are not registered for use under the federal *Pest Control Products Act*. The provincial act regulates aerial and commercial application of pesticides. The *Health Act* regulates the location of livestock facilities.

Quebec

The Ministry of the Environment of Quebec regulates various aspects of the location, construction and operation of livestock facilities (Ministère de l'Agriculture, des Pecheries et de l'Alimentation du Quebec, various years; Ministère de l'Environnement et de la Faune du Quebec, 1997). In contrast to the trends in other provinces, the balance of responsibility for the environmental regulation of agriculture in Quebec is moving toward a greater provincial and a diminished municipal role. In June, 1997, the National Assembly of the province of Quebec passed into law Bill 23, Quebec's version of right to farm legislation. The act is called the *Law and Protection of Agricultural Territory*. The *Law* stipulates separation distances, building codes and other regulations related to reducing emissions from agriculture. The emphasis is on reducing the incidence of nuisance problems related to air and water quality. Contemporaneous changes to Quebec's *Law of Planning and Urbanism*, the province's municipal act, and its *Law on the Quality of the Environment* clarified the role and authority of local municipalities to regulate agricultural operations within their boundaries. At the same time, the provincial Ministry of Agriculture, Fisheries and Food released a Code for the reduction of pollution from agriculture to define standards for reasonable care under the right to farm legislation. The Code emphasises manure and nutrient management to mitigate risks to water and air quality. Cost sharing is available to producers to offset up to 70% of the costs of construction of manure storage structures and for manure application equipment.

Like Ontario, Quebec has recently developed a provincial programme to encourage reductions in pesticide use. The goals of the programme are to reduce pesticide use in the province by 50% and to have 70% of the agricultural land in the province under Integrated Pest Management (IPM) by the year 2003.

Ontario

In Ontario, four provincial statutes, the *Environmental Protection Act*, the *Ontario Water Resources Act*, the *Pesticides Act*, and the *Environmental Assessment Act* authorise the environmental regulation of agriculture (Willson, 1992). The *Environmental Protection Act* sets standards for the release of any contaminant into the environment which may cause injury to humans, flora or fauna. Animal waste is exempt from the provisions of the Act, providing that waste has been handled according to reasonable farming practices.

The *Ontario Water Resource Act* gives the provincial Ministry of the Environment extensive powers to regulate activities that might present a risk to water quality. Standards for water quality for various water uses have been produced by the Ministry of the Environment. As in the Atlantic provinces, Ontario's *Pesticides Act* regulates the storage, distribution and application of pest control products and requires the licensing of commercial applicators. The Ontario *Farm Practices Protection Act* (1988), is the province's version of right to farm legislation. Along with the agricultural code of practice, it defines a standard of reasonable action for application in nuisance actions.

Protection of wetlands in Ontario is authorised under the *Planning Act* and is guided by a Provincial Policy Statement approved in 1992 (Ivy, 1996). The Policy Statement directs local municipalities, in the course of the development of local official plans, plans of subdivision, consents, zoning by-laws, minor variances and other local planning activities to protect provincially significant wetland areas. The aim of the Policy Statement is no net loss in provincially significant wetland area in the province.

Several environmentally related agricultural programmes have been initiated in Ontario in recent years. The Food Systems 2002 programme aims to reduce pesticide use in Ontario agriculture by 50%, relative to 1987 levels, by the year 2002. The primary means of achieving this goal are research, technology transfer and other forms of pest management education and the encouragement of more widespread use of IPM. The CURB programme, undertaken in cooperation with regional conservation authorities, provided financial assistance to support projects to reduce surface water quality problems. Projects receiving support included fencing livestock out of streams and improvements in manure storage facilities.

The Western provinces

Environmental regulations pertaining to agriculture are the shared responsibility of the provincial and municipal governments in Manitoba as well. The livestock waste regulations promulgated under the *Control of the Environment Act* are intended to protect air and water quality from the use, handling and storage of livestock manure in intensive livestock operations. A permit is required to construct, modify or expand a manure storage facility. Manitoba's version of right to farm legislation is the *Farm Practices Protection Act*. Farm practice guidelines have been published describing the normal agricultural practices protected from nuisance actions under the *Act*. A Farm Practices Review Board has been appointed to offer initial mediation over nuisance disputes with neighbours. Provincial regulations were updated in 1997 to accommodate recent growth of intensive livestock operations.

Until 1996, four provincial statutes, the *Pest Control Products Act*, the *Environmental Management and Protection Act*, the *Pollution by Livestock Control Act* and the *Fisheries Act* regulated agricultural activities in Saskatchewan (Ruggles, 1992). The provincial *Pest Control Products Act* requires that all pest control products used in Saskatchewan be registered for use under the federal *Pest Control Products Act*. It also prohibits the application of pesticides over open bodies of water and the cleaning or locating of equipment or containers near open water. The provincial Act requires that commercial pesticide applicators be licensed. Storage and transport of fuel, pesticides and fertilisers are regulated under the *Environmental Management and Protection Act*. Pollution of surface and groundwater by livestock is regulated under the *Pollution by Livestock Control Act*. A permit is required to construct a livestock facility larger than 300 animal units. Design and construction must meet standards to prevent water pollution and adequate provisions for manure disposal must be made. The provincial Department of Agriculture has prepared guidelines and criteria for runoff control, manure storage requirements and land requirements for manure disposal. The provincial *Fisheries Act* requires that a permit be obtained for any agricultural operation that impacts in stream flows or other aspects of fish habitat.

Several policy measures have been developed in Saskatchewan in recent years to respond to environmental regulatory issues associated with the growth in intensive livestock operations in the province. The provincial Department of Economic Development set a goal of doubling hog production in Saskatchewan between 1996 and 2000. This goal has been controversial for environmental reasons and also because of its implications for the promotion of large scale farms. In addition, low prices in 1997 and 1998 discouraged new farm investments in the industry, so the goal has not been achieved. The *Agricultural Operations Act (1996)* and the *Environmental Assessment Act* currently serves as the regulatory framework for intensive livestock operations in Saskatchewan. The Act combines aspects of right to farm legislation used in other provinces with requirements for approval of new or expanded intensive livestock facilities. Guidelines have been developed to describe generally accepted farming practices to be applied in nuisance actions and in municipal by-laws and permitting processes. An Agricultural Operations Board has been established to review nuisance complaints in light of these guidelines.

Agriculture is regulated with respect to environmental issues under the *Public Health Act*, the *Clean Air Act* and the *Clean Water Act* in Alberta. Beginning in 1972, the provincial Department of the Environment formed the Intensive Livestock Operations Committee to study the status of waste management practices in the industry. The committee published a Code of Practice in 1973. The Code contained guidelines for the siting, construction and operations of confined livestock production facilities. It was revised in 1982 to include minimum separation distance formulas. These formulas are intended to reduce problems of nuisance in the form of odour, insects and water pollution. A new *Code of Practice for the Safe and Economic Handling of Animal Manures* was developed in 1995 (Alberta Agriculture, Food and Rural Development, 1995). While the provisions of this code have no formal legal power, they were intended to be used and are in fact being used by municipalities as part of the local permitting process to regulate construction of livestock facilities.

Expansion of the cattle feedlot sector in Alberta since the mid-1980s reflects an adjustment to patterns of regional comparative advantage in feedgrain production, grazing and cattle feeding. Several factors have contributed to this expansion. First, demise of grain transportation subsidies in the prairie provinces has reduced feedgrain costs in the region. In addition, growth in demand for meat in the western USA has provided a ready market for western Canadian production. Finally, favourable climatic conditions have meant that particularly southern Alberta was well positioned to respond to changing market opportunities in cattle feeding in the increasingly integrated North American markets for cattle and beef. With the increased number and average size of cattle feedlots in the province, however, has come an increased number of complaints related to water pollution problems from cattle feeding operations. According to a report prepared by the Commission:

The beef cattle industry faces two clear alternatives in trying to resolve the problem: voluntary action by the industry and individual producers; or increasing regulatory controls initiated by others. The Alberta Cattle Commission endorses the concept of voluntary action by producers both at the individual and industry level (Alberta Cattle Commission, Undated, p. 1).

Waste disposal in general is regulated under the *Environmental Protection and Enhancement Act* in the province of Alberta. The act applies to the release of any substance capable of causing an adverse effect on the environment. Activities related to cattle production that could be considered offenses under the act include:

- Bedding, feeding or watering directly on/from a river, creek or lake where the activity is causing an impact.
- Allowing manure to enter directly into a river, creek or lake or on to public or private property.
- Allowing surface water run-off to become contaminated with manure and allowing that run-off to enter a river, creek or lake or on to public or private property (Alberta Cattle Commission, Undated, p. 4).

Penalties under the act range from $100 to 1 million $, depending on the severity of the offence.

Agriculture is subject to environmental regulation under the *Waste Management Act* in British Columbia. Waste from agricultural operations are regulated under Class C provisions of the act, which limits regulatory initiative if wastes are managed and applied in a reasonable manner from traditional farming operations. According to Nagpal (1992), a Code of Agricultural Practices for Waste Management has been developed for British Columbia that defines accepted agricultural practices for the purposes of the Agricultural Waste Control Regulation of 1992. Farming operations in compliance with the Code are exempt from environmental permitting procedures, but actions that contravene the Code are subject to prosecution under the provincial *Waste Management Act*.

Summary of provincial policies

Comparsion of provincial environmental policies with implications for agriculture reveals variation in approaches, perhaps reflecting regional differences in the nature of environmental problems, but certain themes are evident. Probably for jurisdictional reasons arising out of Canadian constitutional law, provincial governments have generally taken the initiative in this area. Most provinces now have agricultural production codes of practice and right to farm statutes. All have provincial environmental protection legislation and regulation with provisions that apply to the effects of agriculture on water and air quality. Emerging concerns regarding the operation of intensive livestock operations and increased political pressure in the aftermath of the Walkerton crisis suggest that this will be an active area of provincial policy making in the near future.

The role of municipal governments has been emphasised in the discussion of provincial environmental policies and regulations related to agriculture. Responsibility for regulation of the water and air quality problems associated with agricultural production is generally shared between the provincial and municipal levels of government. On the surface, it might appear that hundreds if not thousands of local standards might be in use across Canada. Generally, however, standards are set at the provincial level and these standards are enforced through municipal by-laws, zoning policies and building permit and approval processes. In light of the importance of interprovincial trade in hogs and cattle, especially in the western provinces, some efforts have been made at the provincial level to set compatible if not harmonised standards.

Municipal policies, voluntary initiatives and civil liability

Municipal governments play a pivotal role in the implementation of environmental policy directed toward agriculture in that they are generally required by provincial planning statutes to incorporate environmental considerations into local planning and zoning processes. Designation of land as agricultural or recreational has implications for the types of activities that may be carried out on that land. Most municipalities regulate severances of farm properties for residential or commercial development. A variety of interests are served by such measures. Typically, the rationale offered for such restrictions is to forestall future nuisance complaints by non-farm neighbours.

The Environmental Farm Planning process was initiated by a coalition of Ontario farm groups in the early 1990s. The programme was motivated by concern among members of the farm community over possible increased provincial environmental regulation of agriculture. An Environmental Farm Plan is a confidential review of a farm operation conducted by the farm operator following guidelines prepared by the Ontario Farm Environmental Coalition. An Action Plan is developed to address important environmental problems identified in the review. A peer review process is used to monitor progress on the Action Plan. Small provincial grants of up to $1,500 have been made available to participants to offset out of pocket costs associated with the preparation of a plan. Over 10,000 producers have participated in the programme since its inception. The Ontario Farm Environmental Coalition is currently leading in the development of a provincial Nutrient Management Strategy.

In addition to federal, provincial and municipal environmental regulation, agriculture is subject to common law tort remedies against trespass and public nuisance in Canada. The trespass remedies are primarily directed at damage to property from straying livestock. However, nuisance actions and actions based on riparian rights do play an important role in encouraging producers to take into account the effects of production decisions on local air and surface water quality. Both negligence and strict liability standards have been applied in such cases. Provincial guidelines for generally accepted agricultural practices often set the standard for negligence based liability. Release of livestock waste into watercourses, pesticide drift, odour, noise or dust and runoff contaminated by displaced sediment or excess phosphorus or nitrogen from cropland are all potentially subject to civil litigation.

ENVIRONMENTAL STANDARDS

Separation distances, manure storage and application standards

Under the Newfoundland *Environmental Assessment Regulations*, clearing of land for agricultural purposes in excess of 50 ha or in protected water supply areas is subject to a permitting and approval process. Specific guidelines for agricultural production include prohibition of the spreading of manure within 30 m of a watercourse or within 90 m of any well or public water supply, a requirement that a 15 m buffer zone be maintained between agricultural cultivation and adjacent water bodies and soil testing before extensive manure application is undertaken. The provincial *Environmental Guidelines/Stipulations for Agricultural Development* include a standard of a minimum separation distance of 610 m between any proposed livestock facility and existing residences, a requirement for covered manure storage to mitigate odour, a 30 m buffer strip requirement between agricultural operations and adjacent watercourses and minimum manure storage capacity of 6 months.

A new provincial irrigation policy was introduced in Prince Edward Island in 1995 and new manure management guidelines and separation distance standards were introduced in 1998. New draft regulations regarding exclusion of livestock from watercourses are pending. Separation distances are set at the municipal level in Nova Scotia. There is considerable variation in the requirements across municipalities. A detailed summary of the relevant standards is available in Nova Scotia Department of Agriculture and Marketing and Nova Scotia Department of the Environment (1997).

Currently, 31 watersheds have been designated as protected areas in New Brunswick. Within protected areas, a minimum setback of 75 m from protected watercourses was initiated in 1990. The setback prohibits the establishment of any new agricultural land use within the 75 m setback area, that no agricultural activity occur at all within 30 m of the designated watercourse, that fences prevent grazing livestock from approaching within 30 m of the watercourse and that tillage operations be managed to prevent surface runoff from cropland from entering the watercourse. Up to 80% of the costs of additional fencing and forgone income from cropland within the 30 m zone were offset by a joint federal-provincial subsidy. The New Brunswick Watercourse Alteration Regulation requires any person wishing to disturb the ground within 30 m of a watercourse to obtain a permit from the provincial Department of the Environment. Activities regulated under this provision include the installation of an irrigation system, the construction of ponds, the installation of bridges and culverts and even harvesting trees within 15 m of a watercourse. Several exemptions for agricultural activities are being considered under this regulation. The provincial *Health Act* prohibits the development of a livestock facility on marshy land or land subject to flooding or land located less than 90 m from a waterway or neighbouring dwelling. Air and surface water quality issues are the primary motivation for these regulations. The manure management guidelines for New Brunswick specify minimum separation distances between livestock facilities and other land uses, soil analysis procedures to evaluate sites for manure storage facilities, standards for the construction of such facilities, including a requirement for 210 days storage capacity, land base requirements and application rate standards for manure application to cropland as well as standards for dead animal disposal and milkhouse waste management. The aim of these guidelines is to reduce the incidence of nuisance from odour and to reduce the risk of groundwater and surface water contamination from excess plant nutrients, particularly nitrogen. The guidelines also serve as the basis for evaluating compliance with the provincial Certificate of Compliance.

New regulations came into effect in the province of Quebec in 1997. Livestock facilities larger than a minimum size threshold are required to maintain manure storage facilities with at least 250 days storage capacity. Location, design and construction of manure storage is regulated to minimise risk to surface or groundwater. Generally manure cannot be applied to farmland between 1 October and 31 March or on frozen or snow covered ground. Manure, farm compost and fertiliser application is prohibited within set distances from watercourses, wells, ditches and wetlands. More stringent regulations apply on land which has been determined to be rich in phosphorus. A requirement that producers prepare an approved nutrient management plan is being phased in between 1997 and 2001, depending on the type of operation. *Inter alia*, a plan includes a record of fertiliser, manure and farm compost application rates since 1997, an estimate of the nitrogen and phosphorus content of those substances, soil test results and a listing of parcels of farm land determined to be rich or excessively rich in phosphorus.

Beyond relatively low minimum size thresholds, for example 10 cattle, approval of the Ministry of the Environment is required to build, to expand or to modify a livestock production facility. Information required in an application includes the estimated number of animal units to be housed in the facility, the estimated weight of those animals at the beginning and at the end of the production cycle, a description of the equipment to be used to move and to store livestock waste, a map of the location of the facility and its surroundings, the anticipated volume of livestock waste and its estimated nitrogen and phosphorus content. A four-year nutrient management plan is required. Fines for failure to comply with regulations range from $1,000 to $500,000, depending on the nature of the offence.

The Ontario Code of Practice (Ontario Ministry of Agriculture, 1976) contained guidelines and standards for separation distances, manure storage requirements, the adequacy of land base for manure disposal, guidelines for the timing of manure application and other environmental management standards. The Code was replaced in 1995 by the Guide to Agricultural Land Use, the Minimum Distance Separation I (MDS I) and the Minimum Distance Separation II (MDS II) (Ontario Ministry of Agriculture, Food and Rural Affairs, 1995). The Guide provides recommendations for manure storage facilities and application. At least 200 days storage capacity is recommended. The MDS documents outline calculations to determine the recommended distance between various types and sizes of livestock operations and neighbouring land uses, property boundaries and roads. The MDS I calculations relate to the construction of new residential or other rural land uses in the vicinity of existing livestock facilities and the MDS II calculations relate to the development or expansion of a livestock facility near an existing neighbouring land use or watercourse. The standards are enforced through municipal zoning and permitting processes. Local municipalities can and do enforce local by-laws that exceed the provincial standards. For example, several municipalities are in the process of drafting by-laws that would require the use of a nutrient management plan for intensive livestock operations within their jurisdictions. The by-laws could require, in addition to the provincial separation distance standards, minimum land requirements for manure disposal.

The Guide to Agricultural Land Use stipulates standards for the incorporation of livestock manure. Solid manure applied within 200 m of an incompatible use should be incorporated within 24 hours. A similar time period is indicated for liquid manure applied within 300 m of an incompatible use.

Agriculture is subject to municipal noise control by-laws in Ontario. Enforcement takes place at the municipal level. A negligence standard is generally applied. The standard of normal hearing and the judgement of a reasonable person are used as criteria (Ontario Ministry of Agriculture, Food and Rural Affairs, 1995). These measures apply to noise from ventilation systems, bird scaring devices, grain and hay drying equipment and general mechanical farm equipment.

The separation distance formulas applied to new or expanded livestock facilities in Ontario (MDS II) are based on the degree of expansion, the type of livestock being produced, the type of operation and the nature of the neighbourhood as reflected in local zoning. Different separation distances are applied for residential, recreational or commercial areas and for property boundaries and roadways. Separate calculations are required for buildings and for proposed manure storage facilities. Different types of manure storage structures have different separation distance requirements. To illustrate, consider the case of the expansion of a feeder hog facility from a capacity of 1,000 hogs per year to 2,000 hogs per year. Assuming a liquid manure system, the MDS II separation distance formulas for such a proposed expansion would require a separation distance of 584 m from the nearest neighbour's dwelling. A separation distance of 1,168 m would be required from areas zoned as residential, institutional, active recreational or commercial urban areas. The facility would have to be located at least 117 m from the nearest property boundary and 146 m from the nearest road or road allowance. The land area required increases with the square of the separation distance. The implied land area for the minimum separation from the nearest neighbour's dwelling is 107 ha. The implied land area needed to satisfy the minimum separation distance from land zoned as residential is 426 ha. This implies a stocking density for the proposed facility of slightly less than 19 feeder hogs ha^{-1} year^{-1} in the first instance and slightly less than 5 feeder hogs ha^{-1} year^{-1} in the second instance. The goal for inventory turnover in these types of operations in Ontario is 3 times per year, so the implied stocking rates in terms of animals present on the farm at one time, versus annual capacity, would be a little over 3 animals ha^{-1} in the first instance and 1.67 animals ha^{-1} in the second instance.

The separation distance formula is non-linear with respect to scale and stocking density is permitted to rise slightly with scale. The above example represents a moderate sized hog finishing operation. If a 2,000 hog operation were to be expanded to 4,000 hogs per year capacity, the minimum separation distances would be, for the same assumptions as above, 722 m to the nearest neighbour's residence (and a land area of 160 ha), 1,445 m to the nearest land zoned as residential (and a land area of 650 ha), 144 m from the nearest property line and 181 m from the nearest road allowance.

Approval of a new intensive livestock operation or of an expansion of an existing facility in Saskatchewan considers the proximity of the facility to watercourses, soil, geological and hydrological conditions of the site, the adequacy of the proposed manure handling and storage system and the nutrient management plan including the adequacy of the land base to which manure is to be applied. Separation distances from neighbouring land uses as a function of the proposed size of the facility are also set out in the guidelines. The permitting and approval process can potentially involve the provincial agriculture department, the municipal government for the community in which the proposed facility will be located, the Saskatchewan Water Corporation, as well as provincial departments of environmental and resource management, the highways and transportation department, the provincial health department and the provincial department for municipal affairs. Currently at least half of the municipal governments in Saskatchewan have by-laws regulating intensive livestock operations.

In Alberta, the 1995 *Code of Practice for the Safe and Economic Handling of Animal Manures* (Alberta Agriculture, Food and Rural Development, 1995) replaced the 1982 *Confinement Livestock Facilities Waste Management Code of Practice*. The aims of the *Code* are to reduce conflicts arising from the operation of new livestock facilities through appropriate siting of those facilities and to assist producers in selecting alternative manure storage and use practices that minimise the incidence of nuisance and other environmental problems. The *Code* was written for producers as well as for municipalities and planners concerned with the siting and the management of livestock facilities. Thresholds are defined for the minimum sizes of livestock operations considered to be intensive livestock operations (Alberta Agriculture, Food and Rural Development, 1995). For a beef feeder cattle operation, this is a capacity of 300 head. Minimum distance separation guidelines were developed to provide guidance for the location of intensive livestock operations to mitigate nuisance problems arising from odour. For example, the minimum distance separation for a 10,000 head beef feedlot ranges from 883 m to 1,472 m, depending on the type of adjacent land use. Guidelines for reducing the risk of contamination of groundwater from inappropriate manure storage facilities or practices are included in the *Code*. Many rural municipalities in Alberta have incorporated or are in the process of incorporating the *Code* into municipal by-laws and land use planning procedures.

It is helpful to compare the minimum separation distances in Alberta to the examples for Ontario presented earlier. The Alberta separation distances are specified in terms of animal numbers present at one time, not annual facility production capacity as in Ontario. An operation of 750 feeder hogs, slightly larger than the 2,000 hog capacity operation in Ontario, requires a separation distance of 343 m for Category 1 (low sensitivity neighbours, e.g. a single residence), 456 m for Category 2 (moderate sensitivity neighbours, e.g. multi-parcel country residential, low use recreational) and 571 m for Category 3 (high sensitivity neighbours, large scale country residential development, high use recreational). A 1,500 hog operation, again, slightly larger than the 4,000 hog annual capacity facility in Ontario, requires a separation distance of 441 m for Category 1, 588 m for Category 2 and 735 m for Category 3. Separate calculations are performed for minimum land base under the Alberta *Code*. These requirements vary with soil type. The minimum land area for a 700 head swine finishing operation ranges from 29 to 55 ha of non-irrigated land. The range for a 1,500 head operation is 64 to 118 ha. The implied stocking rates for the smaller operation is about 13 hogs ha^{-1}. For the larger operation, the implied stocking rate is about 24 hogs ha^{-1}, suggesting that the stocking rate does vary somewhat with scale.

Animal welfare

Animal welfare issues are addressed through a series of voluntary codes of practice in Canada, apart from abusive treatment of animals that trigger criminal sanctions. The codes provide comprehensive treatment of a wide range of animal welfare issues, including housing, access to feed and water, protection from extremes of temperature, cleaning of pens, handling and transport as well as slaughter. To illustrate the standards in the voluntary codes, minimum cage sizes for chickens housed in cages are 410 cm^2 for an adult bird weighing 1.8 kg and 450 cm^2 for an adult bird weighing 2.2 kg. Agriculture Canada (1989) also provides guidelines for floor space requirements for broiler chickens, which are typically not raised in cages. These standards are voluntary and the degree of compliance is not known.

Human health

Bovine Somatotrophin has not been approved for use in the dairy industry in Canada. Animal health concerns were cited as the reason that approval was not granted. An early estimate by Oxley *et al.* (1989) indicated that the lost annual income to producers in Ontario from not having access to rBST was about $50 million (Cdn, constant 1983 $).

ENVIRONMENTAL COMPLIANCE COSTS

Estimating the costs to producers of complying with environmental regulations is a daunting task. As the previous section indicated, agriculture is subject to a complex regime of regulatory instruments in Canada. Many of these instruments reinforce one another. For example, municipal permitting processes, common law remedies and provincial and federal water quality legislation all contribute to protecting surface water quality. So it is not possible to attribute the costs of complying with surface water quality standards to any one policy or programme. And many of the compliance costs are unobservable. In the case of livestock operations, compliance with municipal standards is generally required before a new facility can be constructed or an old one can be modified. If approval is not granted, punitive fines can be applied if a producer proceeds to develop a facility anyway. So non-compliant operations cannot generally be observed to inform cost comparisons with compliant ones.

There are a limited number of what we would call synthetic studies that can be used, however, to glean some evidence on the magnitude of environmental compliance costs in Canadian agriculture. These studies have typically used farm planning models based on techniques developed by production economists coupled with environmental simulation models. The combination of tools has been used to characterise the costs of meeting different levels of environmental performance for representative farm types. These studies provide most of the available information on abatement costs for agriculture in Canada. Our purpose in comparing these studies is to get a general sense of the size of environmental compliance costs in Canadian agriculture as a proportion of production costs in total.

Groundwater and nitrogen

Giraldez and Fox (1995) used the Chemical, Runoff and Erosion from Agricultural Management Systems (CREAMS) model to simulate the impact of varying farm nitrogen applications on the rate of nitrogen leaching into groundwater. They conducted a case study of the village of Hensall in south-western Ontario. Nitrate contamination of groundwater had been identified as a problem in the village water supply, obtained from a community well. Grain maize is assumed to be the only crop grown at the study site. The on-farm abatement costs are determined by the loss of profits from reducing the nitrogen application. The profit-maximising level of nitrogen application was found to be 147 kg ha^{-1}. A reduction in average nitrogen application from 147 kg ha^{-1} year^{-1} to 140 kg ha^{-1} year^{-1} would result in a 16% reduction in nitrogen leaching, which would lower nitrate contamination levels to 10 mg per litre. This reduction would result in an additional farm cost of $1.81 ha^{-1} year^{-1}, or $0.25 per metric tonne of grain maize per year. This amounts to an abatement cost to reduce nitrates to the level of the Canadian drinking water standard of about 0.2% of average annual gross revenue from maize production. The cost of further reduction in nitrogen fertiliser was estimated to increase at an increasing rate.

Van Ham (1996) studied alternative policy instruments for reducing the amount of nitrogen from manure and fertiliser leaching into groundwater from dairy farm operations in Ontario. A non-linear optimisation model was used to determine optimal production and to calculate the costs of compliance. The model is based on an above-average sized hypothetical dairy farm in Ontario with 100 milking cows and 80 replacement animals. The farm size was assumed to be 140 ha and the choice of crops are grain maize, maize silage, soybeans, lucerne (alfalfa) silage and lucerne hay. A nitrogen budget is used to calculate the annual nitrogen surplus.

Three different manure handling systems were considered. The impacts of a limit on fertiliser purchases, a tax on fertiliser, a limit on the excess nitrogen and a tax on the excess nitrogen were studied. In the absence of any policy measures, the nitrogen surplus from slatted floor, flush, and scraper systems was 21.91 mg litre^{-1}, 16.47 mg litre^{-1}, and 21.2 mg litre^{-1} respectively. The maximum acceptable limit for groundwater nitrogen is 10 mg litre^{-1}. An excess nitrogen tax of $1.00 kg^{-1} N for a slatted floor system, of $0.47 kg^{-1} N for a flush systems and of $0.85 kg^{-1} N for a scraper system reduced the nitrogen surplus to 10 mg litre^{-1} or less. An excess nitrogen limit on all of the systems also resulted in a nitrogen leachate level of 10 mg litre^{-1}, but compliance costs varied between $0.44 kg^{-1} N for the slatted floor system and $0.23 kg^{-1} N for the flush system. A $0.70 fertiliser tax on nitrogen induced a significant shift in land allocation from maize to lucerne production. This caused an increase in nitrogen fixation by the lucerne which resulted in an increase in the groundwater nitrate level when a nitrogen tax was applied to commercial fertiliser purchases. The 5,000 kg fertiliser limit generated a nitrogen leachate level higher than that calculated under a 7,000 kg fertiliser limit. This is because with a 5,000 kg fertiliser limit there was an increase in symbiotic nitrogen fixation that occurred due to an increase in the amount of land allocated to lucerne production and there was also a reduction in nitrogen output due to a reduction in maize sales. Several policy instruments were capable of getting the excess nitrogen level to within the 10 ppm standard. The cost of these measures ranged from $92 to $2,943 per year at the farm level. Even at the highest estimated cost, this would amount to less than 0.7% of annual average farm gross revenue for a dairy farm of the size considered in this study.

Yiridoe and Weersink (1998) examined the on-farm abatement costs of alternative crop production systems in the Delhi region of south-western Ontario and the groundwater nitrogen pollution from the application of fertiliser. They characterised abatement costs for excess nitrogen for the various crop production systems, with a view to meeting the drinking water standard for nitrogen contamination (10 mg N litre^{-1}). The crops considered in the study are maize, soybeans and winter wheat. Conventional tillage and no-till systems were analysed with four different rotations involving the three crops (maize/maize, maize/soybean/wheat, maize/maize/soybean/wheat and soybean/wheat). A model of a representative cash crop farm was used in conjunction with the CENTURY biophysical simulation model to predict crop production and nitrogen leachate loss to groundwater. The practice that yielded the highest average profits over the entire rotation was continuous maize with conventional tillage. This was designated as the base practice. Nitrogen leaching was always less with conventional tillage than the corresponding no-till system for each crop rotation. The abatement cost of reducing excess nitrogen from 28.06 kg ha^{-1} year^{-1} under the base practice to 17.83 kg ha^{-1} year^{-1} with a soybean wheat rotation with conventional tillage was estimated to be a little over 40% of annual gross revenue per ha relative to continuous maize. But even this level of excess nitrogen would exceed Health Canada's Maximum Contaminant Level (MCL) of 15.2 kg ha^{-1} year^{-1}. A level of 15.2 kg ha^{-1} year^{-1} is estimated to give a nitrate concentration in groundwater of 10 g litre^{-1}. The costs of maintaining an average limit of 15.2 kg ha^{-1} year^{-1} varied substantially across production systems. For the case of a maize/soybean/wheat rotation with conventional tillage, this limit could be reached with a compliance cost amounting to 6.2% of average annual gross revenue per ha. A continuous maize system under no-till faced an abatement cost of 31.7% of annual gross revenue. Soybean/wheat rotations faced virtually no compliance costs to reach the 15.2 kg ha^{-1} year^{-1} limit. The average abatement costs under a peak nitrogen-leachate restriction for the no-till maize/maize rotation and the no-till soybean/wheat rotation had the highest and lowest abatement costs, respectively, for the MCL of 15.2 kg N ha^{-1}.

Several farm level studies have explored the implications of yield and price risk on the economic viability of alternative tillage systems in crop production in Ontario. Weersink *et al.* (1992a and 1992b) used stochastic dominance to compare conventional and reduced tillage practices. Three farm sizes and two soil types were considered. Reduced tillage options like ridge tillage and no-till outperformed conventional tillage for lower levels of risk aversion. Higher levels of risk aversion favoured conventional tillage, however. This study did not characterise differences in erosion rates or off-site sediment deposition under different tillage systems. Weersink *et al.* (1998) studied the relationship between crop prices and abatement costs for excess nitrogen for a representative 162 ha crop farm in south-western Ontario. Crop prices can affect the intensity of input use and the choice of crops. This in turn will affect a farmer's compliance cost for meeting groundwater nitrate standards. Abatement costs will also depend on the risk preference of the individual farmer. It was assumed that most farmers are slightly risk-averse. The crops included in the study were maize, soybeans and wheat. Three different prices were examined: the average market price for the period 1975-1993, in 1993 dollars, the price under the Gross Revenue Insurance Programme (GRIP) for the same time period, and the effective 1993 price, which was the maximum of GRIP or market prices for that year. GRIP is the agricultural support programme for Ontario crop farmers that is currently in place. The authors studied two different policy methods (design- and performance-based instruments). Under the GRIP and the effective 1993 prices, the producer will begin to shift to more wheat production and idle land as farm nitrogen restrictions are more limiting. Conversely, under the average market price, the producer would not change his production system and would not incur any abatement costs because he is already applying less than the maximum amount of nitrogen. With a tax on nitrogen use, there was a cost to the producer, but no reduction in groundwater nitrogen concentration under the GRIP price. A nitrogen tax did cause a reduction in groundwater nitrogen for the other two price regimes. For the average market price and the 1993 price, nitrate concentrations in groundwater were below the 10 mg litre^{-1} standard. Under the GRIP price, a nitrate concentration of 14 mg litre^{-1} occurred for the unconstrained solution, but a total farm nitrogen restriction of 8,000 kg was more than enough to improve groundwater quality to meet the standard and the compliance cost was less than 3% of annual average gross revenue per ha.

When the amount of excess nitrate was restricted directly, there was no effect on the producer under the average market price because the leachate level was already low due to the 29.9 ha of idle land. With the other two prices, a producer would shift from maize and soybean production to maize, soybean and wheat production with more idle land as the restrictions tighten. A nitrate tax of $200 mg^{-1} N $litre^{-1}$ under the GRIP price generated a cost to farmers without any change in nitrogen leachate level. A tax on excess nitrate was able to reduce concentrations to more than meet the standard for a compliance cost of less than 4% of annual gross revenue per ha. The authors conclude that GRIP creates increased abatement costs by increasing the incentive to grow maize, a crop with a relatively high leaching potential.

Erosion and surface water

Fox and Dickson (1990) studied the on farm profitability and off farm costs, the latter due to sediment deposition in waterways, of alternative crop tillage practices in Ontario. Three watersheds in the Thames River basin of south-western Ontario were considered; the Big Creek, Newbiggen Creek, and the Stratford/Avon watershed. In these watersheds, fall moldboard ploughing is the conventional tillage system and the common crops are maize, soybeans and lucerne. The conservation practices considered in the study are autumn chisel plough, ridge planting, and no-till. The Soil Conservation Economics (SOILEC) model was used to estimate the on-farm impacts of the various management practices. The Guelph Model for Evaluating the Effects of Agricultural Management Systems on Erosion and Sedimentation (GAMES) was used to estimate sediment deposition to streams in the watershed. For each of the autumn chisel plough, ridge tillage and no-till methods, the erosion soil rate, gross erosion volume and sediment delivery volume are less than for mouldboard plough. Ridge planting reduced the gross erosion volume by about 66% and the sediment delivery volume by approximately 76% compared to mouldboard plough in the Big Creek and Newbiggen Creek watersheds. The on-farm costs of adoption of ridge planting ranged between $34 and $121 ha^{-1} $year^{-1}$. In the Stratford/Avon watershed, no-till practices reduced gross erosion by 62.6% and sediment delivery volume by 79.7%, and on-farm costs of adoption were approximately $10 ha^{-1} $year^{-1}$.

Autumn chisel ploughing produced lower adoption costs than ridge planting or no-till practices, but the reduction in gross erosion and sediment delivery volumes was only about half of what was realised with no-till or ridge planting. The authors conclude that conservation tillage can be an economically-feasible method of reducing sediment damages due to agriculture in Ontario.

Van Vuuren et al. (1997) evaluated the SWEEP, which operated between 1985 and 1993 in Ontario. The purpose of this programme was to reduce the phosphorus loading from non-point agricultural sources in the Lake Erie Basin. This paper examines single-contaminant reduction policies and the degree of water quality improvement obtained. They also studied the linkages between contaminant loading reduction and the benefits of water quality improvements. The area studied is the Kettle Creek watershed in south-western Ontario. The focus was on-farm conservation methods (alternative crop rotations, tillage practices, and 4.5 m buffer strips) that reduce sediment and phosphorus loading.

The resultant stream delivery of suspended solids and phosphorus was estimated using the GAMES model. The SOILEC model was used to determine the 20-year financial outcomes. With and without a buffer strip, all of the maize/maize, maize/soybean/wheat and maize/soybean rotations had negative abatement costs, meaning that the model indicated that farm income would increase with the practice that reduced sediment deposition. This result could reflect a lack of technical information on the returns available using alternative tillage practices among crop farmers in the region, or possibly that the modelling failed to consider all of the relevant costs that producers would face in adopting these practices. The greatest reduction in loading resulted from hay in the crop rotation, but these practices generated abatement costs on the order of 23% of per ha annual gross revenue.

Erosion and productivity

Summer following has been a common practice in the prairie provinces, but it can cause an increase in soil loss from water erosion. One solution to this problem (Seecharan *et al.*, 1990) is to plant lentils instead of fallow with a wheat rotation to sustain soil productivity. An obstacle to implementing this practice is the variability in productivity, prices, and income. Seecharan *et al.* (1990) assessed the average expected returns for commonly used fallow/wheat rotations versus lentil/wheat rotations. Conventional and reduced tillage were also compared for a 1-year and 25-year period for the brown soil zone in Saskatchewan. The authors used the SOILEC model. In both the short-run and long-run analyses, substantial reductions in gross erosion rates could be achieved with apparently trivial compliance costs.

Turvey and Weersink (1991) examined the on-site costs associated with soil loss and sediment loading constraints for continuous maize or maize-soybean rotations using moldboard plough or no-till on a 352 ha watershed in southern Ontario. The GAMES model was used to derive gross soil loss and sediment loading to surface water. Using a two-period linear programming (LP) model, the profits are maximised with the optimal rotation and tillage system for each constraint combination. The base solution gives a maximum annual profit of $143,484, an annual soil loss of 1683.5 metric tons and an annual sediment loading of 60.85 metric tons for the watershed. The cost of meeting increasingly stringent environmental standards was found to increase at an increasing rate. If one constraint is held constant, the marginal cost increases for increases in the other constraint. A typical crop rotation in the region considered in this study would average between $500 and $700 in gross revenue ha^{-1} year^{-1}. The highest abatement cost value reported of $168 ha^{-1} year^{-1} would amount to 24% to 34% of annual average gross revenue per ha. However, many of the less stringent soil loss and sediment load targets could be achieved at a much lower cost.

Deloitte and Touche Management Consultants (1992a, 1992b) evaluated the cost of reducing sediment and phosphorus run-off into surface water and reducing soil degradation for the Kettle Creek watershed in south-western Ontario. The GAMESP was used to estimate sediment and phosphorus loading and soil loss. The economic impact of different environmental quality constraints was determined using a 3-year profit maximising LP model. The model was allowed to choose one of three rotations; continuous maize, soybean/maize/soybean or soybean/wheat/maize under conventional tillage or no-till conservation tillage.

Without any environmental constraints, the profit-maximising solution was to plant continuous maize using no-till, which gave a 3-year profit of $456,011 for the 412 ha watershed. The model was first run with conventional tillage and continuous maize being restricted from switching to no-till as soil loss restrictions were imposed. The second version of the model was run in which there was no constraint on tillage system, and no-till became the optimal solution when soil loss restrictions were imposed. For each tillage and rotation combination, the model responded to increasingly stringent environmental standards by removing land from maize production. Overall, the results of this study indicate that substantial reductions in soil loss, sediment load and phosphorus levels could be achieved at compliance cost levels below 5% of annual average gross revenue per ha.

Yiridoe *et al.* (1993) examined how the use of cover crops and conservation tillage might increase the financial viability of alternative crops such as beans and wheat as alternatives to tobacco, while also reducing soil degradation and soil erosion, on sandy loam soils in southern Ontario. The on-farm profitability of alternative soybean, white bean, kidney bean and winter wheat cropping systems with conventional tillage are compared to no-till systems with four types of cover crop. Yield was measured, along with the impact of conventional tillage and no-till on soil erosion. The impacts on soil loss are determined using the Universal Soil Loss Equation (USLE). The estimated on-farm costs of different cover crops (with no-till) versus the base (no crop cover and conventional tillage) are estimated for a representative 80 ha farm. None of the alternatives had a cost that was statistically different from the base in each of the three crop cases. Within each bean-wheat rotation, no-till with any of the cover crops generated a negative cost, indicating an increase in net returns relative to the base, except for the kidney bean-wheat rotation with a stubble wheat cover crop. With the kidney bean and white bean rotations, a volunteer wheat cover crop gave the highest net returns. With soybeans, a rye cover crop produced the best net returns. This study only considered the short-term (4 years for beans, 3 years for wheat) and long-term net returns may be different. A comparison of soil loss estimates for conventional tillage and no-till showed that approximately 2 to 10 metric tonnes ha^{-1} year^{-1} of soil is saved with no-till depending on the topography's susceptibility to erosion.

Stonehouse and Bohl (1993) studied the impacts of the use of regulations and taxes for Ontario farms as a means of reducing soil erosion. They used a farm level multi-period LP model that assumed that profit maximisation is the primary goal of the farmers. The two policy alternatives considered are a soil loss regulation and a tax imposed on soil eroded. The most profitable practice without regulation or a tax/subsidy is a 4-year maize/maize/soybean (mouldboard plough)/wheat (no-till) rotation. The introduction of a 4-year soil-loss regulation policy caused a change in the optimal production system. The first switch was to conservation tillage (chisel plough) with maize, soybean and wheat production, but a further decrease in soil loss allowance prompted the production of continuous maize with zero tillage. As taxation rates increased above $0.20 tonne^{-1} year^{-1}, net cash flow greatly declined, while the effects on the erosion rate were marginal. Taxation also induced a change to maize, soybean and wheat production with conservation tillage.

Multiple emissions

Lintner and Weersink (1999) developed a multi-contaminant model to study the effectiveness of policy instruments for achieving reductions in nitrogen and phosphorus levels in ground and surface water. The study considered a sub-watershed of the Lake St. Clair drainage basin in Essex County in south-western Ontario. The principle crops in this area are soybeans, maize and wheat. This study examined 3-year rotations with conventional tillage and no-till, along with two different nitrogen application. Results were obtained for actual fertiliser application levels and for efficient application where the marginal value product of nitrogen is equal to its cost. The Agricultural Non-Point Source pollution model (AGNPS) was used to simulate emissions of sediment-bound and soluble nitrogen and sediment-bound phosphorus to surface water, along with emissions of soluble nitrogen to groundwater. The results showed that the profit-maximising choice for farmers (base practice) was to grow a maize/soybean/soybean rotation with conventional tillage and the efficient application of nitrogen. The management system with the highest abatement costs, relative to this base practice, was maize/maize/maize with conventional tillage and efficient use of fertiliser. This system resulted in a slightly lower emission of sediment-bound phosphorus and nitrogen to surface water, but the amount of soluble nitrogen moving towards surface and groundwater increased, revealing that in multiple agricultural emission problems, there is often a trade-off between emission types or media with alternative management systems. A reduction in one type of emission often results in the increase of another emission compared to the base. Only in the case of the continuous hay rotation with no-till and the actual fertiliser application was there a reduction in all five emissions and media combinations.

Water quality is characterised by the concentration of nitrogen and phosphorus in the surface water and the concentration of nitrogen in the groundwater. In all of the management scenarios considered in the Lintner and Weersink (1999) study, the concentration of nitrogen in groundwater was below the drinking water quality guideline of 10 mg nitrate N litre^{-1} of water. Water quality objectives were violated with respect to the phosphorus concentration in surface water because all were above the eutrophication guideline of 0.01 mg litre^{-1}. Lintner and Weersink arbitrarily chose a 20% reduction in the phosphorus concentration from the base as the optimal level for abatement. The abatement cost of reducing the concentration of phosphorus in surface water was $1,415 per mg of phosphorus per litre of water.

The possible control instruments considered by Lintner and Weersink (1999) were cost-effective abatement, where marginal benefits of abatement equals marginal abatement cost, mandatory switch to no-till, a nitrogen ceiling of 0 kg N ha^{-1} and 29 kg N ha^{-1}, and an ambient tax/subsidy to meet the 20% phosphorus reduction goal. If no-till was mandatory, farms in the area would switch to a no-till maize/soybean/wheat rotation. This causes an increase in groundwater nitrogen but results in the lowest phosphorus concentration of all the policy instruments.

The option of a zero nitrogen ceiling would reach the 20% reduction in phosphorus concentration, but would cause a 59% reduction in profits (an abatement cost of $102 ha^{-1} year^{-1}), and thus it would be extremely costly and inefficient.

Pesticide regulations

Two previous comparisons of the Canadian and the US pesticide regulatory regimes (Deen and Fox, 1991, 1992; McEwan and Deen, 1997) have concluded that while there are differences in some requirements for registration in the two countries and, for these and other reasons, differences in the availability of some products, the overall effect of the two regulatory regimes does not seem to indicate a systemic advantage to producers in either country. Of course, the regulatory regime for pesticides is undergoing potentially significant change in Canada and the United States. The Canadian pesticide regulatory system was restructured in 1995. The 1996 USA *Food Quality Protection Act* promises to change fundamentally the operation of the process of pesticide approval and regulation in that country, but the *Act* has confronted significant resistance in application and it is not yet clear what the final form of its implementation will be.

CONCLUDING REMARKS

The available evidence suggests that surface water quality degradation from erosion, groundwater quality degradation by bacteria and nitrate and air quality problems from odour from livestock production are the most important environmental problems associated with agricultural production in Canada. However, the overall incidence of these problems seems low relative to the situations in the EU or in the USA. We attribute this, in part, to the relatively extensive structure of production in agriculture in Canada, especially in the production of grains, beef and pork, given the historically low level of production subsidies that farmers have received for these products in Canada relative to their counterparts in the EU or even the USA. Like the USA, however, location of large scale intensive livestock operations is becoming more controversial (Prairie Farm Rehabilitation Administration, Undated; and Prairie Provinces Committee on Livestock Development and Manure Management, 1998), especially in more densely populated regions (Marchand and McEwen, 1997). Most provinces, are in the process of developing or updating guidelines and standards for the regulation of these types of facilities (Alberta Agriculture, Food and Rural Development, 1999; Centre for Studies in Agriculture, Law and the Environment, 1996; Nova Scotia Department of Agriculture and Marketing, 1999; New Brunswick Agriculture and Rural Development, 1997; New Brunswick Interdepartmental Committee on Waste Application, 1996; McCormack, 1998;

Saskatchewan Agriculture and Food, Undated, 1997a, 1997b, 1997c, Technology Transfer Working Group, 1998). Nutrient management by-laws like those developed by Evans (1998) and best management practices such as those developed for potatoes in Prince Edward Island (Prince Edward Island Departments of Fisheries and Environment, and Agriculture and Forestry, Undated) are under consideration in many areas. A recent NAFTA challenge that claimed that Quebec was not enforcing its own environmental laws with respect to its provincial swine industry was thrown out on a technicality, but the fact that the case was brought to the Commission for Environmental Cooperation in the first place is likely a harbinger of things to come. The Walkerton crisis of May 2000 has increased the political pressure for provincial governments to act in this area.

Historically, the mitigation of the environmental effects of agriculture in Canada has been accomplished through a complex and interdependent set of institutions. Federal and provincial legislation, municipal policies, civil liability and industry lead voluntary measures have all played a role. It is important in international comparisons of environmental policies, and especially in investigations of the impacts of those policies on competitiveness, to recognise the full range of the institutions that are involved. Comparison of, for purposes of illustration, only federal legislation and regulations, would give a partial and potentially seriously misleading perspective.

The estimation of environmental compliance costs in agriculture is in its infancy. The scope of coverage of published studies of abatement costs in Canadian agriculture is limited and uneven in its coverage. The available evidence suggests, however, that the conjecture made by Ervin and Fox (1998) that environmental compliance costs in agriculture, like those in other sectors, are not particularly high, seems to be confirmed by our analysis. When producers are allowed flexibility in selecting the means by which they can achieve various environmental quality targets, the Canadian literature indicates that compliance costs are generally less than 3% of gross revenue.

REFERENCES

Agriculture Canada (1989) *Recommended Code of Practice for the Care and Handling of Poultry from Hatchery to Processing Plant*, Publication 1757/E, Communications Branch, Ottawa.

Albert, R. (1996) Ground water contamination in rural New Brunswick. *Proceedings of the Agriculture and Groundwater Quality Workshop*, Agricultural Advisory Committee on the Environment, Fredericton, pp. 8-17.

Alberta Agriculture, Food and Rural Development (1995) *Code of Practice for the Safe and Economic Handling of Animal Manures*. Publishing Branch, Alberta Agriculture, Food and Rural Development, Edmonton.

Alberta Agriculture, Food and Rural Development (1999) *A Proposed Regulatory Framework for Livestock Feeding Operations in Alberta*. Alberta Agriculture, Food and Rural Development, Edmonton.

Alberta Cattle Commission (Undated) *Water Quality and Cattle Production*. Alberta Cattle Commission, Calgary.

Canada-Alberta Environmentally Sustainable Agriculture Agreement (1998) *Agricultural Impacts on Water Quality in Alberta: An Initial Assessment*, Publications Office, Alberta Agriculture, Food and Rural Development, Edmonton.

Centre for Studies in Agriculture, Law and the Environment (1996) *Expanding Intensive Livestock Operations in Saskatchewan: Environmental and Legal Constraints.* University of Saskatchewan, Saskatoon.

Chang, C. and Entz, T. (1996) Nitrate content in the groundwater under long-term feedlot manure application. *Proceedings of the Irrigation Research and Development Conference*, Water Resources Institute, Saskatoon, pp. 339-356.

Deen, B. and Fox, G. (1991) *An Overview of the Proposed Federal Pest Management Regulatory System*, Discussion Paper DP91/03, George Morris Centre, University of Guelph, Guelph.

Deen, B. and Fox, G. (1992) *A Revised Federal Pest Management Regulatory System: Implications for Canadian Farmers*, Discussion Paper DP92/02, George Morris Centre, University of Guelph, Guelph.

Deloitte and Touche Management Consultants (1992a) *Field Level Economic Analysis of Changing Tillage Practices in Southwestern Ontario.* Prepared for Agriculture Canada for the Soil and Water Environmental Enhancement Programme, Deloitte and Touche Management Consultants, Guelph.

Deloitte and Touche Management Consultants (1992b) Watershed Level Economic Analysis of Tillage Practices in Southwestern Ontario. Prepared for Agriculture Canada for the Soil and Water Environmental Enhancement Programme, Deloitte and Touche Management Consultants, Guelph.

Eastern Canada Soil and Water Conservation Centre (1997) *Policy Instruments for Environmental Protection in Agriculture: Analytical Review of the Literature*, University of Moncton, Edmonston.

Eastern Canada Soil and Water Conservation Centre (1998) *Acts and Regulations with an Impact on Water Quality in Atlantic Canada*, University of Moncton, Edmonston.

Ecobichon, D.J., Hicks, R., Allen, M.C. and Albert, R. (1996) Groundwater Contamination in Rural New Brunswick. *Proceedings of the Agriculture and Groundwater Quality Workshop.* Sponsored by the Agricultural Advisory Committee on the Environment, April 11.

Ervin, D. and Fox, G. (1998) Environmental policy considerations in the grain-livestock subsectors in Canada, Mexico and the United States. In: Loyns, R.M.A., Knutson, R. and Meilke, K. (eds) *Economic Harmonization in the Canadian/U.S./Mexican Grain-Livestock Subsector.* Texas A&M University and the University of Guelph, pp. 275-304.

Evans, S. (1998) Model by-law to Regulate Nutrient Management for Intensive Livestock Operations in Townships. Economic Development, County of Middlesex, August.

Fox, G. and Barber, D. (1991) Crop Production, Input Use and Pest Management Under a Regime of Declining Product Prices. Invited paper, *Proceedings of the Brighton Crop Protection Conference - Weeds - 1991*, British Crop Protection Council, Brighton, pp. 103-112.

Fox, G. and Dickson, E. (1990) The economics of erosion and sediment control in southwestern Ontario. *Canadian Journal of Agricultural Economics,* 38, 23-44.

Fox, G. (1992) Soil Erosion at a Crossroads. Discussion Paper DP92/01, Department of Agricultural Economics and Business, University of Guelph, Guelph.

Fox, G., Umali, G. and Dickinson, T. (1995) An economic analysis of targeting soil conservation measures with respect to off-site water quality. *Canadian Journal of Agricultural Economics,* 43, 105-118.

Giraldez, C., and Fox, G. (1995) An economic analysis of groundwater contamination from agricultural nitrate emissions in Southern Ontario. *Canadian Journal of Agricultural Economics,* 43, 387-402.

Goss, M., Barry, D. and Rudolph, D. (1998) Contamination in Ontario farmstead domestic wells and its association with agriculture: 1. Results from Drinking Water Wells. *Journal of Contaminant Hydrology,* 32, 267-293.

Government of Nova Scotia (1990) *Nova Scotia Farm Well Water Quality Assurance Study,* Phase 1, Final Report, Nova Scotia Department of Environment, Nova Scotia Department of Agriculture and Marketing and Nova Scotia Department of Health and Fitness, Halifax.

Harker, D., Bolton, K., Townley-Smith, L. and Bristol, B. (1997) *A Prairie-wide Perspective of Non-point Agricultural Effects on Water Quality: A Review of Documented Evidence and Expert Opinion.* Prairie Farm Rehabilitation Administration, Prairie Resources Division, Agriculture Canada, Regina.

Henry, J. and Meneley, W. (1993) *Nitrates in Western Canadian Groundwater.* Western Canadian Fertiliser Association, Regina.

Henry, J. (1995) Nitrate in the groundwater of Western Canada. *Proceedings from the International Association of Hydrogeologists,* Edmonton, Alberta.

Ivy, M. (1996) *Compensation for Regulatory Takings of Private Property: The Case of Wetland Protection in Southern Ontario and the United States,* M.Sc. Thesis, University of Guelph, Guelph, Canada.

Kerr, W., Fox, G., Hobbs, J.E. and Klein, K.K. (1991) *A Review of Studies on Western Grain Transportation Policies.* Working Paper 6/91, Policy and Grains and Oilseeds Branches, Agriculture Canada, Ottawa.

Khan, H. (1992) Agricultural Development and Environmental Control Legislation in Newfoundland. In: Milburn, P., Nicholaichuk, W. and Topp, C. (eds) *Agricultural Impacts on Water Quality: Canadian Perspectives.* National Workshop Proceedings, Agriculture Canada Research Station, Fredericton, pp. 119-138.

Klein, K., Fox, G., Kerr, W.A., Kulshreshtha, S.N. and Stennes, B. (1991a) *Regional Implications of Compensatory Freight Rates for Prairie Grains and Oilseeds.* Working Paper 3/91, Policy and Grains and Oilseeds Branches, Agriculture Canada, Ottawa.

Klein, K., Fox, G., Kerr, W.A., Kulshreshtha, S.N. and Stennes, B. (1991b) *Summary of Regional Impacts of Compensatory Freight Rates for Prairie Grains.* Working Paper 4/91, Policy and Grains and Oilseeds Branches, Agriculture Canada, Ottawa.

Lapp, P., Madramootoo, C., Enright, P., Papineau, F. and Perrone, J. (1998) Water quality of an intensive agricultural watershed in Quebec. *Journal of the American Water Resources Association,* 34, 427-437.

Lintner, A. and Weersink, A. (1999) Endogenous transport coefficients: implications for improving water quality from multi-contaminants in an agricultural watershed. *Environmental and Resource Economics,* 14, 269-296.

Marchand, L. and McEwan, K. (1997) *The Impact of Township Zoning By-laws on Ontario Swine Farms.* Ontario Pork Producers Marketing Board, Etobicoke.

Masse, L., Prasher, S., Khan, S., Arjoon, D. and Barrington, S. (1994) Leaching of metolachlor, atrazine and atrazine metabolites into groundwater. *Transactions of the American Society of Agricultural Engineers,* 37, 801-806.

McCormack, K. (1998) *Water Management Guide: For Livestock Production, Water Quality and Wildlife Habitat, Version 2.* Environmental Youth Corps Programme, Ontario Cattlemen's Association and the Ontario Federation of Anglers and Hunters, Guelph.

McEwan, K. and Deen, W. (1997) *A Review of Agricultural Pesticide Pricing and Availability in Canada.* Ridgetown College, University of Guelph, Ridgetown.

McRae, T., Smith, C. and Gregorich, L. (eds) (2000) *Environmental Sustainability of Canadian Agriculture: Report of the Agri-Environmental Indicator Project.* Research

McRae, T., Smith, C. and Gregorich, L. (eds) (2000) *Environmental Sustainability of Canadian Agriculture: Report of the Agri-Environmental Indicator Project*. Research Branch, Policy Branch and Prairie Farm Rehabilitation Administration, Agriculture and Agrifood Canada, Ottawa.

Milburn, P., Nicholaichuck, W. and Topp, C. (eds) (1992) *Agricultural Impacts on Water Quality: Canadian Perspectives*. Proceedings of a National Workshop Sponsored by the Canadian Agricultural Research Council, Ottawa and distributed by the Agriculture Canada Research Station, Fredericton.

Milburn, P. (1996) Leaching of Nitrates from Atlantic Cropping Systems. *Proceedings of the Agriculture and Groundwater Quality Workshop*. Agricultural Advisory Committee on the Environment, Fredericton, pp. 18-23.

Milburn, P., Leger, D.A., O'Neill, H., Richards, J.E., MacLeod, J.A. and MacQuarrie, K. (1995) Pesticide leaching associated with conventional potato and maize production in Atlantic Canada, *Water Quality Research Journal of Canada*, 30, 383-397.

Ministère de l'Agriculture, des Pecheries et de l'Alimentation du Quebec (Undated) *Ensemble, cultivons le bon viosinage*. Publication No. 97-0148, Ministere de l'Agriculture, des Pecheries et de l'Alimentation du Quebec, Quebec City.

Ministère de l'Agriculture, des Pecheries et de l'Alimentation du Quebec (Undated) *Gestion des odeurs en milieu agricole: Epandage des engrais de ferme*, Publication No. 98-0108, Ministere de l'Agriculture, des Pecheries et de l'Alimentation du Quebec, Quebec City.

Ministère de l'Agriculture, des Pecheries et de l'Alimentation du Quebec (Undated) *Du nouveau pour la Protection des Rives en Milieu Agricole*, Publication No. 92-0060, Ministere de l'Agriculture, des Pecheries et de l'Alimentation du Quebec, Quebec City.

Ministère de l'Agriculture, des Pecheries et de l'Alimentation du Quebec (Undated) *Strategie phytosanitaire*, Publication No. 98-006, Ministere de l'Agriculture, des Pecheries et de l'Alimentation du Quebec, Quebec City.

Ministère de l'Agriculture, des Pecheries et de l'Alimentation du Quebec (Undated) *Plan d'Action Saint-Laurent Vision 2000 - Phase 3, Agriculture*, Ministere de l'Agriculture, des Pecheries et de l'Alimentation du Quebec, Quebec City.

Ministère de l'Agriculture, des Pecheries et de l'Alimentation du Quebec et Agriculture et Agroalimentaire Canada (Undated) *Saint-Laurent Vision 2000*, Publication No. 3923-98-05, Ministere de l'Agriculture, des Pecheries et de l'Alimentation du Quebec, Quebec City.

Ministère de l'Agriculture, des Pecheries et de l'Alimentation du Quebec (1997) *Impacts pour le Monde Agricole et Municipal de la Loi Relative a la Protection et au Developpement des Activities Agricoles*, Publication No. 97-0117, Ministere de l'Agriculture, des Pecheries et de l'Alimentation du Quebec, Quebec City.

Ministère de l'Agriculture, des Pecheries et de l'Alimentation du Quebec (1998) *Agrifood Enterprise Assistance Programme*, 1998 Edition, Ministere de l'Agriculture, des Pecheries et de l'Alimentation du Quebec, Quebec City.

Ministère de l'Agriculture, des Pecheries et de l'Alimentation du Quebec (1998) *Programme d'Aide a l'Investissement en Agroenvironnement*, Publication No. 98-0065, Ministere de l'Agriculture, des Pecheries et de l'Alimentation du Quebec, Quebec City.

Ministère de l'Environnement et de la Faune du Quebec (1995) *Qualite de L'eau en Milieu Agricole: La culture du mais et les Pesticides*, Publication No. 95-3083-10, Ministere de l'Agriculture, des Pecheries et de l'Alimentation du Quebec, Quebec City.

Ministère de l'Environnement et de la Faune du Quebec (1997) *Respecting the Reduction of Pollution from Agricultural Sources: A Summary*, Ministere de l'Agriculture, des Pecheries et de l'Alimentation du Quebec, Quebec City.

Moerman, D. (1997) *Environmental Regulations Handbook for Nova Scotia Agriculture, Edition 1*. Nova Scotia Department of Agriculture and Marketing and Nova Scotia Department of the Environment, Halifax.

Moerman, D. and Briggins, D. (1999) *Nova Scotia Farm Well Water Quality Assurance Study*. Nova Scotia Department of Agriculture and Marketing and Nova Scotia Department of the Environment, Halifax.

Nagpal, N.K. (1992) Impact of Agricultural Practices on Water Quality: A British Columbia Perspective. In: Milburn, P., Nicholaichuk, W. and Topp, C. (eds) *Agricultural Impacts on Water Quality: Canadian Perspectives*. National Workshop Proceedings, Agriculture Canada Research Station, Fredericton, pp. 185-192.

New Brunswick Agriculture and Rural Development (1997) *Manure Management Guidelines for New Brunswick*, New Brunswick Agriculture and Rural Development, Fredrickton.

New Brunswick Interdepartmental Committee on Waste Application (1996) *Guidelines for Issuing Certificates of Approval for the Utilisation of Wastes as Soil Additives*. New Brunswick Departments of Agricultural and Rural Development, Health and Community Services, and the Environment, Frederickton.

Nova Scotia Department of Agriculture and Marketing (1999) *Siting and Management of Hog Farms in Nova Scotia*. Nova Scotia Department of Agriculture and Marketing, Halifax.

Nova Scotia Department of Agriculture and Marketing and Nova Scotia Department of the Environment (1997) *Environmental Regulations Handbook for Nova Scotia Agriculture*. Edition 1, Nova Scotia Department of Agriculture and Marketing and Nova Scotia Department of the Environment, Halifax.

Ontario Ministry of Agriculture (1976) *Agricultural Code of Practice*. Queen's Printer for Ontario, Toronto.

Ontario Ministry of Agriculture, Food and Rural Affairs (1995) *Minimum Distance Separation II (MDS II)*. Queen's Printer for Ontario, Toronto.

Oxley, J., Fox, G. and Moschini, G. (1989) An analysis of the structural and welfare effects of bovine somatotropin in the Ontario dairy industry. *Canadian Journal of Agricultural Economics*, 37(3), 393-406.

Prairie Farm Rehabilitation Administration (Undated) *Assessment of the Environmental Factors Relevant to Hog Expansion in Saskatchewan*, Prairie Farm Rehabilitation Administration Regina.

Prairie Provinces Committee on Livestock Development and Manure Management (1998) *Proceedings of the Sustainable Intensive Livestock Development through Science-based Standards Fall Workshop*. Saskatchewan Agriculture and Food, Regina.

Prince Edward Island Departments of Fisheries and Environment, and Agriculture and Forestry (Undated) *Best Management Practices: Soil Conservation for Potato Production*, Prince Edward Island Departments of Fisheries and Environment, and Agriculture and Forestry, Charlottetown.

Pupp, C., Grove, G. and Betcher, R. (1995a) *Groundwater in Manitoba: Hydrogeology, Quality Concerns, Management*. Environmental Science Division of the National Hydrology Research Centre, Environment Canada, Saskatoon.

Pupp, C., Grove, G. and Betcher, R. (1995b) *Groundwater in Saskatchewan: Hydrogeology, Quality Concerns, Management*. Environmental Science Division of the National Hydrology Research Centre, Environment Canada, Saskatoon.

Racz, G. (1992) Effects of agriculture on water quality: Previous and ongoing studies related to sediments, nutrients and pesticides - Manitoba. In: Milburn, P., Nicholaichuk, W. and Topp, C. (eds) *Agricultural Impacts on Water Quality: Canadian Perspectives*. National Workshop Proceedings, Agriculture Canada Research Station, Fredericton, pp. 55-66.

Ruggles, R. (1992) Agriculture and Water Quality - Saskatchewan Legislation. In: Milburn, P., Nicholaichuk, W. and Topp, C. (eds) *Agricultural Impacts on Water Quality: Canadian Perspectives*. National Workshop Proceedings, Agriculture Canada Research Station, Fredericton, pp. 172-174.

Runge, F. and Fox, G. (1999) Issue Study 2, Feedlot Production of Cattle in the United States and Canada: Some Environmental Implications of the North American Free Trade Agreement, in *Assessing Environmental Effects of the North American Free Trade Agreement (NAFTA)*, Commission for Environmental Cooperation, Montreal, pp. 183-258.

Saskatchewan Agriculture and Food (Undated) *Information Guidelines for the Approval of Intensive Hog Operations*, Saskatchewan Agriculture and Food, Regina.

Saskatchewan Agriculture and Food (1997a) *Guidelines for Establishing and Managing Livestock Operations*, Saskatchewan Agriculture and Food, Regina.

Saskatchewan Agriculture and Food (1997b) *Fact Sheet: Current Review and Approval Process: Intensive Livestock Operations*, Saskatchewan Agriculture and Food, Regina.

Saskatchewan Environment and Resource Management (1997) *Interim Guidelines: Information Required for Proposed Intensive Hog Operations; Feeder Barns, Manure Storage and Disposal*, Saskatchewan Environment and Resource Management, Regina.

Seecharan, R., Narayanan, S. and Biederbeck, V. (1990) An economic evaluation of the impact of water erosion on selected alternative crop rotation systems under risk and uncertainty in Prairie agriculture. *Canadian Journal of Agricultural Economics*, 38, 953-963.

Somers, G. (1998) Distribution and trends for the occurrence of nitrate in Prince Edward Island groundwaters. *Nitrate- Agricultural Sources and Fate in the Environment - Perspectives and Direction, Prince Edward Island*. Workshop proceedings, Eastern Canada Soil and Water Conservation Centre, Charlottetown, pp. 19-26.

Sparrow, H. (Chairman) (1984) *Soil at Risk: Canada's Eroding Future*, Report on Soil Conservation by the Senate Standing Committee on Agriculture, Fisheries and Forestry, The Senate of Canada, Ottawa.

Stonehouse, D.P. and Bohl, M.J. (1993) Selected government policies for encouraging soil conservation on Ontario cash-cropping farms. *Journal of Soil and Water Conservation*, 48, 343-349.

Technological Transfer Working Group (1998) *Agro-Environmental Plan for the Hog Industry: Summary of the Evaluation Report on Liquid Hog Manure Management and Treatment Technologies*. Federation des producteurs de porcs du Quebec, Quebec City.

Turvey, C. and Weersink, A. (1991) Economic costs of environmental quality constraints. *Canadian Journal of Agricultural Economics*, 39, 677-685.

Van Ham, M. (1996) Economic Instruments in the Regulation of Nitrate Emissions from Dairy Farms in Ontario: A Farm Level Approach. M.Sc. Thesis, University of Guelph, Guelph, Canada.

Van Vuuren, W., Giraldez, J.C. and Stonehouse, D.P. (1997) The social returns of agricultural practices for promoting water quality improvement. *Canadian Journal of Agricultural Economics*, 45, 219-234.

Vanderlaan, P. (1992) Regulations Affecting Agriculture in New Brunswick. In: Milburn, P., Nicholaichuk, W. and Topp, C. (eds) *Agricultural Impacts on Water Quality: Canadian Perspectives*. National Workshop Proceedings, Agriculture Canada Research Station, Fredericton, pp. 154-161.

Weersink, A., Dutka, C. and Goss, M. (1998) Crop price and risk effects on farm abatement costs. *Canadian Journal of Agricultural Economics*, 46, 171-190.

Weersink, A., Walker, M., Swanton, C. and Shaw, J. (1992a) Economic comparison of alternative tillage systems under risk. *Canadian Journal of Agricultural Economics,* 40, 199-217.

Weersink, A., Walker, M., Swanton, C. and Shaw, J. (1992b) Costs of conventional and conservation tillage. *Journal of Soil and Water Conservation,* 47, 328-334.

Willson, K. (1992) Agriculture and environment: A review of Ontario's legislation and guidelines. In: Milburn, P., Nicholaichuk, W. and Topp, C. (eds) *Agricultural Impacts on Water Quality: Canadian Perspectives.* National Workshop Proceedings, Agriculture Canada Research Station, Fredericton, pp. 162-171.

Yiridoe, E. and Weersink, A. (1998) Marginal abatement costs of reducing groundwater-N pollution with intensive and extensive farm management choices. *Agricultural and Resource Economics Review,* 27, 169-185.

Yiridoe, E., Weersink, A., Roy, R. and Swanton, C. (1993) Economic analysis of alternative cropping systems for a bean/wheat rotation on light-textured soils. *Canadian Journal of Plant Science,* 73, 405-415.

Zebarth, B. (1992) Water Quality Issues and Research in British Columbia. In: Milburn, P., Nicholaichuk, W. and Topp, C. (eds) *Agricultural Impacts on Water Quality: Canadian Perspectives.* National Workshop Proceedings, Agriculture Canada Research Station, Fredericton, pp. 84-91.

Zebarth, B., Hii, B., Liebscher, H., Chipperfield, K., Paul, J., Grove, G. and Szeto, S. (1998) Agricultural land use practices and nitrate contamination in the Abbotsford aquifer, British Columbia, Canada. *Agriculture, Ecosystems and Environment,* 69, 99-112.

Australia

Randy Stringer and Kym Anderson

INTRODUCTION

Australia is the driest inhabited continent. Most of its soils are shallow and infertile. While 60% of the country's land area is used for agriculture, only 5% of that agricultural land has sown pasture grasses and less than 5% is cropped. The rest is arid leased land used for grazing cattle and sheep on native grasses for meat and wool production. Even so, most sheep and cattle are grazed in the higher-rainfall, mixed-farming zones, with only a small proportion in the arid and semi-arid range lands that cover the majority of the continent. Overall, stocking rates are low compared with other countries because of the low quality and quantity of grass feed available as a result of low rainfall, relatively expensive irrigation water, and poor soils.

During the past four decades, Australia's total agricultural land use has changed considerably: the area has increased by more than one-third in aggregate; the average farm size has doubled in area; and the area under irrigation has expanded sixfold. As well, the volume of farm chemicals used per year has trebled. Between 1950 and 2000, the volume of farm production rose by more than 250%. However, the real gross value of farm production rose by only 25%, because of falling real prices for farm products (ABARE, 1997a, 1997b, 1998, 2000; Chisholm, 1997).

Agriculture is more important to the economic prosperity of Australia than to most other advanced industrial countries. While primary agriculture's contribution to national production is not large, at around 3% of Gross Domestic Product (GDP), the entire agricultural and food processing industry contributes 12% of GDP and 7% of employment (DFAT, 1999; ABS, 2000a). It is expected to continue playing a significant role in regional economic development and as a foreign exchange earner.

Four-fifths of Australia's agricultural production is exported, the value of which accounts for one-fifth of all goods and services exports. While that one-fifth share is much smaller than it was in earlier decades, any downturn in agricultural exports still has important macroeconomic implications for Australia.

That plus the high dependence of farmers on exports means Australia is unusually sensitive to policies and practices at home and abroad that affect its agricultural competitiveness.

An important trade development since the 1950s has been the shift in Australia's export markets away from traditional trading partners in Europe and North America towards Asian countries. Asia as a whole now takes about 60% of Australia's merchandise exports compared with 46% just a decade ago. Food exports to Asia account for 56% of all food exports. In 1998-1999, the major destinations for Australia's food exports were Japan (21%), other Asia destinations (36%), and the United States (9%).

Australian agriculture would be less likely to continue its relative decline in economic importance over the next decade if the Asia-Pacific Economic Cooperation (APEC) process delivers on its promise to achieve free trade by 2010 for industrial countries and by 2020 for developing countries in the APEC region. If agriculture is included as part of that liberalisation, as promised, exports of farm products from Australia, particularly to East Asia, would grow substantially (Anderson et al., 1997a,b).

Since the early 1970s, Australian governments have pursued a gradual transition away from farm price supports, statutory marketing arrangements and trade protection, policies that insulated food and agricultural producers from both domestic and international market signals. The effective rate of government assistance to the sector was close to 30% of value added in 1970, whereas over the past two decades it has averaged around 10%. Protection to manufacturing has dropped even more, which has improved the capacity of farmers to compete for mobile resources such as labour (PC, 2000).

What remains of the agricultural sector's assistance is delivered through a wide range of programmes and policies. Statutory marketing and regulatory arrangements are more important than budgetary assistance (research and development, adjustment assistance and tax concessions) or tariffs on competing imports. Quarantine restrictions also incidentally provide economic assistance to some agricultural commodities, while tariffs on inputs offset assistance for others. The effective rate of assistance for agriculture was less than 10% in the late 1990s, and fell even further after June 2000 with reforms to the domestic milk market (PC, 2000).

Australia's agricultural policy objectives are to increase international competitiveness, encourage sustainable agricultural practices, and promote social and economic opportunities for rural communities (DPIE, 1998). They aim to facilitate market responsiveness and self-reliant risk management. An important consequence of improving market responsiveness is further integration of food and agriculture production with downstream processing industries. The government plans to continue to reduce and eventually remove remaining impediments to efficient markets and trade, including excessive quarantine restrictions at state/territory and national borders, so as to make the sector even more market responsive and better integrated with the rest of the world.

One of the aims of Australia's agriculture policy strategy is to enhance the country's reputation as a 'clean' place to grow and process food. The country's 'clean and green' reputation is seen as a way to gain an edge in markets that are increasingly concerned about food quality and human health. Promoting a 'clean' agricultural image is a response to consumer food safety and health concerns as well as domestic concerns about the impact of agricultural production on the environment. Some 15 food quality assurance projects began during the late 1990s to help document food safety and provide validity to the 'clean' image. Food quality projects are affecting such industries as dairy, wheat, red meat, pork, seafood and fruit and vegetables. Quality assurance systems on farms and along the supply chain are encouraged, as is the use of generic marketing to support brand promotion of the quality and safety of Australia's foods (DPIE, 1998).

Specifically, the long-term objectives and goals for the agricultural sector are (DPIE, 1998):

- *Sustainable agriculture*, through the expansion of whole farm planning and catchment management at local and regional levels.
- *Microeconomic reform*, key areas targeted for reform being the waterfront, land transport, energy supply, and manufacturing including food processing (including through labour market reform).
- *Trade liberalisation,* the objective being to eliminate all production and export subsidies in the next WTO round.
- *Taxation reform*, where the burden imposed on export industries by the cascading impact of indirect taxes (such as wholesale sales tax and fuel excise) along the supply-chain has been lowered as part of the overhaul of Australia's entire tax system.
- *Telecommunications,* where the federal government is working with farmers and service providers to ensure rural, remote and regional areas have access to competitively priced, high-standard telecommunications services.
- *Research and development,* the priority areas being sustainable land and water use, biotechnology, pest and disease control/eradication, opportunities for value adding and market development (leverage for which comes from the federal government's matching of industry levies up to 0.5% of the gross value of production among participating industries).

The federal government has been undertaking comprehensive reform of the country's environmental law regime in an attempt to promote greater certainty and minimise state-federal governmental duplication (DPIE, 1998). Fundamental to the reform package is the integration of environmental, economic and social considerations. The precautionary principle and the principle of inter-generational equity also are expressly recognised. Two other guiding principles are the need to maintain and enhance international competitiveness in an environmentally-sound manner and the adoption of cost-effective and flexible policy measures.

ENVIRONMENTAL AND HUMAN-HEALTH ISSUES

Australia's 19 million inhabitants are heavily concentrated in urban centres along the south-eastern and south-western coastlines, with most people living and working in cities. More than 80% of the entire population occupy just 1% of Australia's land surface. The country's eight state and territory capital cities account for two-thirds of the population and more than two-thirds of employment (SEAC, 1996).

Australia's geographical location and relative isolation provide it with several environmentally-related food producing and trade advantages. For example, the southern hemisphere, which consists mainly of water, is markedly less affected by human activities than the northern hemisphere. The fact that Australia is an island continent means that transborder pollution is not as significant an issue as in other parts of the world. Australia faces less population pressure than most other countries and its rural sector is generally free of polluting industrial activities.

Drinking water quality has never been a major concern in Australia (Maher *et al.*, 1997). While occasional contaminations of rural catchments by agricultural chemicals do occur, evaluations and surveys of drinking water quality conclude that the health risks to humans and livestock presented by organic chemical contaminants from human activity are either not present or extremely rare (NHMRC/ARMCANZ, 1996). Similarly, naturally occurring nitrates are not a frequent concern, with no cases of the nitrate-induced methaemoglobinaemia having been recorded in Australia (Maher *et al.*, 1997).

Around 1800 water monitoring programmes are in place throughout the country to evaluate water quantity, quality and environmental indicators. These programmes are supplemented by community initiatives such as Waterwatch, which began in 1993 and now involves more than 50,000 rural people monitoring nearly 5000 sites (OECD, 1998).

Animal welfare issues are of some concern in Australia, including live transport, surgical husbandry procedures without anaesthetics, and the consequences of drought, bushfire and poor shelter (Wirth, 1998). Abattoirs are usually located on the coast, making livestock transport problems such as prolonged journey time and distances inevitable. Each state and territory has a Prevention of Cruelty to Animals Act to control all uses of animals.

Fertiliser use is comparatively low because low rainfall limits the returns on fertiliser expenditure. So, despite its low soil fertility, Australia uses much less fertiliser than comparable countries: 30 kg per ha of arable land and 130 kg per ha fertilised, which is one-quarter of the amount used in the USA and Spain and one-twelfth of that used in the United Kingdom (SEAC, 1996).

Over time, however, the long-term trends in agricultural expansion, intensification and specialisation have brought about extensive changes to the use of natural resources. While crop productivity growth has been sustained by more intensive land use practices, by-products have been soil structure decline, wind and water erosion, and increased levels of soil salinity and acidity. Dairy and beef activities (especially cattle feedlots) are becoming more intensive and more dependent on purchased feeds, causing waste disposal and contamination problems. Pig, poultry and egg production uses little farmland directly and is very dependent on purchased feeds, and produces waste disposal and contamination problems.

These changing agricultural technologies and management systems, not to mention farm chemicals, are altering soil, landscape, vegetation and water resources, in ways that can result in pollution, contamination, resource degradation and habitat loss. In the case of intensive livestock operations and pesticide sprays, they also contribute to air pollution.

Although agricultural chemical use in Australia is low by OECD standards; data indicate that producers use 2500 types of farm chemicals and 2000 animal health products containing some 500 active ingredients to control 5000 significant pests (ABS, 1996). They also treat some 15 million ha of land with herbicides, 3 million ha with insecticides, and close to 1 million ha with fungicides, out of a total cropped area of 23 million ha.

The area planted to fruit and vegetables represents only around 1% of the country's cropped area, but accounts for more than 15% of fertiliser use. Where such high rates of fertiliser are applied, nutrient run-off may have potentially serious impacts on waterways and coastal zones. Pesticides are also used more intensively on horticultural crops. It is generally sprayed directly onto crops, raising the risk of contaminating non-target areas, entering the soil or washing into river systems and marine environments. Since a large proportion of fruit and vegetable production occurs near rivers or within 100 km of the coast, the potential for environmental damage caused by pesticide spraying is serious, particularly given the country's high propensity for run-off.

The major agriculturally-related environmental concerns receiving policy and programme attention are water, salinity, soil erosion, and biodiversity loss. An overview of the impacts of farming on those aspects of the environment is presented in Table 7.1, along with policy and programme responses to those pressures.

Water-related issues

The four primary agricultural activities causing water-related environmental problems in Australia are:

- allowing runoff from fields to carry sediments, nutrients, organic matter and agricultural chemicals;
- extracting too much water for irrigation, resulting in severe impacts on aquatic ecosystems, water quality and groundwater supplies;
- inverting the natural pattern of river flows in southern Australia (high demands for irrigation during summer when river flows are low and low demand for irrigation during winter when river flows are high); and
- clearing and using land and water in ways that result in rising water tables and salinity (Cullen and Bowmer, 1995).

Table 7.1   Impacts of agriculture on the environment in Australia

Pressures	Impact	Response	Effectiveness of responses
Agricultural land use	Loss or fragmentation of most of native vegetation, exposing areas to many of the soil issues listed below and contributing to greenhouse gases	Legislation and controls on tree clearing; monitoring; One Billion Trees programme; Landcare	Effective in many areas but not implemented in others; the efficacy of the recovery programmes is not tested
Pastoralism	Changes in the density and species composition of vegetation; widespread establishment of weeds and feral animals	Research and extension; legislation; leasehold conditions (animal stocking rates, soil status); National Strategy for Rangeland Management; inventory and monitoring; Landcare; structural adjustment programmes; multiple use policies	Only limited success through lease administration; localised successes in weed and feral animal control but little progress in many areas; variable trends in vegetation on a regional basis
Salinisation due to tree clearing for agricultural and grazing land; over irrigating	Rising groundwater levels; in some areas salt is clearing of agricultural and grazing land; over irrigating	Landcare and catchment planning, Murray–Darling Basin Commission; salt quotas and Saltwatch; monitoring; expenditure on salinity management	Only minor and localised successes through planting and changed management. Regional-scale responses are inadequate
Soil structure decline, especially tillage systems that lead to loss of soil organic matter	Reduce permeability and increased run off and erosion; poor root vigour leading to reduced productivity	Soil conservation research and extension especially into improved farming systems; Landcare	Increased awareness of the issue, but problem is still widespread

Soil erosion: agricultural land use leading to the loss of soil cover	Loss of soil depth; loss of nutrients; offsite effects such as saltation	Land management research and extension; expenditure on structural works; Landcare	Uptake of advice still inadequate in most areas
Soil nutrient decline: agricultural land use, excessive cultivation	Decline in soil organic matter and major nutrients (N and P)	Research and extension on cropping systems, rotations and fertiliser use; promotion of N utilisation, fertiliser applications rate	Varies from successful farming systems, incorporation legume rotations, to reliance on artificial fertilisers to little effective response
Soil nutrient loss due to pastoralism	Wind and water erosion leading to redistribution and loss of nutrients	Improved stock management, fencing and water distribution	Varies from property to property
Soil nutrient accumulation from intensive horticulture	Accumulation of nutrients with risk of water pollution	Guidelines and regulations for effluent discharge and drainage in some area, improved irrigation techniques; soil conservation; education, industry restructuring, water reforms	Locally effective but often problem transferred further downstream; guidelines often based on poor biological knowledge; too early to judge impacts

Soil acidification from fertiliser use and removal of plant products and natural processes	Increased soil acidity (low soil pH) and the release of toxic levels of aluminium and manganese in some soils; poor root growth. Some calcareous soils benefit from the higher acidity	Research and extension; liming and changed fertiliser practice	Very poor uptake of appropriate measure
Soil acidification from agriculture	Impacts include cadmium contamination from phosphate fertilisers; increased herbicide use in minimum tillage systems; pesticide pollution in cattle and sheep dipping sites	Reduction in cadmium levels in fertilisers and use of cultivars that restrict its uptake; regulations on pesticide and herbicide use; education	Cadmium levels are generally low by world standards; more work on identifying and rectifying other contamination problems
Food quality due to hormones, pesticide and fertilisers	Food contamination	Legislation; new farming practices, integrated pest management; reduction in pesticide use	Successful in most cases, but risks of accumulation of residues poorly known
Groundwater	Overuse of water and contamination	Bore metering and licensing; regulations and tradeable water rights	Limited effect
Farm dams	Proliferation has reduced streamflow particularly during dry conditions	Local restrictions and moratoriums	Farm dams only recently considered in water resources planning

Irrigation systems	Major user of stored water with salinisation, waterlogging, nutrients, pesticides	Water pricing, reform and restructuring of industry, demand management; improved technology	Still minimal application
Catchment pollution	Most water bodies in areas of agriculture affected by fine and coarse sediment, elevated nutrients loads, and increased volume and rate of run off in some areas	Strategic revegetation and farm forestry; clearing bans; drainage; soil conservation and fertiliser management; tree planting to reduce salinity; stream bank stabilisation, catchment management and Landcare	Landcare working in some areas, Stream bank stabilisation is costly and partially successful

Source: Adapted from SEAC, 1996.

Agriculture is Australia's largest water user, accounting for some 70% of the country's water consumption (McLennan, 2000). The 2.3 million ha of irrigated crops and pastures represent less than 0.5% of the total agricultural land and about 12% of the total area of crops and pastures (Table 7.2). The value of irrigated production fluctuates between 25% and 30% of Australia's gross value of agricultural output (Cape, 1997). Irrigation supports the production of all rice, most vegetables, milk, fruit, cotton and significant amounts of soybeans and sugar, while its contribution to meat, cereal, pulse and oilseed production is relatively minor (DPIE, 1998). Historically the price of water has been set so low as to not be a critical factor in the choice of crop under irrigation (Smith, 1998). At present, however, water policy reforms are increasing significantly the price of irrigation water, resulting in the introduction of more water saving technologies and land use is shifting from pasture to higher-value crops. In addition, profitable export opportunities are driving the expansion of irrigation intensive crops, including vegetables, fruits and wine grapes.

Table 7.2   Area of crops and pastures irrigated in Australia, 1999 (1,000 ha)

	New South Wales	Victoria	Queensland	South Australia	Western Australia	Tasmania	Northern Territory	Total
Annual pasture	199	267	28	18	12	10	0	534
Perennial pasture	115	251	21	41	5	19	0	452
Rice	148	0	0	0	0	0	0	148
Other cereals	163	17	30	5	1	2	0	219
Cotton	256	0	117	0	1	0	0	375
Sugar cane	0	0	153	0	3	0	0	156
Vegetables	16	21	26	11	8	15	0	97
Fruit (incl. nuts)	23	22	23	15	6	3	2	94
Grapevines	25	32	1	45	5	0	0	108
All other crops	29	9	20	7	2	9	0	77
Total	974	619	417.4	138	42	58	2	2,251

Source: ABS Agriculture, Australia, 1998-99 (Cat. 7113.0).

The water and irrigation-related environmental pressures are greatest in the river system contained in the Murray-Darling Basin (MDB). Roughly equal to the geographical area of France, the MDB covers one-seventh of Australia's landmass, and crosses four states (New South Wales, Queensland, South Australia and Victoria). Known as the nation's 'food bowl', the MDB contains more than 70% of the country's irrigated area and produces 90% of the value of irrigated food crops. In addition, it supports more than 50% of the country's cropland and produces more than 40% of the country's total agricultural output value (ABS, 1996). Some 95% of MDB water diverted for human use goes to agriculture and the flows of all but one of the rivers within the system have been modified to support the growth of agricultural industries.

Extensive use of superphosphate fertiliser in the MDB has been singled out as the primary source of phosphorus, which led to the world's largest toxic blue-green algal bloom in the summer of 1991-92. The algae expanded over a 1,000 km stretch of the Darling River, resulting in the closure of water supplies and major disruptions to local communities. Local, state and national governments responded by establishing the Murray-Darling Initiative. This environmental crisis is credited as the single most important impetus in generating cooperative action and the development of best-practice management activities to address water issues in Australia (ABS, 1996).

Water use inefficiencies, environmental damage and competing demands for agricultural, industrial, recreational, domestic and environmental water are exerting enormous pressures on MDB irrigators. Additional pressures are stemming from the fact that the MDB is rich in Aboriginal cultural heritage sites and includes some 140 conservation areas and numerous internationally-recognised wetlands.

Additional environmental pressures on the country's inland and coastal water systems include salinity, nutrients, toxicants, eutrophication and sediment (DWR, 1992; MDBC, 1994, 1998; Crabb, 1997; Smith, 1998). For example, coral reef degradation in Australia's Great Barrier Reef is attributed primarily to agriculture. A catchment study quantifying the principal sources of sediment and nutrients discharged to coastal waters off Queensland estimates that grazing lands contribute approximately 80% of nutrients and that sugarcane areas contribute 15% (Moss et al., 1992; Crosser, 1997). Nutrients lost from grazing lands are largely those naturally present in the soil and not originating from added fertiliser. The principal causes of the nutrient loss are land clearing and overgrazing, both of which increase the soil's susceptibility to erosion (Gardiner et al., 1988; Beckman, 1991). Sugarcane cultivation results in loss of both natural soil nutrients and added fertiliser, with fertiliser addition and loss far more important than in grazing situations (Prove and Hicks, 1991).

Salinity and soil erosion issues

Soil degradation issues are a second major agriculturally-related environmental issue receiving attention in Australia. Around 20% of the country's soil is considered to be highly erosion-prone, and more than half of the remainder is moderately erosion-prone (OECD, 1998). In addition, naturally acid and acidifying soils are extensive, with the rate of acidification increased by some agricultural practices.

Salinity, both dryland and irrigation-induced, is the most pressing soil related environmental problem, however. A recent national audit reported that approximately 5.7 million ha are considered at risk or affected by dryland salinity (NLWRA, 2000). The estimates further suggest that in 50 years this area may increase to 17 million ha (see Table 7.3). Dryland salinity is concentrated in south-western Western Australia, with a reported 70% of the country's problem

(IC, 1998). Much of the remaining affected land is within the MDB.

The on-farm cost of dryland salinity includes lost agricultural production, secondary degradation of saline land; increased salinity and silting of on-farm water supplies; and increased fertiliser requirements. Off-farm costs include:

- damage to buildings and infrastructure such as roads, bridges, sewerage pipes and water supply systems;
- flood damage caused by increased run-off;
- reduced service life of electrical equipment;
- increased water treatment, cooling and steam generation costs;
- habitat decline (on land and in-stream), with consequences for biodiversity; and
- loss of aesthetic, recreational and tourism values (Watson *et al.*, 1997; IC, 1998).

Table 7.3   Area affected by dryland salinity in Australia (in 1,000 ha)

State	1998/2000	2050
New South Wales	181	1,300
Victoria	670	3,110
Queensland	n.a.	3,100
Western Australia	4,363	8,800
South Australia	390	600
Tasmania	54	90
Total	>5,656	17,000

Source: NLWRA (2000).

Around 20,000 km of major roads and 1,600 km of railways are already at risk, estimated to increase to 52,000 km and 3,600 km respectively by 2050. In addition, 630,000 ha of remnant native vegetation and their ecosystems are at risk with an additional 2 million ha over the next 50 years (NLWRA, 2000).

Biodiversity issues

The third major agriculture and environment linkage important to Australia is biodiversity. Most reduction in the biological diversity of the continent is due to destruction or disturbance of natural habitats and the introduction of non-native species of plants and animals for use in agriculture. Land degradation and water mismanagement are considered the most serious problems affecting the country's terrestrial environment.

Australia is one of the few countries in the world with mega-biodiversity. The continent contains about 10% of the world's biodiversity, with more than a million species of plants, animals and micro-organisms, though only 15% have been described (SEAC, 1996). A large proportion of these, around 85% of the flowering plants, mammals, reptiles and inshore temperate zone fish, are endemic (SEAC, 1996). Unfortunately, these ecosystems and biodiversity are under a range of agricultural pressures from grazing, land clearing and irrigation.

The extent and intensity of pressures leading to habitat loss and modification for both terrestrial and aquatic ecosystems continue to present an extremely serious threat to Australian biodiversity, with a very high number of threatened and endangered species (OECD, 1998). The status of some marine species, including mammals, reptiles and fish, is of particular concern. In the MDB, an estimated 80% of the median river flow is extracted, mostly for agriculture. In parts of the Basin and in eastern coastal regions where water rights are also over-allocated, aquatic environments are under severe stress (IC, 1998). An OCED review of Australia's biodiversity and conservation efforts concluded:

> The coverage and management of protected areas may not be adequate to deal with the pressures involved. In the system of reserves, some areas of poor biodiversity are better protected than areas with high biodiversity. Outside of protected areas, while there has been progress in conservation of natural resources (land, soil and water), progress in conservation of biodiversity (habitats and species) has been extremely limited (OECD, 1998, p. 21).

Over the past decade, communities, agricultural firms and governments have recognised that biodiversity loss, dryland salinity, and irrigation-related soil and water degradation problems are costing the country in lower yields, higher production costs, increasing expenditure on repairing infrastructure and rehabilitation projects and emerging threats to Australia's market advantage as a producer of 'clean and green' goods.

ENVIRONMENTAL AND HEALTH POLICY INITIATIVES

For years, domestic policies inducing Australia's agricultural transformation were guided by socio-economic objectives that seldom included explicit valuation of the environment, much less the concept of ecologically sustainable agriculture. Growth in production, enhanced producer income, stable prices for producers and consumers, and increased net exports of agricultural products tended to be the dominant objectives of farm policy in Australia. Policy instruments used to achieve these objectives included generous tax concessions for land clearing, publicly-funded capital works to expand irrigation, and large fertiliser price subsidies.

Until recently, Australia's agricultural producers have been largely insulated from the demands of environmental groups and from government restrictions on their activities. During the past decade, however, increasing international economic integration and heightened environmental concerns abroad, as well as changing attitudes at home, have forced Australian producers, agribusinesses and policymakers to rethink the role agriculture can play in the economy, society and environment.

The current trend at both local and national levels is to aim for more emphasis on sustainable agricultural development, recognising its tripartite goals of economic efficiency, environmental protection and social equity (intra- and intergenerational). Agriculture is broadening its role beyond increasing productivity, expanding export earnings and enhancing rural incomes to include also evolving concepts of managing ecological processes and protecting environmental resources. In this, it is being accompanied by similar trends in other OECD countries and, increasingly, in middle-income countries as well.

Various local, national and international pressures are forcing policymakers to address agricultural environmental and health-related issues and to examine ways to promote more sustainable farming practices. Domestic concerns include consumer expectations, health considerations, occupational health and safety issues, environmental considerations, trade-related issues, residue in primary produce, sustainable land and water use, wetland and coastal protection, Genetically Modified Organisms (GMOs) and the overall regulatory environment.

International developments include United Nations Conference on Environment and Development (UNCED) follow-up obligations, post-Uruguay Round World Trade Organization (WTO) obligations, environmental treaties on climate change, biodiversity and desertification and adoption of internationally-accepted codes and practices such as those laid out in the Codex Alimentarius Commission (CODEX), Hazard Analysis and Critical Control Point (HACCP) and the International Standard Organisation (ISO). National initiatives include the Business Plan for Australian Agriculture, the National Landcare Programme, Natural Heritage Trust, national water reform initiatives, and the Supermarket to Asia programme, to name a few.

Australia has responded to this environmental challenge by committing to a process of adapting sustainable development concepts, its research foci, and policies to suit domestic socio-economic conditions, including prevailing agricultural production, consumption and trade patterns. It also has been a keen participant in international agencies such as Codex, the WTO and International Plant Protection Convention (IPPC) whose foci include global food safety and plant and animal health issues.

During the 1990s, the environment became a mainstream political concern in Australia to the point that community participation is now recognised formally in agricultural programmes, initiatives and legislation (Campbell, 1994; Alexandra et al., 1996; Thomas, 1996; EDO, 1996; Sarkissian et al., 1997; Archer et al., 1998). Government agencies and large and small businesses incorporate environmental concerns into their overall management plans. Firms, local communities and environmental organisations are increasingly cooperating in environmental management initiatives (Cato, 1995). For example, the Australian Conservation Foundation and the National Farmers' Federation jointly address land degradation issues. Many environmentally-relevant decisions are made by industry, special-interest groups and individuals, either acting alone or collectively.

Australia's primary approach to protecting water, soil, air, animal welfare and human health and safety is through research, education, voluntary adoption of best practice, and the use of guidelines developed collaboratively between governments, industries and communities. Best management practice manuals and published environmental, health and safety guidelines outlining accepted practices are used by:

- Local councils to determine whether to accept new agricultural activities or allow existing activities to expand.
- Courts to determine how to litigate property damage or human injury cases.
- Purchasers of commodities, including livestock, grains, horticultural and dairy products, to meet quality assurance objectives.

Australia's major research efforts on sustainable agricultural practices is organised by industry-specific Research and Development Corporations (RDCs). RDCs are funded by industry and the federal government on a dollar-for-dollar cost-sharing basis up to a maximum government level of 0.5% of the industry's gross value of production. Each year, RDCs invest millions of dollars in developing integrated pest management programmes, disseminating training packages for sustainable practices, and promoting reduced chemical use. Best practice research is conducted for almost every crop. The Agriculture and Resource Management Council of Australia and New Zealand (ARMCANZ) produces a series of model codes of practice and guidelines. This Council consists of Australian federal, state/territory and New Zealand Ministers responsible for agriculture, soil and water.

Australian codes of practice are generally industry based, providing guidance to specific agricultural activities on a range of environmental issues such as resource usage, emissions, waste generation and disposal, animal welfare, occupational or health hazards and regulatory standards. Codes of practice aim to improve industry performance by providing:

- relevant information and suggesting management practices and production processes that individual producers can adopt;
- advice about how to implement environmental improvements; and
- a means by which regulatory bodies can work with smaller producers, firms and industries.

Best practice guidelines are used most often as a benchmark for measuring industry progress (EA, 1999).

In addition to research and extension, and voluntary adoption of guidelines, other approaches to environmental and food safety management include a mix of government regulations and market incentives. The emphasis on economic (i.e. price-based) instruments is increasing.

Environmental policy making and regulatory management are not the province of any one sphere of government. Although the federal government has powers to enact laws affecting the environment and sustainable development, the

Australian Constitution does not specifically deal with environmental powers.

Most environmental and food safety legislative responsibilities rest with state/territory governments and most decisions that affect the environment occur at the local level. When international treaty obligations are at stake, however, the Commonwealth takes precedence over state/territory and local powers.

Odour and noise associated with agricultural activities (e.g. intensive livestock, dairy or mushroom production) are examples of nuisances controlled by local councils, often using guidelines provided by national or state agencies, industry bodies and, in some cases, state legislation. Local council planning and building requirements may limit agricultural activities to an isolated area or prohibit it completely. In general, dairy and feedlot operations must be sited, designed, constructed and operated so as not to cause unreasonable interference with the comfortable enjoyment of life and property off-site or with off-site commercial activity. Special consideration is given to odour, dust, flies and noise above appropriate background levels and to off-site transport effects.

The state and territory governments are responsible for administering some 150 separate pieces of environmental legislation, which are broad ranging and cover areas such as pollution of land, water and air, waste disposal, environmental planning and protection, protection of endangered species, forestry, wildlife, water and catchment management, and natural resource usage (OECD, 1998). For example, pesticide drift is addressed at the state/territory level. State/territory legislation prohibits agricultural spraying which injuriously affects any plants or stock outside the target area or any land outside the target area so that growing plants or keeping stock on that land can be reasonably expected to result in the contamination of the stock or of agricultural produce derived from the plants or stock.

The federal government guides, manages and coordinates state/territory and local governments through councils, programmes, inquiries, and mutually-developed policy strategies. Several of these national processes are having a significant influence on food safety and agriculturally-related environmental management[1].

---

[1] *Ministerial councils* and their advisory groups involve federal, state and territory ministers responsible for various common matters. Numerous councils are relevant to the environment, including the Australia New Zealand Environment and Conservation Council (ANZECC), which is the primary ministerial coordination committee related to the environment and consists of federal, state, territory and New Zealand ministers responsible for the environment and conservation. The Commonwealth provides a secretariat through the Department of the Environment and Heritage. The council provides a forum for the exchange of information and experience and develops coordinated policies on national and international environment and conservation issues (ANZECC, 1997). *National inquiries* into agriculture-related environmental issues may be initiated at the political level by Parliament, governments, individual ministers or ministerial councils. *National strategies* are included under the National Strategies for Ecologically Sustainable Development, Greenhouse Response and Biodiversity. All levels of government, in theory, use national strategies to guide their decision making. The MDB Commission is an example of a strategic management arrangement for a drainage basin and ecosystem. Thirty-three different government departments and 268 local governments have a stake in and some responsibility for the river system (SEAC, 1996). *Commonwealth legislation* is the only means by which specific aspects of environmental management can be consistently governed throughout Australia. *Complementary legislation* is

The Council of Australian Governments (COAG), which includes the Prime Minister and the leaders of the state, territory and local governments, is the peak body overseeing closer cooperation on issues concerning clarification of roles and responsibilities in environmental regulations. Among other tasks, COAG has been responsible for overseeing the finalisation of the InterGovernmental Agreement on the Environment (IGAE) and the National Strategy for Ecologically Sustainable Development (NSESD).

The IGAE was established in 1992 to avoid disputes about environmental matters between the different levels of government and to establish conditions under which the various governments interact. The Agreement provides a mechanism to:

- define the roles of each level of government;
- reduce intergovernmental environmental disputes;
- provide greater certainty in government and business decision making; and
- provide better environmental protection.

The IGAE sets out four main principles to inform government policy making and programme implementation:

- The precautionary principle, which stipulates that where the threat of environmental damage is serious or irreversible, lack of scientific proof of damage is not a defence against action to prevent the degradation.
- Intergenerational equity, which stipulates that the health of the environment should not be eroded for the benefit of the present generation at the expense of the future generations.
- Conservation of biological diversity and ecological integrity.
- Improved valuation, pricing and incentive mechanisms such as including environmental factors in valuation of assets and services, and introducing Polluter-Pays Principle and other market mechanisms to maximise benefits.

IGAE outlines three ways for governments to incorporate environmental issues into their decision making processes:

- ensure that environmental issues are considered when formulating policies;
- ensure that the identified environmental issues are properly examined; and
- ensure that measures adopted are cost effective and not disproportionate to the significance of the environmental problem.

---

legislation mirrored in each jurisdiction to ensure a national approach. Federal funding and/or other services are made available to those states/territories and communities enacting complementary legislation or adopting the guidelines, principles and initiatives established through these processes. For example, states/territories adopting COAG water reform policies receive Commonwealth payments to assist in their water policy reform process only if they adhere to a specified implementation timetable. This provides a powerful incentive since state/territory and local governments have limited revenue raising powers.

The NSESD provides the guiding framework for most environmental and natural resource management efforts. The Strategy aims to enhance individual and community well-being and welfare by following a path of economic development that:

- safeguards the welfare of future generations;
- provides for equity within and between generations; protects biodiversity; and
- maintains essential ecological processes and life support systems.

The NSESD promotes research and adoption of sustainable development and best practice techniques and calls for pricing and economic instruments as means for achieving better resource management. Local and state/territory governments are increasingly experimenting with economic policies such as property rights (for example, establishing tradeable water entitlements and tradeable salt discharge rights), as well as imposing full-cost resource use pricing to cover environmental damage.

Ecologically sustainable development is also promoted by the federal government via the Natural Heritage Trust (NHT), which funds programmes important to agricultural producers. NHT represents a practical example of how resource management in Australia has moved away from technical forms of centralised government planning to a more community-based, participatory resource management regime that recognises the plurality of stakeholder interests (Crean et al., 1999).

NHT-funded programmes relevant to agriculture include the National Rivercare Programme, the Murray Darling 2001 Initiative, the National Land and Water Resources Audit, The National Weeds Strategy, and The National Landcare Programme. All these initiatives promote information, education, community participation and dissemination of research. For example, the National Landcare Programme involves farmers working together with the community and governments to address environmental problems too large for individuals to handle.

The Landcare philosophy advocates a 'whole systems' approach to natural resource management, based on the idea that land, water and vegetation are elements of integrated systems, rather than separate entities. It promotes partnerships and integrated action among governments, industries and communities in order to bring about lasting change in land management practices. The Programme's activities are focussed on raising awareness of sustainable farming practices, thereby ensuring the transfer of information about such practices and encouraging their widespread adoption. It also establishes institutional arrangements to develop and implement policies and programmes that support sustainability.

A recent assessment concludes that the Landcare Programme has been successful in using an integrated approach to natural resource management problems and in achieving genuine public ownership (Cullen, 1997). Another study reports that in the 3 years to June 1996, 59% of broadacre and dairy farmers participated in at least one community landcare training activity (Cary and Webb,

2000). Moreover, Landcare is credited with creating a social climate where landowners are comfortable talking about land degradation problems on their own properties and discussing these issues with their neighbours (Griffin, 2000). While Landcare is credited with raising awareness of resource management issues, the adoption of more sustainable farming practices has still been slow, with motivation, financial incentives, skill capacity and the availability of appropriate technologies considered necessary (Cary and Webb, 2000).

In April 1995, COAG laid the foundation for more emphasis on market mechanisms by establishing principles and guidelines for national standard setting and regulatory reforms. These principles and guidelines, which were amended in November 1997, require a review of existing regulations to determine whether effective alternatives to explicit government regulation exist. The implication is that state/territory and local governments must ensure their regulations and quasi-regulations governing the agricultural and food industry meet economic efficiency and cost minimisation objectives by undergoing cost-benefit analyses in the form of regulatory impact statements.

The aim of the COAG principles and guidelines is to reduce the use of government regulations and to encourage 'minimum effective regulation'. Regulations are supported only where a well-defined social or economic problem exists, where other solutions such as market mechanisms or self-regulation are inappropriate, and where expected benefits exceed likely costs. The guidelines do not prescribe what type of regulation should be used in particular circumstances but set out principles and analytical requirements to be followed in the development of regulation.

Over time, the regulatory approach has been complemented by or, in some cases even replaced with, economic incentives, market instruments and property rights systems (Corbyn, 1996; Jones, 1997; Gunasekera, 1997). Market mechanisms, introduced to meet strategic objectives, include implementing the polluter pays, beneficiary pays and user pays principles. This more market-oriented approach also involves assessing total economic values for the costs and benefits attributed to environmental policies and government regulations, as well as the positive and negative externalities generated by agricultural production activities.

Market-oriented water policy initiatives

Water provides one of the best examples of this more market-oriented approach. Through COAG, the state/territory governments agreed that:

- water should be priced to reflect social cost and benefits;
- subsidies and cross subsidies should be removed; tradeable property rights would be established; and
- market mechanisms would be used to encourage more efficient management and use.

COAG aims to sustain and restore ecological processes and biodiversity of water-dependent ecosystems.

In addition to water pricing, entitlements and trading arrangements, COAG set out recommendations covering institutional reforms, environmental considerations, water-related research, taxation issues and consultation and public education. Community participation in the development and implementation of COAG objectives is encouraged either indirectly through Integrated Catchment Management (ICM), which is partially funded by the Commonwealth, or directly through state/territory legislation requiring community consultation before catchment plans are approved.

The COAG water principles have spawned new water legislation in most states and territories. For example, South Australia's Water Resources Act 1997 provides a comprehensive system of transferable water property rights, incorporates principles of ecologically sustainable development, provides holistic water resources management within the context of ICM, establishes water for the environment, and devolves greater responsibility for water resources to local communities through the establishment of catchment management boards.

The New South Wales government established a framework of water reforms aimed at achieving a better balance in the sharing of water between the environment and water users to ensure long-term sustainability consistent with the COAG requirements. The Water Legislation Amendment Act 1997 allows key aspects of the COAG reforms to be implemented, including:

- ecologically sustainable development principles applied to all decisions made under water legislation;
- new opportunities for temporary trading on regulated streams; and
- new powers for managing groundwater.

Likewise, Victoria passed the Water Resources Act 1997 to amend the Water Act 1989 and to provide for similar COAG-induced reforms.

To address problems associated with nutrient loads and eutrophication, increasing water salinity, declining wetlands and declining river health, the Murray Darling Basin Ministerial Council imposed a cap on water diversions in 1997. This involves the complex task of allocating water for environmental needs as well as for agricultural producers and other users. The cap is set according to the volume of water necessary to flow to keep the entire river healthy.

Agricultural and veterinary chemical management initiatives

Protection of water resources from pesticide and fertiliser residues is directly influenced by regulations controlling the latter's registration, sale and use. Assessment and registration of agricultural and veterinary chemicals is the responsibility of the federal government, while sales and use are controlled by the states/territories. The regulatory approach to agricultural chemical use has changed significantly since the late 1980s, reflecting the need for a more consistent national approach and greater protection of the environment, worker safety and public health (EA, 1998). National programmes operate on a full-cost recovery basis.

Key laws controlling the impacts of agricultural and veterinary chemicals on the environment and food safety have been enacted during the past decade. The overall management of these products reflects Australia's export-orientated and highly competitive agricultural sector. A single inappropriate use of an active constituent could potentially result in bankruptcy, loss of trade, ecosystem damage and public and worker health injury. The relatively small number of agricultural and veterinary chemicals (600) and limited number of specific products registered and used (5000) means that the government considers it is both practical and effective to develop and enforce a uniform system of management at the national level (EA, 1998).

Early in the 1990s, the federal and state/territory governments agreed to establish the National Registration Authority (NRA) to manage agricultural and veterinary chemicals and products. This replaced existing state/territory schemes and became fully operative in 1995.

The federal and state/territory governments provided the NRA with extensive powers in 1994 by introducing legislation known as the Agricultural and Veterinary Chemicals Code Act 1994. This legislative network requires that before any chemical product can be supplied or sold in Australia, it must be registered by the NRA. The NRA evaluates, registers and reviews agricultural and veterinary chemical products; controls the importation, manufacture and export of chemical products; and ensures compliance with and enforcement of the Code.

The registration process involves an evaluation of each chemical's effect on the safety of humans and the environment, the safety of non-target plants or animals, and the impact on trade. The NRA also evaluates the chemistry and efficacy of products and the presence of residues in food (EA, 1998). Any product deemed to pose a high risk to safety is not registered or has restrictions placed on its use. When the review is complete, the NRA also approves the manufacturer's label, which describes the product and specifies (through state/territory legislation and enforcement) how a product may be used, including safe handling information.

States and territories are responsible for the use of products beyond the point of sale and administer a range of legislation controlling use. All states and territories have research, extension and education activities aimed at reducing the risks from chemical use and improving the efficiency and effectiveness of such use. The fundamental requirement of state/territory legislation is to enforce label directions specified by the NRA. Other important objectives include:

- licensing of commercial pest control operators, ground and aerial spray operators, and pilots;
- investigating and taking action for adverse incidents such as spray drift and poisoning of wildlife; and
- monitoring programmes for detecting violations of standards for agricultural product residue.

The results of Commonwealth, state/territory and industry-sponsored monitoring programmes indicate that misuse of agricultural and veterinary chemicals is low (Rowland *et al.*, 1997). Most analyses record no residues or residues below established limits (NFA, 1992; ANZFA, 1996). The cost of agricultural and veterinary chemicals generally represents a small, although not insignificant, part of the overall cost of production. Pesticides are commonly used as insurance (Stirling, 1994); and the rate and frequency of applications tend to vary among growers depending on their individual perceptions of risk (Penrose *et al.*, 1994).

The food safety regulatory regime

Australia's food and agricultural production and processing regulations are a complex and often fragmented collection of laws, rules, standards, and procedures involving numerous federal, state/territory and local government agencies and legislation. Some 150 acts and associated regulations control food and agribusinesses, including imported food and food produced for export and domestic consumption. In addition, there are more than 90 separate national food standards (FRRC, 1998). Laws and standards are administered by more than 40 state/territory agencies and departments, more than 700 local governments, and various federal agencies.

An important aspect of food controls and regulation is the monitoring of agricultural chemicals. The Australian Total Diet Survey (formerly the Market Basket Survey) monitors dietary intake of agricultural and veterinary chemicals and heavy metals to determine whether they are within the safe limits set down by the World Health Organisation (WHO). The survey samples 'market baskets' of over 70 food types for about 50 pesticides, arsenic, cadmium, copper, lead, mercury and aluminium. The National Antibacterial Residue Minimisation Programme monitors antibacterial levels in meat and alerts producers to the risks to trade if safe levels are violated. The National Residue Survey (NRS) monitors chemical residues and heavy metals in raw food commodities, largely those destined for export.

The NRS is fully-funded by participating industries. Since December 1996, it has been compulsory for meat from all species slaughtered for export or domestic use in Australia to be tested for some 90 chemicals by NRS.

Concern about cadmium is such that authorities have taken specific steps to limit dietary intake, including:

- Reducing the level of cadmium allowed in phosphatic fertilisers.
- Eliminating the feeding of phosphate supplements with high levels of cadmium to cattle.
- Banning the sale of offal from aged sheep and cattle (there is an age-related increase in cadmium concentrations in animal offal such as liver and kidney).
- Regulating the disposal of industrial wastes into urban sewerage systems to minimise contamination of agricultural lands by cadmium.
- Monitoring cadmium levels in fertilisers, soils and commodities by federal and state/territory agencies.

The NRS manages and coordinates 4 major testing programmes, receiving and collating the results, making payments to state/territory governments, laboratories and abattoirs, and auditing the operational and financial aspects of the programmes. These programmes are the Hormonal Growth Promotants Audit Programme (HGP); the National Organochlorine Residue Management (NORM); the National Antibacterial Residue Minimisation (NARM); and the Chlorfluazuron (CFZ) and Endosulfan Survey.

The NRA introduced the HGP control system to address EU concerns about Australian meat and meat products containing hormonal growth promotants. The control system is the responsibility of the states/territories and involves importers, wholesalers and retailers. HGP supplies must be registered, suppliers and manufacturers must keep records of the treatment and sale of livestock, and audits are carried out regularly by the states/territories. The system is also liable to audit by European Union (EU) auditors at any time.

Australia's food industry legislation and acts have recently undergone a regulatory review process. The Food Regulation Review investigated all food regulatory matters, focusing on regulations administered by agricultural and health agencies and involving the three spheres of government (FRRC, 1998). It included government regulation-making, compliance and enforcement activities in relation to imported food and food produced for export and domestic consumption and covered the whole of the food industry, including primary production, processing and retail.

The Review concluded that the food regulatory system in Australia was complex, fragmented, inconsistent and wasteful. For example, the food industry incurs costs due to duplication of effort between regulatory agencies, overlap of legislation and functional responsibilities, inconsistency of regulatory approaches between jurisdictions and difficulty in dealing with the large number of agencies and laws involved.

Most agrifood businesses employ practices and equipment that match or exceed the standards required by law (FRRC, 1998). Many proprietors and managers have difficulty separating what they do as a natural part of good business practice from what they do solely to comply with food-specific or food-related regulations. Regardless of regulations, most food suppliers have strong incentives to produce safe food of the type consumers want and for which they will pay.

The FRRC study (1998) reviewed firms with annual turnover of less than A$100,000 through to about A$13 million, with up to 100 employees. The average cost of food-related regulatory compliance per firm was just over A$13,700, representing 0.3% of average annual turnover. The main elements of the regulatory costs are the cost of the firms' time (44%), capital expenditure (26%), inspection fees and charges (14%), licence fees (9%) and test fees (7%).

The major reforms adopted as a consequence of that Review include: new simplified national food safety standards; nationally consistent State and Territory Food Acts; and supporting projects to help businesses and government with the implementation of the new measures incorporated in the new standards. The reforms represent a co-regulatory approach to food regulation based on government, industry and consumers working together. Government's aim is to set minimum performance-based standards through consultation, and to give business greater flexibility in how it meets the standards, without reducing business' responsibility for meeting the standards.

Other public concerns: GMOs and animal welfare

Australia is taking a cautious approach to the release and use of GMOs and the adoption of GMO technologies is progressing slowly. Major environmental risks include the potential for herbicide-resistant genes from transgenic herbicide-resistant crops to escape into weedy relatives (Mikkelsen *et al.*, 1996) and excessive use of herbicides to control weeds growing in a tolerant crop, leading to over-exposure of the environment to chemicals, and possibly to the emergence of resistance (Holland and McDowall, 1995).

To date, more than 100 field trials of transgenic crops and 100 extensions to those trials have been approved with research into transgenic crops undertaken by both commercial companies and public research organisations. Herbicide resistance is the trait most frequently tested, followed by insect resistance, product quality and agronomic characteristics. Cotton and canola are the most trialed transgenic crops, accounting for more than half of the trials and extensions. This contrasts with the major Australian agricultural crops, by area, which are wheat, barley, lupins and oats.

To address the GMO-related food safety issues, the government established new labelling rules for GM foods in July 2000. The new food standard requires the labelling of food and food ingredients where novel DNA and/or novel protein is present in the final food. The law also requires labelling of food and ingredients where the food has altered characteristics.

To address animal welfare concern, each state and territory has a Prevention of Cruelty to Animals Act to control all uses of animals. Most legislation lists those acts of commission or omission deemed to be acts of cruelty. Producers are obliged under the Act to provide animals with proper and sufficient food, drink and shelter, which it is reasonably practicable in the circumstances for the person to provide. The general approach to transport is to not allow persons to carry or convey an animal, or to authorise the carriage or conveyance of the animal, in a manner, which unreasonably, unnecessarily or unjustifiably inflicts pain upon the animal.

Aggravated cruelty is said to occur where the animal dies or is seriously disabled. Animal welfare organisations have opportunities to contribute to the content and direction of the Codes. Proven adherence to the provisions of the Codes may assist a person defending a charge of cruelty. In South Australia, codes of practice are regulatory, while Victoria follows the English system of non-regulatory codes proclaimed under the Act. In all other states and territories codes of practice are not recognised by the Prevention of Cruelty to Animals Acts.

Since 1980, responsibility for the administration of each state/territory's Prevention of Cruelty to Animals Act has been vested in the ministers for agriculture, except in South Australia and Western Australia. Australian

Prevention of Cruelty to Animal Acts recognise by statute full-time officers of the RSPCA, members of the police force, and designated officers of the state agencies as delegated by the responsible minister to enforce the Act. Officers of the RSPCA undertake the bulk of prosecutions.

ECONOMIC IMPLICATIONS OF COMPLIANCE WITH ENVIRONMENTAL REGULATIONS

To date, the available evidence on the impact of environmental policies on agricultural activities suggests that compliance costs are increasing. Local communities enforce ever-stricter controls on a range of production activities. State and federal authorities are responding to international and domestic concerns over human health, food safety, ecological and biodiversity. Environmental standards are rising, producers are being forced to avoid polluting activities; and managers are having to pay an increasing proportion of the full social costs of resource use.

For some activities, in some locations, increasingly stringent environmental policies may slow the rate of expansion, while for other activities policies may accelerate their decline. In still other cases, raising environmental standards may have an insignificant effect if competing exporters have similar or higher standards. However, Australia's experience suggests that the major factors influencing agricultural growth, expansion, productivity and location are much more related to output prices, ongoing structural reforms and both domestic and export market opportunities. Among the important explanations for this situation include the following.

First, Australia's export-oriented agriculture has learnt over the generations to adapt quickly to evolving market conditions, including more-stringent environmental, health and safety imposed on its exports. As well, the country's public and private institutions have built up a range of signalling mechanisms to pass on key information to support producers. These signalling mechanisms include: relatively undistorted input and output prices; reliable product information on evolving consumer preferences; easy access to documentation on import requirements by commodity and by country; and research and extension networks that encourage resource conservation practices.

Second, Australia's producers market a great deal of their food production in designated premium markets (that is, in markets with the highest food safety and environmental standards, namely Canada, EU, Japan, Republic of Korea and the USA). Import requirements often vary significantly between these countries, so agricultural and food producers have developed procedures to react quickly to changes within and between markets. The standards and requirements also vary by commodity and level of processing, packaging and labelling. In general, Australian exporters attempt to meet the most stringent requirements whether at home or abroad. This is a demand side argument, i.e. meeting these strict environmental requirements enables Australian exporters access to specific markets.

Third, the predominant view within Australia's agribusiness community is that food safety is the country's most important agricultural issue (FRRC, 1998). The response of processors, input suppliers, retailers, and exporters is to work up and down the food chain with quality assurance programmes. Quality assurance serves the dual purpose of minimising the risk of 'tainted' produce, while the record keeping requirements allow tracing the problem to its source for quick resolution. Beef, dairy, horticulture and grains are all implementing quality control programmes.

Fourth, retailers are responsible for ensuring the safety of the products they sell. Retailers are able to pay producers for the higher costs of providing 'safe' and environmentally-friendly food, because the higher costs are passed on to consumers. Domestic retailers and food exporters tend to focus on ways to leverage non-price factors to improve profit margins and to gain and maintain market shares. These factors include 'clean and green' promotion, counter seasonality, generic and specific branding and retailers' capacity to manage production, processing, packaging, logistics and supply chains.

Fifth, while agricultural producers have become increasingly land and input intensive over the past two decades, dairy, beef, sheep, vegetable and fruit production is still based on relatively extensive production methods. For example, dairy and beef cows rely on pasture for most of their nutrients. In addition, Australia is one of the least densely populated countries in the world: the average population density of OECD countries is eleven times Australia's level. This low population density combined with its highly urbanised population mean that a very high proportion of Australians live and work in locations far removed from agricultural activities. The result is fewer conflicts between urban communities and rural industries.

Sixth, environmental policies do contribute to location shifts but they are unlikely to inhibit output expansion or dictate producer decisions except at the margin. In the longer run, environmental pressures and market forces are likely to impact on those pasture, rice and cotton producers who are pesticide intensive or rely on inexpensive water.

Seventh, the trend in environmental policy is to use more economic instruments that allow individual producers to abate at least cost. Over time, this approach to environmental policy making leads to higher-value uses of scarce natural resources, more efficient producers, and technological choices that minimise relatively expensive inputs. The level of waste and misuse of natural resources in Australia suggests that efficiency gains can offset production and productivity losses.

Case one: wine production and export compliance

Wine grape producers provide an important example of how Australia's export-oriented industries adopt to ever-increasing compliance costs. In addition to an export license and meeting country-specific labelling requirements, all Australian wine exports undergo chemical analysis to ensure the product meets Australian standards for export as well as import requirements of the destination country.

Each wine in an export consignment of more than 100 litres needs an Australian Wine and Brandy Corporation 'Certificate of Analysis'. This analysis includes specific gravity, volatile acidity, alcoholic strength, total acidity, free and total sulphur dioxide, pH, and glucose and fructose. Some countries (e.g. Singapore and the USA) require only the Australian Certificate, while exports to the EU require an additional test, known as the VI Certificate of Analysis Report. A laboratory registered with the National Association of Testing Authorities must perform the VI Certificate test. The level of analysis varies by the type of wine (table wine, bulk wine or late harvest botrytised wine). Exports to Japan can only be tested by a laboratory approved by Japanese authorities. The differences in import requirements are one reason that Australia recommends maximum residue limits at the lowest level set by any one country.

Once certificates are in place, the Australian Wine and Brandy Corporation subjects export wines to a sensory inspection. Wines are examined against prescribed criteria to determine whether the wine is sound and merchantable and is not likely to detract from Australia's reputation as a wine producer (Weeks, 1999). The remaining procedures are paper work intensive. For the wines going to Europe, exporters must complete:

- an application form for each wine type and package size;
- a declaration on company letterhead stating that the wine complies with all required winemaking practices and the Australian Food Standards code;
- a set of front and back export labels that are used on current shipment; and
- a copy of the VI Certificate prepared by the laboratory.

In total, the export license, laboratory costs, export fees and costs associated with processing applications can cost several thousand dollars per consignment. More importantly, while these export markets provide profitable opportunities for grape growers, the importing country requirements directly affect and constrain on-farm management options. This grape production example is also representative of how the export-oriented Australian agricultural sector is so often dependent upon off-shore environmental and health regulations.

Case two: water reforms and compliance costs

Environmental compliance costs related to water, air and chemical use are only a small proportion of total costs, with irrigation-related expenses averaging from 3% to 15% for most irrigated crops. For those cases where input costs are small yet this input is essential, producers tend to find new ways to offset increases (such as taxation options, corporate restructuring, or leasing water rights instead of purchasing water (since payments to lease water are fully tax deductible)).

Australia's water policy reforms are leading to higher water costs per unit of water used as the country moves to adopt the COAG water reforms. Much higher unit costs are expected in the future as further reforms are enacted and as market forces influence the price for water. For example, public irrigation system fees are increasing not only to cover operating expenses and capital costs of storage and supply structures, but also to cover refurbishment and replacement of depreciating infrastructure. Many irrigation structures are now at or near the end of their useful lives (Crabb, 1997). In addition, irrigators are forced to present farm management plans, pay environmental levies on water use, establish salinity prevention plans and to pay salinity fees when purchasing water that amounts to 10% or more of the value of the water.

At the same time, the volume of water used per irrigated crop is likely to decline and, in many cases, so too will the total expenditure on water-related expenses. This is because producers conserve water by using less as its price rises, including through investing in new water-saving technologies. Other market incentives are also influencing water use. For instance, the trend among grape growers is to produce high-value grapes with concentrated juice through lowering yields. Keeping grape yields down means using less water. This trend implies even lower water requirements and higher returns per unit of water used. High-quality grape production is encouraged by wineries through appropriate incentives in their contracts with growers.

There is little evidence in Australia that these higher short-term unit costs are imposing an enterprise-threatening situation, except for pasture. Even there, pasture producers are taking advantage of the higher water prices and selling some of their unused allocation, or selling their properties, which are then put to higher value uses. Moreover, the dairy industry is restructuring due to deregulation and reforms implemented on 1 July 2000.

The evidence suggests that water entitlement transfers between irrigators have been from lower- to higher-valued commodities (Crabb, 1997; Young et al., 2000). Investments in the wine industry, horticulture and the dairy industry in the state of South Australia, for example, are sending strong price signals and banks are insisting that these developments be underpinned by secure water rights. From 1995 to 2000, water prices in South Australia's horticultural areas with access to Murray River water have increased from[2] A$500 to $1,200 per 1,000 m^3. In those areas dependent on groundwater, water prices are as high as A$10,000 per 1,000 m^3. Dairy, rice and cotton producers are facing a greater cost-price squeeze than horticultural producers as overall subsidies to their industries are reduced.

Increased product prices from export demand are the main reason why these increasing expenses imposed by water reforms and cost-increasing environmental policies are not reducing horticultural production in Australia. The real value of both wine-grape production and wine production has grown at more than 10% per annum over the past 12 years (Anderson, 2000). Over the last 9 vintages, grape prices have increased 90% (ABS, 1997, 1998, 1999, 2000b). The volume of wine exports has risen from less than 5% to more than 30% of production, and will soon exceed 50% of sales. Australia is now the world's largest wine exporter after the European Union bloc, having been a net importer of wine as recently as the early 1980s.

Since 1995, grape growers in South Australia with access to River Murray

[2] An Australian dollar was worth 78, 73, 62, 64, and 57 US cents, respectively, over the five years to 2000.

water expanded their vine area by 67%. Grapes, especially premium grapes, are a relatively high value, low volume water user compared with other irrigation activities. Irrigation costs (including fuel for pumping costs and all related fees and levies) can range from 5% to 15% for grape growers. Grape growers can out-bid most other irrigation activities due to higher profit margins compared with other competing agricultural water uses.

A recent study of how irrigation reforms are impacting horticulture farms in the Murray Darling basin concludes that the rate of expansion is likely to slow but production is unlikely to fall as water becomes scarcer (ABARE, 1999). The study suggests that producers will purchase more water, invest in improved irrigation equipment, adopt new irrigation management practices, and change their enterprise mix. The ABARE study also concludes that increases in water charges are unlikely to lead to a fall in water use by horticulture in the Murray Darling Basin since water charges are only a small proportion of total costs, while the returns to water in horticulture are high. Water expenses ranged from 2 to 7% of total cash costs. As water expenses increased from \$40 to \$100 per 1,000 m^3 (with other prices held constant), gross margins fell from around \$6,500 ha^{-1} to \$5,900 ha^{-1} for grapes and from \$3,500 ha^{-1} to \$2,800 ha^{-1} for citrus.

In summary, there are a number of reasons why higher unit water costs and increasing water delivery costs are not imposing problems for other non-pasture irrigators as well. These include:

- Water costs are only a small proportion of total costs. For those cases where input costs are small yet critical, producers have flexibility to reduce costs in other areas (taxation options, corporate structure).
- Alternatives are available to offset higher unit water costs and provide opportunities for cost-saving improvements (as the cost of leasing water rights is tax deductible).
- Much more water is used in production processes than necessary, so many irrigators are able to make management changes to greatly reduce their water consumption.
- All current on-farm irrigation investments are much more efficient in water saving technologies than previous investments, and they utilise management systems such as under-canopy sprinklers, mini-sprinklers, drip systems, centre-pivot systems that produce 'soft rain', laser land forming, soil moisture measuring equipment, and efficient drainage.
- In South Australia, the use of furrow irrigation declined from 50% of the total area in 1976 to 20% by the mid 1990s (while in Victoria's Sunraysia district 70% is irrigated by furrow and flood methods and in New South Wales it is 84% - Crabb, 1997).
- Changes to improve water use efficiency are also taking place in the livestock sectors (such as replacing dams with piped reticulation systems for stock watering).
- Over the past decade, higher water charges have encouraged water transfers out of low-value uses and locations (as with water entitlements used on pasture in South Australia's Lower Murray being transferred to vegetable

production in the Riverland (WRMS, 1994; Crabb, 1997).

- Higher water prices have allowed irrigators to sell off part of their entitlement, invest the money in new water-saving technologies and then sell more of their entitlement (Young et al., 2000).

CONCLUSIONS

The greater awareness and emphasis on environmental issues has required Australian farmers to modify existing ways of doing things so as to reduce their negative environmental consequences and increase their positive effects. This is being encouraged by the greater use of market-based incentives, by designing alternative technologies and less-polluting inputs, by establishing information campaigns, by promoting voluntary best-practice codes, and by redirecting advisory services.

Australia is a highly urbanised, small economy with a low population density and low levels of protection for a largely export-oriented agricultural sector based on relatively low-intensive practices. Only 4% of Australia's agricultural land is sown to improved pastures and less than 5% of agricultural land is cropped. Stocking rates are low compared with other countries, while arid and rugged lands held under grazing licences account for 88% of the agricultural land area, where livestock mainly graze on native grasses. Producers, processors, retailers, input suppliers and exporters not only recognise their comparative advantage in exporting to countries with high environmental and food safety standards, they strive to find ways in which to expand that advantage.

One of the aims of Australia's contemporary agricultural policy strategy is to enhance the country's reputation as having a 'clean' environment that is able to grow and process safe food in a sustainable manner. The country's 'clean and green' reputation is seen as a way to gain an edge in markets that are increasingly concerned about food quality and human health. Quality assurance systems on farms and along the supply chain are not only encouraged, they have become a widespread practice in almost every agricultural activity in the country.

The most common policy responses to land and water degradation problems have been the promotion of information, education, community participation, dissemination of research and the development of codes of practice. Research and extension on sustainable practices dominate Australia's policy response. Full-cost resource pricing, overcoming market failures and addressing policy failures are increasingly encouraged.

Australian codes of practice are generally sector-based, providing guidance to specific industries on a range of environmental issues such as resource usage, emissions, waste generation and disposal, animal welfare, occupational or health hazards and regulatory standards. Codes of practice attempt to improve agriculture's environmental performance by providing: relevant information and suggesting management practices and production processes that individual producers can adopt; advice about how to implement environmental improvements; and a means by which regulatory bodies can work with smaller producers, firms and industries.

In addition to the greater community involvement in policy making and resource management, the regulatory approach has been complemented by or, in some cases even replaced with, economic incentives, market instruments and property rights systems. Market mechanisms are being introduced to meet strategic objectives, including implementing the polluter pays, beneficiary pays and user pays principles. This more market-oriented approach involves assessing total economic values for the costs and benefits attributed to environmental policies and government regulations, as well as the positive and negative externalities generated by agricultural production activities.

Structural change in Australia's agricultural sector is primarily a function on the supply side of technological changes and on the demand side of international price and quality movements, which in turn are affected in part by developments in the farm and trade policies of other countries. Hence unilateral, regional and global (WTO) trade policy reforms are important, as are policy reforms that are trade-related. That is, the quality assurance programmes, designed to ensure Australia's reputation as a high-quality and safe supplier in foreign markets is maintained, are probably going to continue to have at least as much influence on costs of production in Australian agriculture as are domestic environmental and food safety regulations.

REFERENCES

ABARE (1997a) *Australian Commodities: Forecasts and Issues.* Australian Bureau of Agriculture and Resource Economics, Canberra.

ABARE (1997b) *Australian Commodity Statistics.* Australian Bureau of Agriculture and Resource Economics, Canberra.

ABARE (1998) *Australian Commodities.* September quarter, Australian Bureau of Agriculture and Resource Economics, Canberra.

ABARE (1999) Irrigation water reforms: impact on horticulture farms in the Southern Murray Darling Basin, *ABARE Current Issues.* 99.2, April 1999, Australian Bureau of Agriculture and Resource Economics, Canberra.

ABARE (2000) *Farmstats Australia.* Australian Bureau of Agriculture and Resource Economics, Canberra.

ABS (1996) *Australian Agriculture and the Environment.* Australian Bureau of Statistics, Canberra.

ABS (1997, 1998, 1999, and 2000b,) Australian Wine and Grape Industry, Cat no 1329.0, Australian Bureau of Statistics, Canberra.

ABS (2000a) *The Labour Force Australia.* Cat. No. 6203.0, Australian Bureau of Statistics, Canberra.

Alexandra, J., Haffenden, S. and White, T. (1996) *Listening to the Land: a Directory of Community Evironmental Monitoring Groups in Australia.* Australian Conservation Foundation, Melbourne.

Anderson, K., Dimaranan, B., Hertel, T. and Martin, W. (1997a) Economic growth and policy reform in the APEC region: trade and welfare implications by 2005. *Asian-Pacific Economic Review*, 3(1), 1-18.

Anderson, K., Dimaranan, B., Hertel, T. and Martin, W. (1997b) Asia-Pacific food markets and trade in 2005: a global, economy-wide perspective. *Australian Journal of Agricultural and Resource Economics*, 41(1), 19-44.

Anderson, K. (2000) Lessons for other industries from Australia's booming wine industry. *Australian Agribusiness Review* 8. Paper 6, accessed 5 February 2001. http://www.agribusiness.asn.cui/review/2000v8/2000_Anderson/2000_Anderson_wine .html.

ANZECC (1997) *Facts*, accessed 2 March 2001, http://www.environment.gov.au/library/ pubs/fs_anzecc.html

ANZFA (1996) *Australian Market Basket Survey 1996*. Australia New Zealand Food Authority, Canberra.

Archer, M., Burnley, I., Dodson, J., Harding, R., Head, L. and Murphy, A. (1998) From Plesiosaurs to People: 100 Million Years of Australian Environmental History, Australia. State of the Environment Technical Paper Series, Department of the Environment, Canberra.

Beckman, R. (1991) Measuring the costs of clearing, *Ecos*, Authority Workshop Series No. 17, 13-15.

Campbell, A. (1994) *Landcare: Communities Shaping the Land and the Future*. Allen and Unwin, Sydney, 344 pp.

Cape, J. (1997) Irrigation. In: Douglas, F. (ed.) *Australian Agriculture: the Complete Reference on Rural Industry*. Morescope Publishing, Melbourne, pp. 35-62.

Cary J. and Webb, T. (2000) Community landcare, the National Landcare Programme and the landcare movement. Social Sciences Centre, Bureau of Rural Sciences, Canberra.

Cato, L. (1995) *The Business of Ecology: Australian Organisations Tackling Environmental Issues*. Allen and Unwin, Sydney, 159 pp.

Chisholm, A.H. (1997) Policy case of Australia. In: *Agriculture, Pesticides and the Environment: Policy Options*. Organisation for Economic Co-operation and Development, Paris, pp. 35-38.

Corbyn, L. (1996) Pollution reduction strategies in the Hawkesbury Nepean Catchment. In: *Proceedings of the 1994 Fenner Conference on the Environment, Sustainability: Principles to Practice*. Department of Environment, Canberra.

Crabb, P. (1997) *Impacts of Anthropogenic Activities, Water Use and Consumption of Water Resources and Flooding*. Australia: State of the Environment Technical Paper Series (Inland Waters), Department of the Environment, Canberra.

Crean, J, Pagan, P. and Curthoys, C. (1999) *Some Observations on the Nature of Government Intervention in Natural Resource Management*. Paper presented at the 43rd Annual Conference of the Australian Agricultural and Resource Economics Society, Christchurch, New Zealand, 20-22 January 1999.

Crosser, P.R. (1997) *Nutrients in marine and estuarine environments*. State of the Environment Technical Paper Series (Estuaries and the Sea), Department of the Environment, Canberra.

Cullen, P. (1997) The Australian Scene; Visions for Integrating Catchment Management. *Proceedings of the Second National Workshop on Integrated Catchment Management*, Australian National University, Canberra, 220 pp.

Cullen P. and Bowmer, K. (1995) Agriculture, Water and Blue Green Algal Blooms. In: *Sustaining the Agricultural Resource Base*, Office of the Chief Scientist, Department of the Prime Minister and Cabinet, Canberra, 90 pp.

DFAT (1999) *Exports of Primary and Manufactured Products, Australia*, Trade Analysis Branch, Department of Foreign Affairs and Trade, Canberra, 136 pp.

DPIE (1998) *Action plan for Australian agriculture actions, roles and responsibilities 1998-2008*. Department of Primary Industries and Energy, Canberra.

DWR (Division of Water Resources) (1992) *Towards healthy rivers.* Consultancy Report No. 92/44, CSIRO. Canberra.

EA (1998) N*ational profile of chemicals management infrastructure in Australia.* Environment Australia, Canberra.

EA (1999) *Eco-efficiency and cleaner production homepage, An introduction to codes of practice.* Environment Australia, Canberra.

EDO (Environmental Defender's Office, NSW) (1996) *The Bush Lawyer: A Guide to Public Participation in Commonwealth Environmental Law.* Environmental Defender's Office Ltd., Sydney.

FRRC (1998) *Food a growth industry: The report of the Food Regulation Review.* Food Regulation Review Committee, Canberra.

Gardiner, D. J., McIvor, J. and Williams, J. (1988) Dry Tropical Rangelands: Solving One Problem and Creating Another, *Proceedings of the Ecological Society Australia,* 16, pp. 279-286.

Griffin (2000) Agribusiness and Sustainable Agriculture. Agriculture, Fisheries, Forestry Australia, Canberra.

Gunasekera, D. (1997) *Role of Economic Instruments in Managing the Environment,* Invited paper, Industry Economics Conference, Panel 3, Environmental Regulation, Melbourne, Industry Commission.

Holland, J. and McDowall, P. (1995) Regulation of herbicide-resistant crops. In: McLean, G. and Evans, D. (eds) *Herbicide-resistant Crops and Pastures in Australian Farming Systems.* Proceedings of a GRDC, CRDC, British Council, RIRDC and BRS Workshop, Canberra, 15-16 March, Bureau of Resource Sciences: Canberra, pp. 140-167.

IC (Industry Commission) (1998) *Full Repairing Lease: Inquiry into Ecologically Sustainable Land Management.* AGPS, Canberra, 524 pp.

Jones, D. (1997) *Environmental Incentives: Australian Experience With Economic Instruments for Environmental Management.* Environmental Economics Research Paper No. 5, Environment Australia, Canberra.

Maher, W., Lawrence, I. and Wade, A. (1997) *Drinking Water Quality.* Australia: State of the Environment Technical Paper Series (Inland waters), EA, Canberra.

McLennan, W. (2000) *Water Accounts for Australia.* ABS Catalogue No. 4610.0, ABS, Canberra.

MDBC (1994) *The Algal Management Strategy for the Murray-Darling Basin.* Murray Darling Basin Commission, Canberra.

MDBC (1998) *Murray-Darling Basin Cap on Diversions Water Year 1997/98 Striking the Balance.* Murray Darling Basin Commission, Canberra.

Mikkelsen, T.R., Andersen, B. and Lorgesen, R.B. (1996) The risk of crop transgene spread, *Nature,* 380, p. 31.

Moss, A.J., Rayment, G., Reilly, N. and Best, E. (1992) *A Preliminary Assessment of Sediment and Nutrient Exports from Queensland Coastal Catchments.* Technical Report No. 4, Queensland Department of Environment and Heritage, Brisbane.

NFA (1992) *The 1992 Australian Market Basket Survey: A total diet survey of pesticides and contaminants.* National Food Authority, Canberra.

NHMRC/ARMCANZ (1996) *Australian Drinking Water Guidelines. National Water Quality Strategy,* DPIE, Canberra.

NLWRA (2000) *Dryland Salinity in Australia, Key Findings.* National Land and Water Resources Audit, Canberra.

OECD (1998) *Environmental Performance Reviews: Australia.* Organisation for Economic Co-operation and Development, Paris, pp. 211.

Penrose, L.J., Thwaite, W.G. and Bower, C.C. (1994) Pesticide use reduction logical decision making. In: *Proceedings of the 47th New Zealand Plant Protection Conference,* Waitangi, New Zealand.

PC (Productivity Commission) (2000) *Trade & Assistance Review 1999-2000,* Annual Report Series 1999-2000, AusInfo, Canberra.

Prove, B.G. and Hicks, W.S. (1991) Soil and nutrient movements from rural lands of North Queensland. In: *Proceedings of the Workshop on Land Use Patterns and Nutrient Loading of the Great Barrier Reef Region.* James Cook University, Townsville, pp. 67-76.

Rowland, P., Evans, G. and Walcott, J. (1997) *The Environment and Food Quality.* Australia, State of the Environment Technical Paper Series (Land Resources), Department of the Environment, Canberra.

Sarkissian, W., Cook, A. and Walsh, K. (1997) *Community Participation in Practice: A Practical Guide.* Institute for Science and Technology Policy, Murdoch University.

SEAC (State of the Environment Advisory Council) (1996), *Australia: State of the Environment,* An Independent report to the Commonwealth Minister for the Environment, Department of the Environment, Sport and Territories, (CSIRO Publishing), Melbourne.

Smith, D.I. (1998) *Water in Australia: Resources and Management.* Oxford University Press Australia, Melbourne, 384 pp.

Stirling, G.R. (1994) *Integrated Management as an Alternative to the Use of Methyl Bromide for Nematode Control.* National Workshop on Alternatives to Methyl Bromide as a Soil Treatment for Temperate and Sub-tropical Horticultural Crops, Institute for Horticultural Development/RIRDC Workshop, 28-29 July.

Thomas, I. (1996) *Environmental Impact Assessment in Australia: Theory and Practice.* The Federation Press, Sydney, 241 pp.

Watson, B., Morrisey, H. and Hall, N. (1997) *Economic Analysis of Dryland Salinity Issues.* Paper presented at Outlook 97, National Agricultural and Resources Outlook Conference, Canberra, 4-6 February.

Weeks, S. (1999) How to maintain your sanity during the export process. *The Grapegrower and Winemaker,* March, No 423.

Wirth, T. (1998) *Animal welfare in Australia.* RSPCA: Canberra.

WRMS (NSW Water Resources Management News) (1994) *Rush to rice lifts MIA water rights,* 1 (2), 2-4.

Young, M., Hatton, D., MacDonald, D., Stringer, R. and Bjornland, H. (2000) Interstate water trading in the MDB. CSIRO, Adelaide.

New Zealand

Anton D. Meister

INTRODUCTION

New Zealand (NZ) lies in the southwest Pacific Ocean approximately 1,600 km southeast of Australia and consists of two main islands, the North and South Island (Figure 8.1), plus several smaller islands. The combined size of the two main islands is 26.8 million ha (MAF[1], 2000a), which is similar in size to Japan or Britain but with a population of only 3.8 million. The country is predominantly hilly and mountainous.

Most of New Zealand's land is used for agriculture. In 1995, 16.6 million ha were used for agriculture, of which 55% being pastoral use, 10% forestry, and the remainder in horticulture and arable farming and other land on farms. Competition for land in New Zealand has seen a major loss of land from the pastoral sector to plantation of forestry and horticulture. This competition is expected to continue over the next decade. Meanwhile, even within the shrinking pastoral sector, there is a steady redistribution of traditional sheep, cattle, goat and crop usage into dairy, deer and other holdings such as 'lifestyle blocks'[2]. Ownership of agricultural land is still predominantly private.

Agriculture's percentage contribution to the national Gross Domestic Product (GDP) in 1998 was 5.3% (this includes horticulture), while the contribution of forestry and logging was 1.2% (MAF, 2000a). In terms of sectoral contribution to New Zealand's GDP, agriculture may not appear to be vital, as is the case with all developed countries. However, contrary to most of its developed counterparts, agriculture's contribution has remained stubborn, and has actually increased over

---

[1] MAF, currently stands for Ministry of Agriculture and Forestry, but before 1997 was Ministry of Agriculture and Fisheries.

[2] Relatively small acreages, often in the peri-urban zone occupied by people earning the majority of their income in the city but living on an acreage in the rural area. Sometimes these people make intensive use of the land while in other cases little use beyond a few horses or sheep and cattle are kept. Houses built on those acreages are often substantial. The control of the development of such land-use conversion has been and still is a major land-use planning issue in New Zealand especially around the larger cities.

the last decade. Between 1986-1987 and 1999-2000, agriculture's contribution to GDP (now including agricultural processing, input supply, transport and wholesale/retail contributions) increased from 14.2% to 16.6%. GDP itself grew by 26% during the same period. This has happened at a time when 17 years of reforms have seen agricultural protection, as measured by PSE (Producer Support Estimate) decline from just over 20% to around 2% (MAF, 2000b).

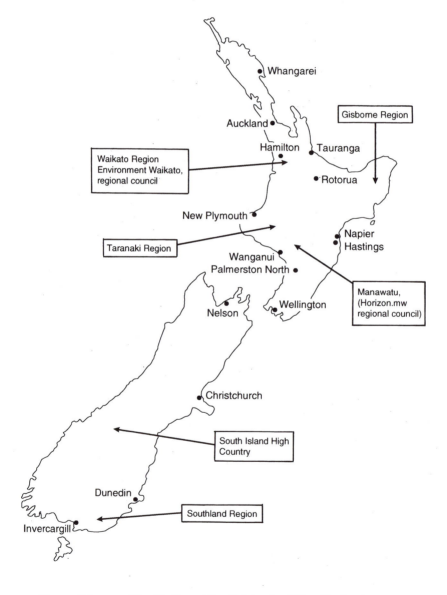

Figure 8.1   The North and South Islands of New Zealand

By this time, all protection and price support has been removed and whatever meagre government support there is, is restricted to agricultural research, extension and pest control activities. Agriculture has thus claimed a bigger slice of an even bigger pie.

Livestock accounts for three-quarters of New Zealand's agricultural production. Over the last 16 years, the sheep population has declined from 70.3 million at June 1982 to stand at around 45.2 million at June 1999. The beef cattle population has declined from a high of 6.3 million to a current low of 4.6 million, while the total number of dairy cattle has risen over the same period to its current level of approximately 4.4 million and deer at 1.7 million (MAF, 2000a).

The horticultural industry represents approximately 13% of total agricultural output. Production has been steadily rising over the years and horticultural exports now comprise approximately 8% of New Zealand's total agricultural based exports (55% of total horticultural export is made up by apples and kiwi fruit). The wine industry is becoming a success story and acreage is increasing rapidly. Arable crops represent 3% of total agricultural output (MAF, 2000b).

Figure 8.2   Agricultural based exports 1993-2000[a, b]

a) Values are NZ$ free on board (years ending June). Values for 2000 are provisional (Source: MAF, 2000b); b) All values quoted in this chapter are in NZ$. The value of the NZ$ has changed rapidly over the last year. By December 2000, it stood at NZ$=US$0.42; however only a few months before that it was US$0.38.

The vast majority of agricultural and horticultural produce is exported. Historically New Zealand depended on Britain and Europe as export market for agricultural commodities. However, since Britain's inclusion in the EEC in 1973, New Zealand has diversified its export markets as well as its products. In terms of export destinations, in order of importance, Australia, USA, Japan, Europe and South East Asia, now form our major export markets.

While historically, New Zealand's external trade has been dominated by agricultural exports, that dominance has declined (see Figure 8.2).

Although agriculture's share of total exports has decreased, it is still New Zealand's dominant export sector. The dependence on agricultural exports reflects also a dependence on natural resources. In order to ensure sustainable economic growth, these natural resources must be carefully managed for the benefit of present and future generations. The image of a green and clean[3] country helps in the export of meat, wool, timber, and fish and attracts foreign tourists. To protect this image as well as the environment, New Zealand needs to take strong measures to curb pollution, manage the landscape and protect its indigenous flora and fauna (OECD, 1996). Those measures will directly affect the agricultural sector.

REFORMS TO THE NATIONAL ECONOMY

New Zealand was, in the 1950s, one of the richest countries in the world on a per capita basis through the sale of agricultural commodities to the British market. In 1973 the United Kingdom joined the European Community and New Zealand lost its guaranteed export market (then 35% of its exports, now 6% (The Economist, 2000)). Exporters were exposed to international competition and the returns from alternative markets were lower than those received from the United Kingdom market.

To maintain the standard of living that New Zealanders had become used to during the 1950s and early 1960s, the government, realising that agriculture would not provide enough jobs for a growing population progressively introduced protection to develop a small manufacturing base. The government also began a programme of borrowing on the international markets. This borrowing was used largely to sustain consumption rather than for investment for further economic growth of the economy (Walker and Bell, 1994). A fixed exchange rate policy masked the impacts controls were having and the overvalued exchange rate disguised the real size of external borrowing. In the late 1970s and 1980s there was also a massive fiscal expansion, including big subsidies for industry and farming, and heavy public investment in industrial projects (The Economist, 2000).

The subsidies to agriculture were to encourage farmers to increase output. Agricultural policies in the 1960s and 1970s reflected an increasing degree of intervention by government in the agricultural sector. The intervention was aimed at guiding farmer-decision making toward investment and production and came in the form of stabilization and supplementary minimum prices schemes, concessionary development loans, tax allowances on development expenditures, etc. The objective was to increase export returns from pastoral production. By

---

[3] Clean and green is an image New Zealand attempts to project of the environment in which agriculture is conducted. It was based on its clean water, air and soil (low use of fertilisers and pesticides and its reliance on all year outside grass feeding), low population density and wide open spaces. This chapter demonstrates that relative to many other countries, New Zealand is greener and cleaner but that it needs to work hard at being able to preserve that image.

1984 government assistance to agriculture was equivalent to over 30% of the final value of output of all pastoral farmer products in the country (Sandrey, 1991), and around 40% of sheep and beef farm income (Walker and Bell, 1994).

By 1984, therefore:

> New Zealand was a highly regulated economy …. Interest rates were tightly controlled by the government and lending policies of banks were directed by the government: the exchange rate was fixed; and wages and prices were subject to a comprehensive 'freeze.' At the same time, several economic indicators showed that the economy was performing poorly. Deficits of the current trade account and the federal government's budget were very high and increasing. Inflation was artificially low due to the price freeze. It was expected to accelerate sharply in 1985 (Sandrey, 1991, p. 25).

New Zealand was achieving low economic growth, and agriculture, still the key export sector, was heavily subsidized and had become isolated from the realities of the (overseas) marketplace (Johnson, 1992). Declining terms of trade, high domestic inflation, a growing public debt, a budget and current account deficit of 8 and 9% of GDP respectively (The Economist, 2000), and a large overseas debt, created a situation in which international lenders were no longer willing to extend credit. The country was on an unsustainable course and a change of direction was required.

In 1984, after a snap election, which brought the Labour Party in power, the new government pursued free market economic policies. The New Zealand dollar was devalued by 20% and floated, there was general market liberalisation in the commercial and exchange sector, export assistance was discontinued, import protection was lowered, and taxes were made more direct (Sandrey and Reynolds, 1990). It had also long been recognized that the level of assistance provided to agriculture was unsustainable in the face of falling world prices. Hence as part of the 1984 reforms, agricultural support was withdrawn, supplementary minimum prices schemes abolished (but it was thought that the 20% devaluation would compensate farmers for this loss), concessionary loans were brought in line with market rates, user fees were announced for agricultural product inspection and farm advisory services, and farm input subsidies terminated (Sandrey, 1991). From the early 1980s when it was one of the most regulated economies in the OECD[4], New Zealand is now, one of the least regulated. According to the OECD (1998:71) the PSE percentage New Zealand's agriculture peaked in 1983 at 36. International comparison of agricultural support over time is provided in Table 8.1.

The reforms in the mid 1980s improved the way price and incentive systems operate in individual markets through more efficient allocation of private resources. In particular, international trade was liberalised and markets deregulated, the taxation system was fundamentally transfigured, the delivery of

---

[4] New Zealand, including industry was highly protected with import quotas, export incentives and tariffs. While agricultural support was high, New Zealand's agriculture was not one of the most protected agricultural sectors in the world, but the country was one of the most protected ones.

income support, health and education was significantly changed, the efficiency of core government departments was extensively improved, and many government trading activities were corporatised and privatised (Pomeroy, 1997).

Table 8.1   International comparison of support to agriculture (year ending 31 December) expressed as Producer Support Estimate (% PSE)[5]

	New Zealand	Australia	Canada	EU	Japan	USA	OECD average
Average 1979-1981[a]	18	9	20	36	60	15	29
Average 1986-1988	11	8	34	44	67	25	40
Average 1997-1999	2	7	17	44	61	20	36
1999	2	6	20	49	65	24	40

a) The average 1979-1981 is not directly comparable with the other averages. The 1979-1981 figures are obtained from MAF (1997), and although the source was quoted as OECD, the numbers do not quite correspond with the OECD numbers provided for % PSE in the MAF (2000b) source from which the other figures are derived. However, the 1979-1981 figures do show relative magnitudes between countries and trends over time.

The removal of supplementary minimum prices, price supports, tax concessions, capital subsidies, input subsidies, low interest loans and free extension services to farmers, in the short term led to reduced farm incomes, falling land values, a decline in farm profitability, and an increase in farm debt (Pomeroy, 1997). The impact of the removal of subsidies was compounded by low international prices for most commodities during the mid-to-late 1980s and increasing interest rates (OECD, 1997). After the initial depreciation of the exchange rate it appreciated and this also hit farmers hard.

The New Zealand government put into place adjustment programmes for farmers who had become marginal and non-viable following the withdrawal of agricultural assistance. A Special Assistance to Farming programme operated between 1986 and 1989. This allowed farmers and their families, after satisfying certain criteria, to receive grants and welfare benefits equivalent to the unemployment rate. An Exit Grant scheme was introduced in 1988 to provide assistance to non-viable farmers to encourage them to leave farming. The rapidly falling land prices since 1984 eroded farmers' equity and the Rural Bank (the major farm loan lender) introduced a loan-discounting scheme. Johnson *et al.* (1989) reported that by 1988 approved applications involved average discounting of 33% of the original debt to the Rural Bank. Many farmers were assisted through creditor meditation meetings with financial and legal experts to agree appropriate action. Walker and Bell (1994) noted that for most, debt restructuring

---

[5] The PSE is an indicator of the annual monetary value of gross transfers from consumers and taxpayers to agricultural on farm production or income. The PSE percentage is expressed as gross transfers as a percentage of gross farm receipts valued at farm gate prices, including budgetary support.

and debt write-off followed, although for some selling was the only option - about 20% of the total debt owed by the farm sector was written-off, and about 5% of farms were sold.

Despite the loss of government assistance, adverse international prices and fluctuating exchange rates, New Zealand farmers have shown remarkable resilience and farm incomes have actually increased steadily (except in the first 2 years of reform). The agricultural sector didn't as such change structurally but responded, rapidly in some cases, to the new market environment and this caused major changes in the mix of agriculture output. There has been a marked trend away from the traditional pastoral farming activities, where most subsidies were once directed. Market forces have led to a shift of resources into a wide range of other activities particularly horticulture and forestry. Farmers and exporters have had to explore and develop new non-traditional markets. Most noteworthy are the rise in exports to Asia and Australia and the fall in exports to Europe.

The conclusion of the Uruguay Round brought new confidence to the agricultural industry. New Zealand vigorously pursued trade in the newly liberalised trade environment under GATT (General Agreement on Tariffs and Trade). Considerable success has been achieved in some sectors (for example dairy). Since the early 1990s, the agricultural sector has somewhat recovered with better overseas prices and a depreciating exchange rate during the late 1990s. Agricultural output has been severely affected in the late 1990s by drought caused by El Niño (two seasons in a row), and export incomes have been affected by the Asian crisis. It has been estimated that between July 1997 and June 1999, the impact of the Asian financial crisis, as a percentage of New Zealand's actual exports and GDP, were about 3.4% and 1.4%, respectively, and that the drought caused a 1.4% loss to GDP (MAF, 2000b).

The more efficient and intensive use of private resources, resulting from the reforms and new market opportunities, while leading to higher productivity, in some areas also led to concerns about environmental quality (Meister, 1995; Meister and Gardiner, 1997). With continued intensification likely in the future, these concerns will increase.

KEY ENVIRONMENTAL AND HUMAN-HEALTH ISSUES

A number of environmental issues have been raised over the past century by the development of New Zealand agriculture. Although farming in New Zealand is generally much less intensive than in other countries, nevertheless, in some areas of New Zealand significant soil erosion problems have resulted from removal of the natural forest cover for pastoral farming. The resulting sediment along with nutrient run-off and discharge of agricultural wastes has also contributed to water quality concerns in some areas. The removal of indigenous vegetation raised the issue of protecting biodiversity. The continual conversion of land to horticultural and dairy production has made water allocation, water quality and minimum stream flows all issues of importance.

In many cases, environmental problems associated with agriculture can be attributed to government policies. Until the mid-1980s, for example, agricultural support programmes encouraged over-intensive use of chemical inputs and other physical resources. Following the removal of subsidies, the number of sheep in New Zealand declined by well over 35% (Statistics New Zealand, 1998). Well over 500,000 ha of pasture have been converted to exotic pine forests over the same period. An even larger area of marginal pasture on steep erodible slopes has been left to regenerate in scrub and native forest (Ministry for the Environment[6], 1997b). While the decline in sheep numbers has continued, this has been partially offset by an increase in the number of dairy cows and other livestock. Fertiliser and pesticide use, which significantly decreased after the economic reforms of the 1980s, increased again when farm incomes started to rise (MAF, 1997).

Farming is intensifying in some fertile downland and lowland areas. This intensification (particularly dairy conversions in lowland areas) has placed water quality and aquatic biodiversity at greater risk. Similarly, continuous intensive cropping in arable areas is threatening soil structure. Urban areas continue to encroach into rural areas and 'rural sprawl' (rural residential development) continues in areas in close proximity to urban centers.

At the national level, the environmental dimension of New Zealand's agriculture has been described by the Environment 2010 Strategy, the New Zealand Sustainable Land Management (SLM) Strategy, the Government's policy on sustainable agriculture and associated legislation, particularly the Resource Management Act (RMA) of 1991 (RMA, 1994).

In October 1994, the New Zealand Government released a strategy outlining environmental priorities for the future. The Strategy's agenda focuses on eleven priority issues that implicitly identify some of the environmental concerns held by New Zealand society and the role agriculture plays. Some of these priorities (the ones more directly related to the agricultural sector) are:

- Managing our land resources - maintaining and enhancing our soils, so that they can support a variety of land-use options.
- Managing our water resources - managing the quality and quantity of all types of water to meet the needs of people and ecological systems.
- Maintaining clear, clean, breathable air - maintaining clean air in parts of New Zealand where it is already clean, and improving its quality elsewhere.
- Protecting indigenous habitats and biological diversity - maintaining and enhancing New Zealand's remaining indigenous forests and other indigenous ecosystems, and promoting the conservation and sustainable management of the diversity of plants and animals.
- Managing pests, weeds and diseases - to protect the diversity of plants and animals in ecosystems, to protect human-health, and reduce risks to the economy.
- Sustainable fisheries - for the benefits of the fisheries resources and all New Zealanders, including for commercial, recreational, and customary use.
- Reducing the risk of climate change - to help address levels of greenhouse gases in the atmosphere, and to meet New Zealand's international obligation under the Framework Convention on Climate Change (FCCC) (MFE, 1995).

---

[6] Ministry for the Environment is often abbreviated as MFE.

The SLM Strategy was adopted in 1996, while a discussion paper on sustainable agriculture was released by the Ministry of Agriculture and Fisheries (MAF, 1993). In 1998, the draft New Zealand Biodiversity Strategy was released by the Department of Conservation (DOC) (DOC, 1998). As much of the remaining biodiversity is on private land, much emphasis in policy circles is currently focused on addressing the effect of private land management on biodiversity and conversely, the effect of managing biodiversity on the landowners (MFE, 2000a)

Current policy directions reflect the increasing concern in New Zealand and worldwide about the effects of some farming systems on the environment. At the same time, it is realized that dealing with these concerns provides opportunities for New Zealand to position its products as coming from a high quality, more sustainable farm environment. The policies that are in place, or are being put into place, acknowledge the need to protect the land resource (including biodiversity and landscape preservation) and address issues such as soil erosion, weeds and pests, and water pollution as some of the physical threats to agriculture, now and for future viability. Further issues of animal health and food safety associated with current agricultural practices and systems are being dealt with under appropriate legislation. The protection of the consumer and the environment is also reflected in the current moratorium on genetically modified organisms and the establishment of an Environmental Risk Management to deal with current and future proposals for the growing or introduction of genetically modified crops or organisms.

In the following sections, the major environmental and human-health issues affecting agriculture are discussed. Where possible, available evidence on the incidence of environmental problems associated with agriculture is presented. The sequence is land, water, air, biodiversity, pests, weeds and diseases, animal welfare and human-health. Before doing so, below is a summary of the changes in environmental quality that have been observed in New Zealand as summarised by the Parliamentary Commissioner for the Environment (PCE) in a recent speech (Morgan Williams and Gebbie, 2000).

Positive signs
- Icon species saved
- Pesticide contamination low
- Methane 5% below 1990 levels
- Over 30% of land in conservation uses
- Forest area increasing

- Most coastal waters healthy

Negative signs
- Biodiversity under great pressure
- Land use more intensive on lowlands
- CO_2, 19.2% above 1990 levels
- Pest pressures rising both on land and in the sea
- Water demand exceeds supply in some areas, and reduced quality in some systems
- Urban amenities value and air quality declining

Many of the issues mentioned by the PCE relate directly and indirectly to agriculture and therefore any attempts to mitigate the negatives will affect agriculture.

Land

New Zealand soils are mostly evolved under forest and tend to be thin and acidic with low levels of nitrogen, phosphorus and sulphur. Before they can grow productive pasture or crops, the soils must be improved with fertilisers. Over the past 100 years, a large area of New Zealand has been converted from natural forests, wetlands or dune land to farm and forestry land (52%). In many areas pastoral agriculture, with its dense grass coverings and fertilizer application, has increased the organic carbon content of the soils leading to improved water retention and nutrient cycling capacity. However in some areas the soil quality has come under pressure from overgrazing or too much cultivation (MAF, 2000a).

On the hilly and steep areas of New Zealand this conversion of forests has caused significant erosion. Over half of New Zealand is affected by slight to moderate soil erosion (approximately 8.3 million ha [Clough and Hicks, 1992]) and of this, 10% is severe (particularly the eastern North Island - the Gisborne region, parts of Taranaki, and the South Island high country). Some of the highest erosion rates in the world have been recorded in New Zealand (OECD, 1996). Most of the erosion-prone land is hilly or drought-prone pasture land. Overstocking and over-cultivation have further contributed to erosion, compaction of soil and contamination of waterways.

Throughout New Zealand some 2.1 million ha of agricultural land have been identified as being susceptible to soil structure degradation (MAF, 1993) and a total of 7.6 million ha of hill country and high country farmland are susceptible to nutrient decline.

The consequences of erosion are seen in agricultural productivity losses, flood damage, destruction of physical infrastructure on and off-farms, siltation of rivers, streams and hydro lakes and recreational and amenity costs. No estimates are available of the cost of these problems to the nation. Also much of the information on the national state of our soils, vegetation and land ecosystems is out of date and/or incomplete.

Water

While New Zealand has substantial rainfall and an extensive lake and river system, water is distributed very unevenly. In upland areas the water is of high quality by international standards; however, in many lowland areas including streams, lakes and groundwater, water quality is poor and biodiversity has been significantly affected.

Water use by agriculture and related industries is three times greater than the consumption by all households and industrial use combined. Agriculture's demand for water is highest during summer months when river flows and groundwater levels are at their lowest. This has led to increased competition for water in some rivers and aquifers, particularly in cropping, market garden and horticultural areas.

Livestock produce some 40 times the organic waste produced by the human population (MAF, 2000a). With the increase in the number of dairy farms, water quality has become an environmental concern in the lowland river system throughout New Zealand. Often nutrient levels are excessive, baseflows are often turbid and the waterways aesthetically degraded. It was noted that:

> the most significant (and most frequent) kind of groundwater pollution from agriculture is the accumulation of nitrate-nitrogen, especially where concentrations equal or exceed 10 g m^{-3} [50 g NO_3^-]. Only shallow groundwater (down to 60 meter) show the effects of agricultural inputs, but the phenomenon is common throughout rural New Zealand - especially in areas where stock densities are high and groundwaters are vulnerable to contamination from surface drainage (Smith *et al.*, 1993, p. 82).

In some of the regions e.g. the Waikato area, it was found that 9% of tested groundwater bores has nitrate levels above acceptable drinking water guideline (Environment Waikato, 1998) while in the Manawatu-Wanganui region the number is 20% (Horizons.mw, 1999). Of course not all nitrate is derived from agricultural sources alone. However, all regional authorities are placing great emphasis on getting farm effluent disposal systems upgraded so those farms (especially dairy farms) no longer are a source for groundwater pollution. In 1998, 70% of all dairy farms were disposing effluent to land and the remainder was, in the main, using best management practices set by regional councils. Therefore, although there is evidence of nitrate levels in some groundwater area exceeding drinking water standard this is often not groundwater used for drinking water purposes, and the problem is still very localised (Lincoln Environmental, 1997).

It is not only the discharge of effluent that is placing pressure on water quality. The expansion of dairy farming means many dairy farmers are now applying larger amounts of nitrogen fertiliser to their soil (MFE, 1997c). Until recently, nitrogen fertiliser was used mostly for crops, while pastures derived their nitrogen from clover, with just a light application of fertiliser in the winter. With the more intensive pasture use, many farmers are now applying 25-100 kg N fertiliser ha^{-1} year^{-1} and some are applying more than 200 kg ha^{-1} year^{-1}.

The continued intensification of agriculture (especially dairying) has, since the publication of the 1993 report mentioned above, led to a mixed result in areas in New Zealand. For example, one region reports a continuing upward trend in the number of bores (groundwater) contaminated with nitrates (Horizons.mw, 1999), while another states that with the increased awareness of agriculture's impact on water quality and measures taken by regional councils, in many streams and aquifers improvements have been noted or at least no further deterioration has occurred since the publication of the above report (Environment Waikato, 1998).

Other pressures on water flows have been from drainage and channelisation (which have reduced wetlands and altered the natural character of rivers including lowland aquatic habitats). Deforestation (which has intensified flooding and sedimentation in steep catchments) has also contributed to land quality and flooding problems.

Overall, however, the quality of our water is high by international standards, except as pointed out in some low-lying rural streams and small lakes, some shallow groundwater and some piped water supplies. Over the last decade responses to water quality have successfully focused on improving point source discharges (sewage, factory and dairy shed outfalls) but the more difficult and pervasive problems of non-point source discharges have yet to be fully addressed and will require changes in land management.

Air

New Zealand is thought to have good air quality by world standards, although very little data is available. The good air quality is largely due to our remote location, relatively low level of industrialisation, and dispersion of local emissions by persistent winds. However, recent monitoring has shown that in some urban locations, air pollution occasionally exceeds air quality guidelines for human-health. Of more significance to agriculture and horticulture are spray drift and odour. Although frequent complaints are received in some regions by the territorial authorities, the overall issue is small and very localised.

Biodiversity and the protection of indigenous habitats

New Zealand's long period of geographic isolation, diverse terrain and ecosystems have contributed to a unique range of habitats and animal and plant species. Biodiversity is vital for our primary industries, providing environmental services and new gene and drug discoveries, and for ecological, cultural and spiritual values.

A recent study suggested that the total annual value provided by indigenous biodiversity could be more than twice that of national GDP (Patterson and Cole, 1999).

For all these reasons, the decline in biodiversity in New Zealand is seen as the most pervasive environmental issue, with 85% of lowland forests and wetland now gone, and at least 800 species and 200 subspecies of animals, fungi and plants considered threatened. Or, as one pertinent quote says:

> We have managed the country's resources largely for the benefit of the
> second community [humans and the species that come with us]
> (including unintended beneficiaries, such as rabbits) while leaving the
> native species to survive as best they could, often in habitat remnants
> and protected areas (MFE, 1997b: 10.5).

While much habitat and species protection has taken place in natural parks and reserves (the conservation estate), the main threat to indigenous biodiversity is insufficient habitat in lowland areas, declining quality of many of the remaining land and freshwater habitats, the impacts of pests and weeds and, for some marine species and ecosystems, human fishing activities.

Much remaining indigenous biodiversity is on private land and is threatened by habitat modifications and weeds and pests, and while:

agricultural pressure on native ecosystems, through large-scale conversion to pasture, is largely a thing of the past, ...however regenerating areas [and protection of remaining remnants] may need to be actively managed to exclude pests and invasive weeds, and in some cases sympathetically restored, if we are to re-establish fully functioning ecosystems (MFE, 2000a, p. 4).

Weeds, pests and diseases

Introduced pests, weeds, and diseases pose a serious risk to biodiversity, agriculture, forestry and aquaculture. The nation's estimated 70 million possums are currently considered to be the nation's most destructive pest. Besides their impact on indigenous vegetations, orchards and cropland, they also affect pastoral agriculture by transmitting the highly contagious bovine tuberculosis bacterium to cattle and deer (MFE, 1997b). Rabbits are another high profile pest, although their greatest impact is limited to the tussock-predominant grasslands of the South Island, where their pressure has combined with that of invasive weeds and grazing sheep to degrade an area of about 1 million ha. Less obvious but more widespread is the combined pressure from many smaller invaders such as insects, parasitic worms, weeds and fungi. These threaten native species, exotic crops, forests and livestock and their pressure grows with each new arrival. New Zealand lives with a significant risk of the introduction of new exotic pests, weeds and diseases as demonstrated by the white-spotted tussock moth, fruit fly, rabbit calicivirus disease and the varroa mite. All of these, if they become established can wreak havoc with our primary industries. Although some success in pest management has been achieved, there is still a long way to go. In many instances, however, it may be neither practical nor economic to attempt to eliminate existing pests, weeds and diseases using current techniques and containment and management becomes the only option (as demonstrated by the latest pest outbreak of the varroa mite [affecting the honey industry]).

Animal welfare

The impacts of farming on animal welfare are coming under close scrutiny from the general public, lawmakers and consumers (including consumers from the markets to which we export). Of concern is protection against deliberate cruelty, including issues such as the lack of farm shelter in winter, early lambing and shearing (which can result in stress and death in cold weather), stressful transport in lorries and on ships, the restrictive confinement of pigs and chickens in 'factory farms', and amputations performed without anaesthetics (e.g. docking cows' tails, removing the highly sensitive velvet from deer antlers, dehorning cattle, castrating pigs, mulesing merino lambs, and trimming the beaks and toes of chickens) (MFE, 1977c). Of all these issues, the treatment of battery hens is particularly controversial, especially since with 60 million birds, poultry is our most abundant livestock.

Human-health

Human-health impacts have come to the fore through a variety of issues such as pesticide residues, animal feeds and environmental quality. The human-health effects of genetically modified foods (GMF) and genetically modified organisms (GMO) are discussed separately.

Pesticide use in New Zealand, in common with other developed countries, has progressed from a past in which persistent and highly poisonous chemicals (such as arsenates) were used, through the era of DDT and other persistent (but less directly toxic) organochlorines, to the present era. Research initiated in 1995 on the levels of organochlorine contamination in the New Zealand environment and in food products purchased from retail outlets reported '..that the level of these contaminants in air, soil and water are 'generally low' and that New Zealanders have one of the lowest dietary intakes of these chemicals in the western world' (MFE, 1998b, p. 1). This confirms earlier results from a 1994 survey (Close, 1996). Similar results have also been found for ambient concentrations in water and sediments.

With regard to animal feeds and hormones, rBST is not registered as an acceptable compound hence its use is illegal in New Zealand. Hormones are used in some beef cattle but the meat is tagged so that it is treated completely separate from other meat.

Animal feed and veterinary products are all controlled under the Agricultural Compounds and Veterinary Medicines Act 1997 (ACVM Act) for oral nutritional compounds. Oral nutritional compounds include both proprietary and non-proprietary products ranging from unharvested grass and crops to complete, balanced foods or feeds, components of feeds and animal diet supplements. In addition there may be additives in oral nutritional compounds that have no specific nutrient value but are incorporated to improve the preservation, digestion, colour, palatability, texture or nutritive value of food, including for example preservatives, antioxidants, enzymes, emulsifiers, stabilisers, acids, no-stick agents, humectants, firming agents, anti-foaming agents, colourings and flavourings, solvents and direct-fed microbials.

Genetically modified organisms (GMO) and genetically modified food (GMF)

With respect to GMFs, until recently there was no legal obligation to notify the Ministry of Health prior to putting a new food on the market and no legal requirements existed for pre-market assessment or testing. There was an interim policy that requested food importers and producers to confine imports of GMFs to those that have been assessed and found suitable by competent overseas authorities. These include the US Food and Drug Administration and the UK Advisory Committee on Novel Foods and Processors. Today GMF is a major issue in New Zealand and is being dealt with by appropriate authorities (jointly with Australia), mainly through food standards and labeling.

GMOs have also become a major issue in the last couple of years. New Zealand has been rather conservative with regard to the application and acceptance of gene technology. So far only 21 field trials have been approved. There has been no commercial release of GMOs, and at the moment New Zealand has a 12-month voluntary moratorium in place on all applications to release and, with some limited exemptions, field test GMOs. The moratorium started on 14 June 2000 and ends on 31 August 2001; 3 months after that date a Royal Commission on Genetic Modification is to report. As stated:

> Not all genetic modification work is subject to the voluntary moratorium. The moratorium provides a balance between the benefits of maintaining research on genetic modification, the risks associated with the work and public attitudes towards GMOs. The Government considers that the controls already imposed by the Environmental Risk Management Authority (ERMA) on approved field tests, the voluntary moratorium, and additional monitoring to be undertaken by the Ministry of Agriculture and Forestry, will ensure that there is a minimal risk to the environment from GMOs during the course of the Royal Commission inquiry (MFE, 2000c, p. 1).

New Zealand, as stated by the NZIER (2000), is 'facing a difficult judgement: does a better economic future lie with securing our status as an 'organic' grower, or do we have to make sure we can match or better the genetically modified attributes of the products of our competitors'? (p. 1). Recent research has revealed a strong lack of support for GM technologies among primary producers, or, as the researchers state, 'In contrast, the GM fad among farm industry leaders and agricultural scientists is not catching on with the grassroots industry. Farmers are basically skeptical about the prospects of GM technologies' (Press release, 2000).

ENVIRONMENT AND HEALTH LEGISLATION

Over the last 15 years, New Zealand has witnessed a major reorganisation of the institutional framework dealing with environmental and animal and human-health issues. While until 1986 government responsible for the environment was highly fragmented, the reorganisation (which has only recently started in the animal and human-health areas) provided a major clarification of function of all departments, reduced central control and integrated the management of New Zealand's natural resources under one piece of legislation, the RMA of 1991. It is this Act, with its purpose of sustainable management, which impacts most significantly on the agricultural sector. Other legislation deals with the animal and human health issues.

The reforms also mirrored the reforms that took place in the political arena, in that greater emphasis was placed on devolution, less government involvement and a greater use of markets (or market-based instruments) as well as voluntary measures with/or without market incentives. With regard to environmental legislation common themes are:

- Sustainability: this is the umbrella principle for the management of natural and physical resources, indigenous forests and fisheries.
- Precautionary principle: where there are threats of serious or irreversible damage, lack of full scientific certainty shall not be used as a reason for postponing cost-effective measures to prevent environmental degradation. The corollary of this principle is that people proposing to undertake activities with potential effects on the environment should carry out environmental impact assessments. In this way the nature and extent of any environmental risk is identified before action is taken. This approach is the basis on which the RMA is implemented, i.e. an effect-based focus.
- User-pays: people using resources that are managed at public costs are required to help pay for a share of the management cost.
- Polluter-pays: the RMA (as well as others such as the Biosecurity Act (discussed below)) requires people conducting activities that damage the environment to account for the environmental costs of their activities and to pay for measures to mitigate, remedy or avoid those effects. The resource consent process under the RMA is an example of the practical application of this principle.

Integrated resource management: the Resource Management Act 1991

The RMA involves several key concepts (OECD, 1996, p. 32):

- The development of comprehensive effects-based legislation for environmental management, requiring examination and consistent treatment of all factors in any issue (removing the need for sector-specific responses).
- The desirability of intervening only where required and clearly justified.
- The requirement of clearly focused outcomes (targets) where intervention is justified.
- The need to use appropriate policy instruments to achieve cost-effective solutions.

The RMA entails a significant shift away from mandating technologies or discharge standards to protect the environment and towards a focus on ambient environmental quality, with flexibility as to how it is to be achieved (OECD, 1996). The reform moved legislation from a prescriptive to an enabling mode or, as Le Heron and Pawson (1996) state, the new planning regime established by the RMA regulates the environmental effects of resource development, rather than controlling land uses directly.

Since the RMA plays a major role in describing the legislation relating to the environment and some animal and health issues, its structure is presented in Figure 8.3. The figure also reflects the reform in local government.

Figure 8.3 The Resource Management Act and implementation structure

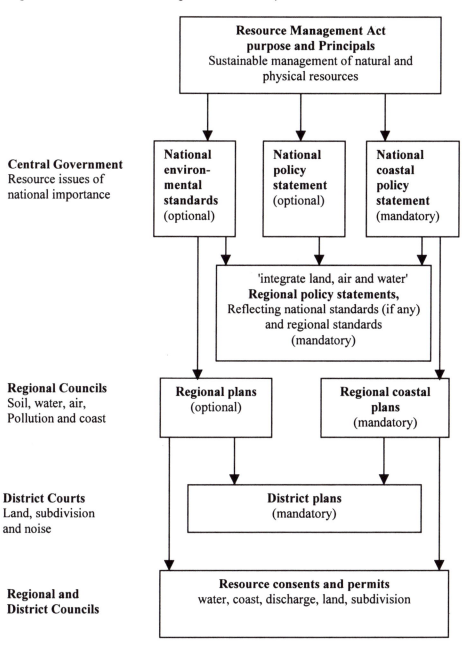

Source: MFE (1997b, 4.12).

As Figure 8.3 shows, regional government has a significant and key role in implementing the RMA. They fulfil this task in two major ways: planning for resource use (regional plans) and setting environmental standards (policy statements), and implementing the plans by granting resource use consents (or permits):

- take, use, dam or divert water;
- discharge contaminants to land, air or water (including marine waters);
- occupy the coastal marine area;
- place structures or carry out other activities in river and lake beds;
- control activities on land for the purpose of soil conservation, the maintenance and enhancement of the quality or quantity of water in a water body, the avoidance or mitigation of natural hazards and the prevention or mitigation of any adverse effects from storage, use, disposal or transportation of hazardous substances.

The standards set by regional authorities can differ from region to region as environmental issues and situations differ. However the RMA does provide guidance and so do materials supplied by the Ministry for the Environment (MFE). Much work is going on by providing regional councils with guidelines for air, water, and other resource quality standards.

Under the RMA, the presumptions for consents are positive in the case of land use. This means that consent is required only if the plan explicitly requires it. Consents for the subdivision of land, for the use of water, coastal marine areas and river or lake beds, and for discharge of any pollutants to land, air or water have the opposite presumption: consents will always be needed unless a plan provides otherwise. All consent applications require an environmental impact assessment.

Territorial authorities must prepare district plans and issue consents for activities under their jurisdiction. These include controlling the effects of land use and subdivision, controlling noise and protecting the surface of rivers and lakes (see Table 8.2). Finally, central government may prepare national policy statements on the management of specific resources and these must be taken into account by local authorities, i.e. they would be binding.

It is this devolution to regional and territorial authorities that determine the legislation that deals with the environmental issues discussed in the previous section. As a result, the majority of resource management programmes relevant to agriculture are now carried out at the regional level. These include soil conservation activities, water quality monitoring and control, pest management, etc. Under the RMA, councils are developing policies, in consultation with their communities, to address these issues. This means that the government, and district and regional authorities are required to identify the environmental risks in their area and develop policy statements and plans containing ways to regulate activities in response to those threats. There is no single set of overriding rules and measures, but legislation and approaches vary from region to region.

The MFE coordinates development of environmental standards and guidelines to help local authorities and resource users implement their responsibilities under the RMA. Standards and guidelines help define the 'environmental bottom line' of sustainable management. The principle followed is that standards and guidelines should prescribe the minimum amount of regulation to best achieve the desired environmental outcome. Further, national standards should only be developed where the advantages of protecting national values or providing national consistency outweigh the advantages of regional resource management.

Central government as yet has not produced any national policy statements on specific resources. National guidelines have been developed (or are being developed) to help people manage water better. These are:

- The draft Australian and New Zealand Guidelines for Fresh and Marine Water Quality (released for comment in July 1999).
- Revised guidelines to monitor beaches and report on public health risks in relation to bacteriological contamination (to be released 2001).
- Guidelines for microbiological fresh water quality guidelines (in development). They will also be used to improve human and stock drinking water guidelines.
- The Ministry of Health's drinking water standard, which the MFE has been closely involved with.
- A number of other technical guidelines to assist water managers to monitor and manage ecosystem health (MFE, 2000b).

Table 8.2   Responsibilities of regional councils, district councils and unitary authorities

	Regional council	District or City council	Unitary authority (district and regional councils combined)
Air	Yes	No	Yes
Water	Yes	No	Yes
Soil	Yes	No	Yes
Coast	Yes	No	Yes
Pollution and discharges	Yes	No	Yes
Land	No	Yes	Yes
Subdivision	No	Yes	Yes
Natural hazards	Yes	Yes	Yes
Hazardous substances	Yes	Yes	Yes

Source: MFE (1999a, p. 6).

Currently much effort is going into preparing indicators for environmental effects which (when adopted) will bring about greater consistency over all regions. However, one of the purposes of the Act was also to allow variety so that different regions could define standards and approaches more suitable to the specific region (see, for example, MFE, 1997c; 1998a, c).

While air, soil, and water are managed under the RMA, weeds and pests are managed under the Biosecurity Act 1993. This Act is implemented by regional councils, and requires them to clearly identify weed and animal pests and determine if they represent a pest, then to develop a pest management strategy, and then to look for the most efficient and effective way of implementing this strategy.

Soil, air and water

Regional council programmes, providing technical advice on soil conservation and other environmental issues, have taken on a renewed importance with the reduction in grants from central government. In many regions, council officials assist land users to develop property management plans dealing with soil conservation and other environmental problems on the farm. In some cases, farm plans have been developed that include protection covenants on natural forest or other significant natural features on their properties. Some councils charge for this service to recover some of their costs, while others provide free advice. Several councils have ratepayer-funded programmes to assist with soil conservation and sustainable land management activities by groups of landowners.

The aim of all these approaches is to collectively deal with land management problems such as erosion, flooding, irrigation, pest and weed control, and also more conventional concerns relating to livestock and financial performance and consumer acceptability of products (i.e. market access).

Through these programmes and landowners groups, farmers are showing a new interest in soil and vegetation monitoring and a greater willingness to resolve the problems themselves. Regional councils provide landowners groups with advice and some funding (in some regions up to $300,000 for the whole region).

Regulation does apply to land, soil and vegetation disturbances (the latter one especially affects logging, felling or harvesting trees and the burning of vegetation). Regional and district plans both contain rules for such activities. The approaches to land disturbance cover a wide spectrum with some councils applying predominantly regulatory approaches and others using only non-regulatory approaches. Morriss and Workman (1998) demonstrated the great variation that existed in the mid-1990s. They also showed that the whole situation is in a fluid state, and regional councils are still finalising land use plans and rules since the transition phase following the RMA. Also, most rules and regulations apply when changes are being made in terms of afforestation, tracking, vegetation removal through burning and uprooting, etc. With grazing being an existing use, all this doesn't affect the farmers too much except in areas, that have been declared sensitive, and the councils carefully control all activities. Those most seriously affected are the forestry companies whose activities nearly all require consents. The overriding philosophy is that regulation should be minimised, but the acknowledgement is there that some form of regulation will be needed in certain cases. It is difficult to summarise the overall situation in New Zealand since there is such a great variation.

Since the report by Morriss and Workman (1998), a less regulatory approach has been adopted by a number of councils. In particular, councils have reduced control on vegetation disturbance on erosion-prone areas, the major concern of forest owners. Many councils now propose using permitted activity conditions to replace the requirement for resource consent (MFE, 1999b). This change has come about in response to the councils' experience of requiring consents for a number of years and in response to submissions from forestry companies on the relevant regional plans.

With regard to water, regional councils set desired quality for the waters in their region in the policy statements. Once receiving water quality standards have been set (the Act provides a set of minimum water quality standards[7] and then provides a schedule of higher water quality classes that can be aimed for), the councils manage the effects of discharges to water in such a way that after reasonable mixing[8], the quality of the water does not fall below the class aimed for.

Any significant point source discharge to water requires a discharge permit from the regional council, usually subject to conditions.[9] In the agricultural sector, this applies primarily to discharges from effluent ponds of dairy and other intensive livestock operations and discharges from agricultural processors and input manufacturers. With regard to non-point sources, councils leave the decision on how these are dealt with up to the land-users, but monitors the effects. All regional councils charge application fees for consents, and these are increasing as councils move to recover all costs associated with processing applications and monitoring compliance. All consents are monitored annually by council inspectors (or delegated authorities) at a cost to the farmer. Councils are trying to reduce regulatory approaches and aim to achieve reduction of effects through education, information, and self-regulation by industry.

The fertiliser industry has produced a voluntary code of practice that specifies a limit of 300 kg N ha^{-1}. The use of this code is a proactive response in

---

[7] This RMA stipulates generic environmental objectives. With regard to discharges, the Act in sections 70(1) and 107 set minimum water quality standards, which prohibit, after reasonable mixing:
- conspicuous oil or grease films, scums or foam, or floatable or suspended materials;
- conspicuous change in the colour or visual clarity;
- objectionable odour;
- fresh water unsuitable for consumption by farm animals; and
- significant adverse effects on aquatic life.

[8] 'Reasonable mixing' is the length of a river reach downstream of a discharge point, or an area around a discharge point in a lake, where standards are allowed to be exceeded without compromising overall policy objectives. Note that reasonable mixing does not necessarily equate with full mixing; or with waste assimilation through natural treatment within the receiving water (pers. comm. Ewen Robertson: Horizon.mw, 19 December 2000).

[9] For example discharges to land (spray irrigation or irrigation from ponds) receive in all areas of New Zealand a permitted consent, i.e. the activity is automatically permitted with some conditions dealing with:
- distance to neighbour's property, well or bore;
- loading rate: N loading rates (by soil type), in some regions P loading rates;
- land area guideline (used in compliance monitoring);
- application interval;
- system design, no rules, and maintenance – checked annually (Taranaki Regional Council, 1998a).

recognition of the need to ensure the environment is managed sustainably. Regional authorities have written guidelines in regional schemes with regard to the use of nitrogen application to land. For example, Environment Waikato (New Zealand's largest dairy region) advises farmers to avoid applying N fertiliser to areas where effluent loadings reach 150 kg N ha^{-1}, and that rates of artificial fertiliser over 200 kg N ha^{-1} year^{-1} will significantly increase environmental impacts. Most farmers (and this mainly applies to dairy farmers) operate well under this level.

Air issues (such as odour and spray) are mainly handled through buffer distances. Guidelines are given for preferred minimum buffer distances between sheds and other features. However, for example the Taranaki regional plan warns that it should be noted that buffer separation is not a substitute for good management of sheds, nor will observation of these distances without consideration of other factors ensure no odour problems off-site. The guidelines provide a general indication of the matters that a regional council may consider and the nature of the conditions that might be attached to resource consent for the discharge of contaminants into the air from piggeries. The material presented must not be considered as a set of rules that will be applied universally. Each individual situation will be considered by the Council on its particular merits and circumstances with regard for the level of environmental protection that is appropriate in that situation (Taranaki Regional Council, 1998b, p. 135).

Pesticide spraying is a permitted activity, but regional plans state that:

- You need a proper spray plan when you intend to carry out widespread application of agrochemicals within 50 metres of your property boundary (this should include chemicals used, how they will be applied, and the GROWSAFE[10] certificate number of the person who will carry out the spraying activity[11]).
- You need to notify your neighbour if he or she asks you to let them know when you are going to spray.

Only approved chemicals are allowed to be used (those approved under the Pesticides Act 1979 or the Hazardous Substances and New Organism Act of 1996). Currently around 900 registered pesticides can be bought, representing 320 active ingredients (pers. com., Pesticide Board).

Biodiversity

Biodiversity was identified, in the State of the Environment Report (MFE, 1997b), as New Zealand's most pervasive environment issue. In response the government launched the New Zealand Biodiversity Strategy (NZBS) in March 2000. The strategy has a 20-year timeframe to implement 147 recommended

[10] Grow Safe Agrichemical Users' Code for the application of pesticides. It is a requirement that applicators have followed this course. The test must be retaken every three years.
[11] Applicators of pesticides need to have followed a course in safe pesticide application. The certificate is evidence that the course has been followed.

actions aimed to conserve and sustainably use and manage biodiversity. Implementation of priority actions identified to deliver greatest gains for biodiversity in the first 5 years has already begun. During this first 5-year period, the Government will spend an extra $187 million towards implementing NZBS. A breakdown of this allocation is highlighted below (DOC, 2000a):

- Controlling animal pests and weeds on public conservation lands ($57 million).
- Increasing the funds available to protect and maintain biodiversity on private land ($37 million).
- Increasing the number of marine reserves around New Zealand ($11.5 million).
- Improving the protection of the marine environment from invasive marine species ($8.9 million).
- Researching New Zealand's marine biodiversity, leading to better management ($14.1 million).
- Development of a comprehensive biosecurity strategy for New Zealand and the assessment of biosecurity risks to indigenous flora and fauna ($2.6 million).

Since most of the remaining biodiversity is on private land, one of the most compelling management issues is how to bring conservation of biodiversity about. Under the RMA of 1991, territorial authorities must recognise matters of national importance, which talk about the ethic of stewardship, intrinsic values, essential characteristics that determine the integrity of ecosystems and maintaining and enhancing high-quality environments. Councils achieve this under the Protected Natural Areas Programme (mainly indigenous scrub, forest, and wetlands). With the help of the Department of Conservation (DOC), significant areas and features can be identified and incorporated in regional or district plans, and these can no longer be converted but need to be protected.

After the DOC has identified significant areas and features, the incorporation of such areas follows consultation between territorial authorities and private landowners. This implementation is a controversial issue in New Zealand since no compensation payments are made and farmers often are faced with having to fence the land. Sometimes regional or district councils will help with some funding, while in other situations some habitat conservation (especially indigenous forests) occurs as voluntary action under a wide variety of schemes. By 1997, 3% of New Zealand's surviving indigenous forests have been voluntarily protected or committed for protection by private owners who have offered them for sale to the DOC or have attached legal covenants to the titles. The covenants may last from 20 years to perpetuity. These voluntary protection arrangements are funded by: the Forest Heritage fund (established by the government in 1990); Nga Whenua Rahui (established for use by Maori forest landowners); and the Queen Elizabeth II National Trust (established in 1977 with funding from both government and non-government sources) (MFE, 1997b; Bayfield and Meister, 1998).

Under the Forests Amendment Act 1993, the MAF must promote sustainable forest management on areas of indigenous forestland. Nearly all indigenous forest harvesting for timber production requires obtaining an approved sustainable management plan or permit from MAF's Indigenous Forest Unit. Some landowners have to go through a costly process of obtaining consents under both the RMA and the Forestry Act (MFE, 2000a).

To see how the NZBS and the Acts work out on private land, the Government put in place a Ministerial Advisory Committee to 'address the effects of private land management on indigenous biodiversity' (MFE, 2000a).

The report came out in 2000. The executive summary nicely summarises New Zealand's approach to management issues on private land:

> The Committee recognises that, ultimately, achieving New Zealand's and New Zealanders' goals for biodiversity will result not from force compliance or from increased public funding alone. Outside public conservation areas it also depends largely on understanding, acceptance and informed decision-making by individuals, companies and regulatory and non-regulatory public agencies. It will take the combined resources and the co-operation of all these to halt the decline in New Zealand's biodiversity (MFE, 2000a, p. iii).

The report concludes the section on property rights by stating there is still a great divergence among people about the issue of property rights and that there is a need for a better understanding of what land use property rights entail in New Zealand and of the established system in place to provide for change (MFE, 2000a).

The Committee calls for a national policy statement under the RMA; a range of non-statutory guidance to assist with implementation of biodiversity aspects of the RMA; a range of other related services including information systems, education and incentive schemes which should be at least partially government-funded; and assistance to local communities with the development of local accords.

All this will impact on landowners since sustaining biodiversity on their land and controlling pests (under the Biosecurity Act) is not a free exercise. As the report points out, 'it is appropriate that the wider community assist the management transition by providing some contribution to the costs of change' (MFE, 2000a, p. 65). The report is up for discussion and was not implemented by the year 2000 by government.

The whole situation is well summed up by the NZBS when it states:

> Habitats are being protected on private land through government and private-funding covenants and other mechanisms (such as reserves, protected private land and resource consents provisions). Individual landowners are also choosing to fence off and maintain remnant areas of bush, riparian margins and wetlands on their land using their own resources (DOC, 2000b, p. 38)...[To date] RMA provisions to promote

the protection of significant indigenous vegetations and habitats have not been effectively implemented across New Zealand. This is due to: difficulties in defining the meaning of 'significant'; the lack of clarity over values to be protected; uncertainty over the right mix of rules and non-regulatory methods; ineffective consultation with landowners; resourcing problems in local authorities; and unresolved issues relating to private property rights, community benefits and cost sharing (DOC, 2000b, p. 40).

Weed and pest control

Weeds and pests are controlled under the Biosecurity Act 1993 (BA). This Act has two principal areas of operation. The first is the exclusion of unwanted organisms from entering New Zealand by way of requiring risk goods to conform to import health standards. The other area of operation relates to unwanted organisms already present in New Zealand. For such organisms the BA allow agencies to plan for, and carry out pest management. Being an enabling law, the BA contains powers but no mandatory obligations for agencies, nor any obligation for any individual to carry out pest control. The decision on whether to prepare a pest management strategy (PMS) and for what reason - mandating the management or eradication of particular organisms - is at the discretion of individual agencies (subject to cost/benefit analysis). When a PMS is implemented, the BA requires that the cost of the PMS is carefully allocated between beneficiaries (private landowners and the general public) and exacerbators.

Animal health and welfare

Animal welfare is managed under the Animal Welfare Act 1999 (which became operative on 1 January 2000). The emphasis of the Act is on punishing acts of cruelty towards animals. The primary focus of the Act is on a proactive and preventative approach. There is a greater focus in the Act on positive obligations towards animals. The onus of care lies with the owner or person in charge of an animal. They have the responsibility to meet their animal's physical, health and behavioural needs in accordance with both good practice and scientific knowledge. 'Physical, health and behavioural' needs are defined in the Act by reference to what is referred to internationally as the five freedoms. These are:

- proper and sufficient food and water;
- adequate shelter;
- the opportunity to display normal patterns of behaviour;
- appropriate physical handling; and
- protection from and rapid diagnosis of injury and disease.

The Act itself does not expand on these obligations but leaves that to 'Codes of recommendations for the welfare of specific animals', which contain minimum standards with regard to care of, and conduct towards animals as well as recommendations for best practice. Currently existing codes are carried over under the new Act, but will be reviewed within the next 3 years after a public consultation process.

Breach of the provisions in a code of welfare will not be an offence under the Animal Welfare Act; rather, any prosecutions will be for failure to meet the obligations in the Act relating to the care of an animal or for ill treatment of an animal. Failure to adhere to the minimum standards set out in a code will, however, be able to be used as evidence to support a prosecution. Recently, for example, two farmers in the Manawatu, have been prosecuted for failing to look after animals (calves on farms) and fined. There is a case pending in court dealing with battery hens, and the local SPCA (Society for the Protection and Care of Animals) would like to see some poultry transporters prosecuted for small cage sizes and the number of dead birds (pers.com., local SPCA 19/12/00). The DOC has been prosecuted for the treatment of wild horses on its estate.

Animal welfare is also handled under the Agricultural Compounds and Veterinary Medicines Act 1997 (ACVM), which will, besides looking at animal welfare, also cover trade, agricultural security and food residue standards. The ACVM Act is primarily responsive to standards and outcomes set under other legislation (Meat Act 1981, Dairy Industry Act 1952, Biosecurity Act 1993, Animals Protection Act 1960 and Food Act 1981). Agricultural compounds will be assessed and controlled under the ACVM Act to ensure that the outcomes and standards set under the above laws or by international agreements, are not compromised. The purpose of the ACVM Act is to manage risks from the use of agricultural compounds to trade in primary produce, animal welfare and agricultural security. For example, the reform taking place with the ACVM will facilitate a move away from prescriptive blanket regulation of all animal remedies and pesticides to a regime where regulation will only be provided where there are demonstrated risks to be managed, and where there is a net benefit in providing regulation (Burdon, 1999).

Human health

Human health is protected by a variety of Acts and standard-setting organisations. For example while the ACVM covers general risks to producers and their production systems, the Hazardous Substances and New Organism Act 1996 (HSNO) covers risks to the environment and human health. The purpose of this Act is to protect the environment, and the health and safety of people and communities by preventing or managing the adverse effects of hazardous substances and new organisms. This purpose is achieved by assessing substances for adverse effects on the environment (including social, cultural and economic effects) and approving the substance if it has a net benefit to New Zealand after taking account of controls.

Food safety is controlled under the New Zealand Food Act of 1981 (amended in 1996). Food standards are set jointly with Australia under an Australia New Zealand Joint Food Standards Agreement. Increasingly, food producers in New Zealand are responding to consumer demands for safe, quality food by applying systematic quality management systems such as the Hazard Analysis and Critical

Control Point (HACCP) system. These systems specify the standards to apply at critical points of each stage in the marketing chain; that is, from the consumer's plate back to the farm or orchard. This had led to the development of several quality assurance programmes, such as (MAF, 1998):

- Game Industry Board's 'Deer QA' programme;
- Pork Industry Board's 'PQIP' programme;
- Meat Board's 'Quality Mark' for the domestic market;
- Kiwifruit Marketing Board's 'Kiwigreen' programme, which focuses on minimum pesticide use;
- Apple and Pear Marketing Board's 'Quality Pack' programme; and
- 'Environmental Choice' - administered by the Testing Laboratory Registration Council (TELARC) and approved by the MFE. TELARC has responsibility for the promotion of good quality assurance practices and the maintenance of a scheme for the registration of suppliers.

With regard to export products, New Zealand has to accept the standards set for food safety by its major trading partners. For example, exports of New Zealand dairy products to the European Community are of great value to the national economy, but are subject to compliance with the national law and some specific technical market access requirements. The NZ/EU Veterinary Agreement of 1996 formally recognises NZ's regulatory system as equivalent to relevant EU law (92/46/EEC). The Agreement sets out reciprocal rights of both parties to audit one another's regulatory systems, to ensure the strength of the official assurances provided (this happens annually when the EC visits New Zealand). The requirement in the New Zealand Dairy industry is that all dairy exports receive a MAF 'Approved for Export' mark. Its use is not required by New Zealand law, but is applied to all dairy products for exports to the European Union. The mark is recognised in Directive 97/132/EC as New Zealand's health mark. MAF permits companies to use the mark on products, which satisfy the conditions of the Dairy Industry Act 1952 and the Dairy Industry Regulation 1990.

Also, annually 300 herds are sampled and their milk tested for 130 residues (dictated by the EC). In all this the New Zealand Dairy Board is trying to achieve equivalence of our approach to that of the EC, i.e. so that the content of the EC directives for food safety and quality are fully taken into consideration.

Similar audits exist for meat export slaughtering plants (regular visit from EC auditors). Japan has tight standards for the production of many fruit and vegetables and meat and dairy. They require ISO 9001 and also conduct audits in the countries they import from.

New Zealand exporters must also abide by the residue standards (MRLs) of the various countries it exports to. MAF maintains a MRL database of these pesticide residue standards established by New Zealand's trading partners.

Genetically modified foods and genetically modified organisms fall under the new HSNO Act. Under the new legislation, starting in 2000, two statutory bodies will assess the safety of genetically modified food. ERMA will assess the safety to people and the environment of all GMFs for use in research or for release into the environment (ERMA, 1997). The Australia New Zealand Food Authority

(ANZFA) is responsible for assessing the safety of all foods derived from gene technology before they are marketed. Both ERMA and ANZFA will use a case-by-case, risk-based assessment of safety, and both allow for public input into their decision-making processes.

An amendment to the New Zealand Food Standard 1996, which recognises changes to the Australian Food Standard Code (AFSC), took effect from 13 May 1999. The AFSC Standard A18 - *Foods Produced Using Gene Technology* - was recognised by New Zealand and declared a mandatory standard by this amendment (ANZFA, 1999). Standard A18 prohibits the sale of food produced using gene technology or using such foods as ingredients in other foods, unless the food has been listed in the standard and complies with any special conditions stated. The standard also specifies labelling requirements.

On 28 July 2000, the Australia New Zealand Food Standards Council agreed to new labelling rules for GM foods. Mandatory labelling will be required for all foods containing GMOs (i.e. zero thresholds). This came following a poll showing more than 90% of the population approved of such a move. Australia and New Zealand will therefore have one of the most rigorous and progressive labelling requirements for GM foods in the world. The labelling will be progressively introduced from September 2001 (ANZFA, 2000).

Currently the MAF is developing a proposal for a standard for oral nutritional compounds prescribed under the ACVM Act 1997 (MAF, 1998).

The proposed standard for oral nutritional compounds to be described under the ACVM Act focuses exclusively on the specific risks that can be managed under the Act. The requirements are designed to ensure that such compounds when used as directed do not:

- Purposefully contain or are contaminated with substances at levels that could produce residues in primary produce that violates either the New Zealand or specified overseas maximum residue limits (MRLs) for those substances.
- Result in toxic reactions (in the form of pain or distress) in animals.
- Result in malnutrition to the extent that an animal suffers unnecessary pain or distress.
- Contain micro-organisms (exotic or endemic) at pathogenic levels.

Given the relatively low risk or oral nutritional compounds in general, MAF considers that the risks specified in the ACVM Act can be managed adequately via a prescribed standard without requiring registration of such products. The proposed standard contains:

- General requirements regarding information provision, labelling, directions for use (all material used must contain no contaminants in excess of limits specified by MAF, comply with the provisions of the HSNO Act 1996, BA of 1993 and the Food Regulations 1994, plus other general requirements).
- Specific requirements, dealing with compound food/feeds (mainly cat/dog feeds).
- Food/Feed additives: to be approved by MAF in schedule of substances generally regarded as safe in oral nutritional compounds.

- Premix (approved ingredients, mixing instructions and danger warning about concentrated form).
- Dietary supplement, such that ingredients need to be fit for purpose. Statement about possible danger of supplements such as copper or selenium.
- Industry standards and codes of practice: compliance with these is required.
- Also a set of proposed standards for fertilisers.

Voluntary measure and approaches

Voluntary approaches have been part of the agricultural sector for quite some time, especially in for example the setting aside of land of significant natural or heritage value under the Queen Elizabeth II Trust. However, the start of farmer-led environmental groups can be traced back to the Rabbit and Land Management Program (RLMP), which ran from November 1989 through June 1995 (OECD, 1998). The RLMP was set up to deal with the complex economic, biophysical, social and institutional issues of the rabbit-prone tussock grassland area in the South Island high country.

Although the RLMP did not originate as a farmer initiative, *per se*, it was an important factor in spurring the formation of farmer groups, and is seen as the progenitor of community based action to deal with problems of sustainability in the rural area (OECD, 1998, p. 73). From this intitial partnership of central government, regional and district councils and landholders, sprung many voluntary groups, in the form of FARMER, landcare groups and the Rural Futures Trust, after the RLMP was concluded.

In 1993 the New Zealand Government released a position paper on sustainable agriculture as part of a wider policy on sustainable land management.

Sustainable agriculture is defined as the use of farming practices that maintain or improve the natural resource base, and any parts of the environment influenced by agriculture, are financially viable, and allow people and communities to provide for their economic and cultural well-being (MAF, 1993). The MAF has undertaken a facilitation programme designed to encourage the adoption of sustainable agricultural practices, and regional councils are also promoting sustainable agriculture as part of their responsibilities under the RMA. Meanwhile, grassroots landcare or community-based groups have been established in many parts of New Zealand.

The government announced in May 1996 assistance for the establishment of a Landcare Trust and the establishment of a national Sustainable Land Management Advisory Group. The Landcare Trust is intended to develop a network of trained, landcare or community group facilitators, and to provide support for small community-based projects. The Sustainable Land Management Advisory Group will help to co-ordinate information needs and delivery systems, encourage the adoption of better land management practices, and provide feedback to government on policy initiatives and associated land management issues.

Landcare groups are formed by land occupiers (often including other community stakeholders), which have a common interest in environmental issues relating to land and water resources and in developing practical approaches to address those issues.

There are several other industry-led initiatives on sustainable agriculture in New Zealand. Project FARMER (Farmer Analysis of Research, Management and Environmental Resources) has farmers helping to train other farmers in the use of computer-based decision support systems, including better monitoring and analysis of environmental and farm performance data. The project also has helped identify knowledge gaps and worked with researchers to fill them. The leaders of this project helped establish the Rural Future Trust as a vehicle for funding FARMER and other farmer-led sustainable land management initiatives. This project reflects the way many current and future issues pertaining to agriculture and the environment will be addressed in New Zealand (OECD, 1998).

The industry-owned Dairy Research Institute has funded a project in the Waikato region to address farming impacts on water quality. In co-operation with the regional council and other research institutes, dairy farmers have developed an operational definition of sustainable dairy farming, compiled a list of sustainable management practices, and designed subjective indicator scales to monitor their performance. The current phase of the project aims to convert the subjective scales into objective criteria.

The New Zealand Meat Research and Development Council (also industry-owned) uses 21 privately owned farms to monitor and demonstrate to sheep and beef farmers how farm business planning and monitoring can improve performance. The Council has incorporated environmental objectives into two of the monitor farms on a pilot basis, with some funding from regional councils and the MAF. These monitor farms have associated 'sustainable farming community groups' that discuss issues and possible management approaches, and follow the progress of the farm.

In 1997 ENZA (fruit brand as used by the New Zealand Apple and Pear Marketing Board) introduced the Integrated Fruit Production - Pipfruit programme (IFP-P). The programme's aim is to ensure that fruit is grown in an environmentally safe way with minimal use of pesticides. IFP-P bridges the gap between organic and conventional growing methods. Growers are required to monitor their crops for pests and diseases and, where possible, use non-chemical methods to control where problems occur. The programme prioritises fruit production methods that are safest for the environment and human health. Chemicals used in the programme aim to leave beneficial insects unharmed so as to naturally assist the grower with pest control. All ENZA - supplying growers will be IFP producers by 2000 (Statistics New Zealand, 1998).

Codes of practice have been developed by the pork industry, the logging industry, the kiwifruit industry, and by an agrochemical education trust initiated by leaders in the horticultural industry. The fertilizer industry has developed guidelines for responsible fertilizer use, and grazing guidelines are under development by leaders in the pastoral sector. In many of these cases, farmers are motivated not just by the desire to do the right thing or the possibility of regulatory pressure if problems are not addressed, but also by market considerations. Being self-produced, codes are usually better-targeted methods of encouraging self-change in management practices. Examples of some codes of practice are:

- Farmers in the Waikato and Hawke's Bay are developing Pastoral Codes of Practice that seek to reduce the adverse effects of pastoral farming on water quality.
- The NZ Fertilizer Manufacturers' Research Association has prepared a Code of Practice for the use and handling of fertiliser.
- There is an Agrichemical Users' Code of Practice.
- Others are the Dairy Farmers Code of Practice, the Pig Farming Code of Practice and the New Zealand Forest Code of Practice.

Farming leaders sense that consumers in New Zealand and in overseas markets are increasingly interested in how a product is produced, in addition to traditional quality concerns. They are therefore supporting efforts to establish systems to ensure that their production practices are sustainable and that this can be demonstrated to consumers. Thus, in a variety of ways, the environmental costs and benefits of sustainable agriculture are being internalised to the production process.

Regional and district councils currently try to find the right balance between encouraging voluntary approaches to environmental issues and sustainable land management (SLM) and regulation. The impetus of many sustainable land management groups was very much to avoid regulation and to bring about more sustainable land management on their own terms, rather than through the use of force. This has led in some regions and with some groups to excellent outcomes. However, with other groups little monitoring had been done and environmental benefits were hard to identify. In a recent survey of voluntary responses to achieve sustainable land management, the following conclusions were drawn (MAF, 1999):

- The survey shows that SLM initiatives will be limited in their effectiveness unless a clear link between environmental and economic benefits can be made.
- Environmental sustainability was not always the overriding aim of the group and it was found that land-holders outside the group where just as likely to undertake SLM practices as those within the group. However, the SLM groups are effective in fostering the adoption of practices to improve economic sustainability.
- Where monitoring and enforcement are impractical, the voluntary approach may still have a role to play in changing behaviour. The survey showed that financial incentives are the mechanism likely to be most favourably received by land users in the short-to-medium term.

There is little doubt that voluntary action among land-users (whether individually, through landcare groups or through industry-led initiatives) will continue to play an important role in New Zealand when it comes to dealing with issues such as water quality, biodiversity, soil erosion, animal welfare and human health. This is not to say that sustainable land management can be fully achieved by voluntary action. Rather as the MAF (1999) report concludes, a mix of voluntary, regulatory and incentive programmes is needed to achieve SLM goals.

IMPACT OF LEGISLATION ON AGRICULTURE

The new ethical and ecological dimensions in the government's strategy for Environment 2010 are reflected in such Acts as the RMA of 1991, the BA of 1993, the Fisheries Act of 1996, the 1993 amendment to the Forests Act, the HSNO Act 1996 and Animal Welfare Act 1999. Much has changed in New Zealand since the introduction of those acts and environmental as well as food safety and animal welfare have become integral part of decision-making.

Looming in the background is the issue of climate change or greenhouse gas emissions. With an agricultural industry dependent on energy inputs (although relatively a lot less than, say northern hemisphere agricultural producers), the New Zealand Government's response towards unilateral implementation of the Kyoto Protocols may impose new burdens on the agricultural industry.

When we consider the impact on the agricultural sector of the changing emphasis with regard to the environment and the objective to achieve sustainable farming systems, one has to say that the main impacts have already occurred. Aspects of farming that caused environmental problems and were not sustainable have been changed, either through regulations or through voluntary action. Examples here are the change in effluent disposal facilities on dairy farms (from ponds to discharge on land), land management on erodable lands by planting trees, the protection of some area of significant natural and heritage value, the management of riparian zones, and the implementation of Codes of Welfare.

Some have implied capital outlay (e.g. on dairy farms) and others have implied increases in variable costs. The whole regulatory process with its consents (for land, water, air and subdivision), inspections on farm and in processing has, under the user-pays philosophy of the RMA, imposed extra costs on farmers.

Changes with regard to animal health and food safety are currently all in the 'melting pot', with new legislation having only taken effect in 2000, and some of the details of the HSNO Act still being worked out. These latter legislative changes are on the whole not expected to bring great changes to the way things are currently being done. In some cases the legislation brings more flexibility to situations where previous outdated standards caused inefficiencies. However, there is concern about compliance costs. There is increasing concern (and some evidence) that the compliance costs under the HSNO Act and the ACVM Act for agricultural chemicals may be detrimental to the environment, to human and animal health and to trade outcomes (pers.comm. Burdon, 5 October 2000. MAF Policy). It is, however difficult at this stage to be definitive about any of this.

It is for the above reasons that it is very difficult to draw any conclusions or make predictions of the impact of recent legislation and future legislation on costs and prices in the agricultural sector. There is little research available in New Zealand that considers this issue. Some research did start in the year 2000 and will present some results shortly on food safety and environmental regulatory compliance costs facing farmer and the food processing industry. Some research findings on identifiable impacts, or directions of impact, of environmental and food safety regulations will be discussed below under the appropriate resource topic.

Water

The major area of importance is the concern about water quality, mainly caused by non-point discharges from dairy farms in the low-lying regions of New Zealand, such as the Waikato and Southland. The particular concerns are nitrate and nutrients.

The RMA 1991 sets the standards for water quality by controlling abstraction from and discharges into surface and groundwater. As a result, the trend nationwide has been to require dairy farms to cease discharging their effluent directly to waterways in favour of land-based effluent disposal systems. The legislation provides regional councils some flexibility in how they administer and enforce the Act, thus differences in consent costs and on-going monitoring costs are observed between regions. There is also variability across the regions in the percentage of dairy farms which are now discharging their farm dairy effluent to land, as for some regions the transition from water discharge to land discharge is slow. Nevertheless, approximately 70% of New Zealand dairy farms are now discharging effluent to land.

To estimate the cost imposed on the dairy sector of policy measures for water quality, the assumption has been made that all New Zealand dairy farms will utilize a land-based effluent disposal system. Depending on climate, soil type, and terrain, this will either be a traveling irrigator system or tanker-spread effluent from pond storage systems. Total net annual costs for all New Zealand dairy farmers to convert to a land-based effluent disposal system have been calculated as $39.4 - $67.8 million (these cost represents a one-off capital cost, annualized as a loan repayment for 15 years, the consent cost and the annual increase in variable costs minus any benefits from effluent as fertiliser). Therefore, with some 30% of farmers still having to comply (1998), further cost to the dairy sector of complying with environmental regulations is estimated to be between $11.8 and $20.3 million. The cost to the dairy farmer of these policy measures has been estimated as between 2.0% and 3.2% of the farmer's total costs (Cassells and Meister, 2000b).

Further analyses were conducted using a Global Trade Analysis Project (GTAP) model to look at the impact on the competitiveness of New Zealand's dairy industry in the world trade market. The model showed that if only New Zealand were to fully enforce compliance with water regulations, there would be a loss in international competitiveness for the New Zealand dairy exporting sector and a change in the pattern of dairy trade across the main dairy exporters. If, however, all four principal dairy exporters (Australia, EU, New Zealand and USA) enforce their water quality regulations and internalise their compliance costs, New Zealand will be better off. The dairy sector is predicted to expand, gaining global market share, and again there will be a realignment of international trading patterns (Cassells and Meister, 2000a).

Soil

The removal of subsidies from the agricultural sector has removed much pressure from some of the more fragile hill and high country and has helped to reduce soil erosion and dissolved nutrient and pesticides entering water bodies. Also the conversion of some of this fragile land to forestry has also helped. Soil erosion control will continue to remain an issue and farmers will be required to either adopt stricter soil conservation practices or to change the use of fragile land to forestry or let it revert. There is nothing new in the legislation that will impose new and additional costs on farmers in relation to this particular issue.

Air

Air quality is not a great issue in relation to the agricultural industry. The rules and regulations as found in the code of practices of the various industries that could create air quality problems have been exercised for a quite some time and the industry has adjusted where it needed to adjust. The only area of new legislation that will impose costs on some growers is the removal of methyl bromide by 2005. This will in the main affect the strawberry industry, and currently this issue is being researched (i.e. substitutes and costs) with a grant ($150,000) from the MFE's Sustainable Management Fund.

Nature conservation

It is hoped that the new Biodiversity Strategy will preserve a lot of remaining biodiversity on private farms. This is of some concern to farmers as in the past this was often achieved through legislation and rules that forced farmers to fence off sections of their farms without any compensation. It is expected that the Biodiversity Strategy, however, will achieve most conservation by voluntary means. There is still uncertainty about how this is going to be achieved and where some of the funds are going to come from, or:

> Regulation alone is not a preferred option to protect remnant natural areas on private land. Many landowners actively manage remnant habitats now and want to be acknowledged for, and assisted in, what they are doing. Landowners generally don't react positively to being told what to do on their land; therefore regulation is likely to be counterproductive and also risks losing many private 'conservators' across the country. Nor is it possible to monitor and enforce a regulation-based regime on the scale that would be necessary. Securing the willing and active participation of landowners is therefore pivotal to sustaining indigenous biodiversity on private land (DOC, 2000b, p. 41).

While most farmers in the agricultural sector are to some extent affected by all of this, most affected would be the sheep and beef farmers whose land ownership is very large and other farmers with significant riparian strips on their farms. Preservation of riparian strips has become an important issues in resources management both from a biodiversity and water quality point of view. The DOC would like to see most riparian strips fenced off. However, the RMA does not allow compulsory taking (only when land is subdivided) and so most of the fencing off will have to come through voluntary means. However, the President of Federated Farmers of New Zealand (pers. comm.) expressed real concern that more than voluntary measures may be used (as well as to set aside areas of natural bush) and that this may cost the farming community up to $30 million in terms of lost production and fencing costs.

Animal health

As indicated before, the whole animal health regulation is being revamped, and the new acts do gradually become operational. The new Act, however, will continue the changes put into place and will continue to use the existing codes of practice. Although changes are bound to occur when these codes of practices are reviewed, it is hard to predict where the changes will be and what the possible cost implications may be.

Human health

The same comment made for animal health holds here. Changes are everywhere but without clear cost implications. The Food and Beverage Industry complains about the high compliance costs of environmental and food and safety regulations, but there is little evidence to support this. In 1998 they commissioned a report on regulation affecting the food and beverage industry (New Zealand Business Round Table, 1998). This report indicated some high transactions costs. There was also some duplication, and conflict, between different agencies (MAF and the Ministry of Health) in the implementation of rules. Research is currently ongoing with a reporting date in 2001 which will document the excess cost burden that food safety regulations are imposing on the industry.

As explained earlier however, most of the regulations, standards and rules, derive from New Zealand's major trading partners. Hygiene, food safety, maximum residue levels etc. are all those that are acceptable to New Zealand's trading partners; hence even though they impose costs on the primary industry, they are simply costs of being able to do business.

Perhaps a telling newspaper clipping shows how New Zealand must live by overseas standards and how that affects farmers.

> Tough on-farm food product safety regulations being planned by the
> Agricultural Ministry could force dairy farmers to keep written records
> of each cow. ...the need for the tough measures was highlighted this
> week by the European dioxin scare in which food containing Belgium
> dairy products was removed from shelves in many countries. The

regulations will include a requirement for all water used in cleaning dairy sheds to be up to human drinking standards. It is likely that each animal will need to be given an individual identity to allow for food safety traceback in the event of contamination. ... European Community inspectors are due in New Zealand in a few months to check product monitoring systems. ...The Europeans have a standard which they use to audit us and we have to comply or demonstrate that we have got equivalent standards to them (The Dominion, 1999, p. 16).

GMO and GMF

There is much debate in New Zealand about this country becoming GMO- and GMF- free. This would fit with our 'clean and green' status. As explained in this chapter, New Zealand is only just now finalising legislation dealing with GMFs.

CONCLUSION

Most of the adjustments to environmental legislation have been completed by the agricultural sector in New Zealand. With changing societal values and attitudes, farmers will continually be required to change in line with a demand by society for higher environmental quality levels and more preserved natural environment. This will continue to place greater costs on farmers. However, as this chapter has indicated, most change will be achieved through voluntary means (although in areas guided by legislation). Compensation is something that was vigorously debated by the farming community when the RMA was reviewed in 1999. The decision has not been made, although the draft review report hardly mentions compensation. The proposed amendments to the Act call for a redefinition of the environment with more emphasis on biophysical and less on the social and economic consideration; it suggests a refinement of consents and consent processing with emphasis on timeframes, costs, appeals and categories; and it proposes a greater emphasis and urgency in the development of national policy statements and national environmental standards (MFE, 1998d). The select committee was to report back to Parliament on these proposals in October 2000. Hence it is uncertain which proposal will finally be accepted and when to form the new amended Act.

The changing philosophy by the New Zealand Government has also caused more of the costs of food production, meat slaughter, border inspection and weed and pest control to be paid by the farming community (a user-pays and beneficiary-pays philosophy).

The Biodiversity Strategy worries many in the agricultural sector, especially the demand of some environmental groups for the countryside to be returned to pre-European arrival flora and fauna. Although this probably will not occur, it does indicate the increased pressure on the farming community to preserve whatever native vegetation remains on their farms.

With increasing demand on the nation's waters, the cost of water will continue to increase, thereby imposing another cost on those requiring water for irrigation or other purposes.

The food industry is under continuing pressure to achieve standards set by the country's major trading partners and to satisfy the demand and desires of domestic consumers (especially with regard to GMOs and GMFs). This industry believes the compliance cost the government imposes on them is very high and that they are losing competitiveness because of it. This, however, has not been proven, and much of the evidence presented is anecdotal. However, the New Zealand Government has recognised that the current institutional set up for food safety and health does suffer from overlaps, duplication and high administrative costs. Some of this was highlighted in a report prepared for the Food and Beverage Industry (New Zealand Business Roundtable, 1998) and a newspaper article dealing with proposed restructuring of the MAF and its responsibility for food regulation and biosecurity and the Ministry of Health dealing with food safety. As Edlin (2000) wrote, '..officials unanimously favour rationalisation of food regulation under the administration of MAF. Such rationalisation would be welcomed by food processors keen to reduce the compliance costs they incur in dealing with both MAF and Health official' (p. 2). The restructuring however, did not take place and late 2000 no further decisions regarding reorganisation have been made.

Therefore, even though New Zealand by many measures is a 'clean and green' country, this 'clean and green' has been under stress and one of the main culprits has been (and still is) the agricultural industry. However, these pressures have been recognised and especially since the early 1990s the country has started to deal with them. The removal of all agricultural and other subsidies aided in bringing about a much more sustainable agriculture. By now, most of the major impacts of legislation and regulations have worked themselves through the cost structures of the farming industry. As indicated above, costs will continue to increase both due to environmental demand and food and safety demands, as the population continues to increase and the values and demands of society change. Predicting the impacts of those costs increases is difficult as they all will be small and gradual.

Acknowledgements

I am grateful to Dr Shamim Shakur and Sue Cassells for help in collecting material for this chapter.

REFERENCES

ANZFA (1999) Food produced using gene technology. *Food Standards*, April, edition 4. http://www.anzfa.gov.au/FoodStandard/Latest_GMO_Update.htm.
ANZFA (2000) Labelling genetically modified foods. *ANZFA Fact Sheet*, http://www.anzfa.gov.au/document/fso36.asp

Bayfield, M.A. and Meister, A.D. (1998) *East Coast Forestry Project Review.* Report to Ministry of Agriculture and Forestry, Wellington, New Zealand.

Biosecurity Act (1993) New Zealand Government, Wellington, New Zealand.

Burdon, B. (1999) Risk Management for Agricultural Compounds and Veterinary Medicines. Paper presented at the Australian Agricultural and Resource Economics Society Annual Conference, Christchurch, January 1999 (unpublished).

Cassells, S.M. and Meister, A.D. (2000a) *Cost and Trade Impacts of Environmental Regulations: Effluent Control and the New Zealand Dairy Sector.* Discussion Paper No. 8, Department of Applied and International Economics, Massey University, Palmerston North, New Zealand.

Cassells, S.M. and Meister, A.D. (2000b) *Cost of Land-Based Effluent Disposal to the New Zealand Dairy Sector.* Discussion Paper No. 10, Department of Applied and International Economics, Massey University, Palmerston North, New Zealand.

Close, M.E. (1996) Survey of pesticides in New Zealand groundwater 1994. *New Zealand Journal of Marine and Freshwater Research,* 30, 455-461.

Clough, P. and Hicks, D. (1992) *Soil Conservation and the Resource Management Act. Summary.* MAF Policy, Wellington, New Zealand.

Department of Conservation (DOC) (1998) *New Zealand's Biodiversity Strategy. Our Chance to Turn the Tide.* A Draft Strategy for public consultation. Department of Conservation and Ministry for the Environment, Wellington, New Zealand.

Department of Conservation (DOC) (2000a) *New Zealand Biodiversity Strategy,* DOC Science Publications, Wellington, New Zealand.

Department of Conservation (DOC) (2000b) *New Zealand Biodiversity Strategy* (February). http://www.doc.govt.nz/cons/biodiversity/pdfs/entire_doc.pdf

Edlin, B. (2000) New government reprieves boards and MAF bureaucrats. *The Independent,* 26 January, p. 2.

Environment Waikato (1998) *Waikato State of the Environment Report 1998.* Environment Waikato Regional Council, Hamilton, New Zealand.

ERMA (1997) *Consultation Document. A Proposed Methodology for the Consideration of Applications for Hazardous Substances and New Organisms under the HSNO Act 1996.* Environmental Risk Management Authority, Wellington.

Horizons.mw (1999) *Measures of Changing Landscape. State of the Environment Report Manawatu-Wanganui Region 1999.* Manawatu-Wanganui Regional Council, Palmerston North, New Zealand.

Johnson, R. (1992) Policies for the agricultural sector. In: Birks, S. and Chatterjee, S. (eds) *The New Zealand Economy: Issues and Policies,* Dunmore, Palmerston North, pp. 73-91.

Johnson, R.W.M., Schroder, W.R. and Taylor, N.W. (1989) Deregulation and the New Zealand agricultural sector: A review. *Review of Marketing and Agricultural Economics,* 57, 47-71.

Le Heron, R. and Pawson, E. (eds) (1996) *Changing Places. New Zealand in the Nineties.* Longman Paul, Auckland.

Lincoln Environmental (1997) Nitrogen Inputs at Land Surfaces and Groundwater Quality. Notes from a Christchurch Workshop on 15 August 1997, Report No 2776/3, Lincoln Venture Ltd, Lincoln College, Canterbury.

Meister, A.D. (1995) Environmental legislation in Europe. The European Union nitrate directive. Lessons for New Zealand. *Natural Resource Discussion Paper No. 18.* Dept of Applied and International Economics, Massey University, Palmerston North: 58pp.

Meister, A.D. and Gardiner, P.B. (1997) New Zealand's agriculture and environmental issues. In: Meister, A.D. and Rae, A.N. (Eds) *Environmental Constraints to Pacific Rim Agriculture: Further Evidence.* Centre for Applied Economics and Policy Studies, Massey University, Palmerston North, pp. 108-123.

Ministry of Agriculture and Fisheries (1993) *Sustainable Agriculture Policy Position Paper*. Ministry of Agriculture and Fisheries, Policy Paper 73/2, Wellington, New Zealand.

Ministry of Agriculture and Forestry (1997) New Zealand: The environmental effects of removing agricultural subsidies. In: OECD, *Environmental Benefits from Agriculture. Issues and Policies. The Helsinki Seminar. Country Case Studies*. OECD/GD (97)110, Paris.

Ministry of Agriculture and Forestry (1998) *The Role of On-Farm Quality Assurance and Environmental Management Systems (QA/EMS) in Achieving Sustainable Agriculture and Sustainable Land Management Outcomes*. MAF Policy Technical Paper 98/2, Policy Information Group, Ministry of Agriculture and Forestry, Wellington, New Zealand.

Ministry of Agriculture and Forestry (1999) *Evaluation of the Focus Farm and Orchard Programme*. Technical Paper 99/6, MAF Policy and Ministry for the Environment. Wellington, New Zealand.

Ministry of Agriculture and Forestry (2000a) *Agriculture and Forestry in New Zealand. An Overview*. Ministry of Agriculture and Forestry, Wellington, New Zealand.

Ministry of Agriculture and Forestry (2000b) *Situation and Outlook for New Zealand Agriculture and Forestry*. Information Bureau, Ministry of Agriculture and Forestry, Wellington, New Zealand.

Ministry for the Environment (1995) *Environment 2010 Strategy: A Statement of the Government's Strategy on the Environment*. Ministry for the Environment, Wellington, New Zealand.

Ministry for the Environment (1997a) *Reducing the Impacts of Agricultural Runoff on Water Quality. A Discussion of Policy Approaches*. Ministry for the Environment, Wellington, New Zealand.

Ministry for the Environment (1997b) *The State of New Zealand's Environment 1997*. Ministry for the Environment, Wellington, New Zealand.

Ministry for the Environment (1997c) *Environmental Performance Indicators: Proposals for Air, Fresh Water and Land*. Ministry for the Environment, Wellington, New Zealand.

Ministry for the Environment (1998a) *Flow Guidelines for Instream Values. Volumes A and B*. Ministry for the Environment, Wellington, New Zealand.

Ministry for the Environment (1998b) *Reporting on Persistent Organochlorines in New Zealand*. Ministry for the Environment, Wellington, New Zealand.

Ministry for the Environment (1998c) *Environmental Performance Indicators: Summary of proposed indicators for terrestrial and freshwater biodiversity*. Ministry for the Environment, Wellington, New Zealand.

Ministry for the Environment (1998d) *Proposals for Amendment to the Resource Management Act*. Ministry for the Environment, Wellington, New Zealand.

Ministry for the Environment (1999a) *A Guide to Preparing a Basic AEE*. Ministry for the Environment, Wellington, New Zealand.

Ministry for the Environment (1999b) *The Way Forward: Forestry and the RMA. Managing vegetation disturbance activities under the Resource Management Act*. Ministry for the Environment, Wellington, New Zealand.

Ministry for the Environment (2000a) *Bio-What? Addressing the Effects of Private Land Management on Indigenous Biodiversity*. Preliminary report to the Ministerial Advisory Committee, Ministry for the Environment, Wellington, New Zealand.

Ministry for the Environment (2000b) *We're making every drop count... and it's more than just a drop in the bucket*. Ministry for the Environment, Wellington, New Zealand.

Ministry for the Environment (2000c) *Guide to the voluntary moratorium* (June). Ministry for the Environment, Wellington, New Zealand.

Morgan Williams, J. and Gebbie, E. (2000) Aotearoa - A land of great sailors, but are we setting a sustainable course? Address at the Conference of the International Society for Ecological Economics, July, Canberra, Australia.

Morriss, S. and Workman, M. (1998) Regional council and unitary authority management of land disturbance and vegetation removal activities under the Resource Management Act 1991. Ministry for the Environment Working Paper (unpublished), Ministry for the Environment, Wellington, New Zealand.

New Zealand Business Round Table (1998) Regulation of the Food and Beverage Industry. Report prepared by Credit Suisse First Boston, for the New Zealand Business Round Table and the New Zealand Food and Beverage Exporters' Council, Wellington, New Zealand.

New Zealand Institute of Economic Research (NZIER) (2000) Is business green around the gills? *Update,* August, Wellington, New Zealand.

OECD (1996) *Environmental Performance Reviews. New Zealand.* Organisation for Economic Co-operation and Development, Paris.

OECD (1997) Agriculture and the Rural Economy - New Zealand Case Study: Agriculture Reform. Committee for Agriculture, Working Party on Agricultural Policies and Markets, Organisation for Economic Co-operation and Development, Paris.

OECD (1998) *Co-operative Approaches to Sustainable Agriculture.* Organisation for Economic Co-operation and Development, Paris.

Patterson, M. and Cole, A. (1999) *Assessing the Value of New Zealand's Biodiversity.* Occasional Paper No. 1, School of Resource and Environmental Planning, Massey University, Palmerston North, New Zealand.

Pomeroy, A. (1997) Impacts of recent economic reforms on rural communities. In: *Proceeding, New Zealand Agricultural and Resource Economics Society Conference.* 4-5 July, 1997, Blenheim. Agribusiness and Economics Research Unit, Discussion Paper 145, Lincoln, Canterbury.

Press release (2000) New Research Shows New Zealand Farmers Clearly Favour Organic over GM, http://www.scoop.co.nz/archive/scoop/stories/ad/fc/200009121117.833 fla39.html

Resource Management Act 1991 (1994) (reprinted). New Zealand Government, Wellington, New Zealand.

Sandrey, R.A. (1991) Economic reforms and New Zealand's agriculture. *Choices,* First Quarter, 17-19.

Sandrey, R. and Reynolds, R. (1990) *Farming without Subsidies. New Zealand's recent experience.* A MAF Policy Services Project. GP Book, Wellington, New Zealand.

Smith, C.M., Wilcock, R.J., Vant, W.N., Smith D.G. and Cooper, A.B. (1993) *Towards Sustainable Agriculture: Freshwater Quality in New Zealand and the Influence of Agriculture.* MAF Policy Technical Paper 93/10, Ministry of Agriculture and Fisheries, Wellington, New Zealand.

Statistics New Zealand (1998) *New Zealand Official Yearbook 1998.* Statistics New Zealand, GP Publication, Wellington, New Zealand.

Taranaki Regional Council (1998a) Proposed Regional Fresh Water Plan for Taranaki. Taranaki Regional Council, Stratford, New Zealand.

Taranaki Regional Council (1998b) Regional Air Quality Plan for Taranaki. Taranaki Regional Council, Stratford, New Zealand.

The Dominion (1999) Dairy farmers face new regulations. 11 June, p. 16.

The Economist (2000) Can the Kiwi economy fly? 2 December, p. 83.

Walker, A. and Bell, B. (1994) Aspects of New Zealand's experience in agricultural reform since 1984. Ministry of Agriculture and Fisheries, Wellington, New Zealand, MAF Policy Technical Paper 94/5.

Environmental and Human-health Standards Influencing Competitiveness

Floor Brouwer and David E. Ervin

INTRODUCTION

This chapter is a synthesis, drawing on five previous chapters, of the principal environmental and human health related standards applying to agriculture in the European Union (EU), United States of America (USA), Canada, Australia and New Zealand respectively (see also Brouwer *et al.*, 2000 for a strategic comparison of environmental and health-related standards influencing the relative competitiveness of EU agriculture vis-à-vis main competitors in the world market). It provides a first comparison of the main environmental and health issues causing concern in the five countries, the policy measures and standards that have been put in place and some of the implications at the farm level. On this basis a preliminary assessment is made of the implications of differences in standards for the relative competitiveness of EU agriculture vis-á-vis the other four countries, which are competitors on the world market.

The chapter first seeks to identify where there are clear differences in standards between the countries, at the farm level (i.e. direct, operational constraints upon farming practices). These differences need to be evaluated in the context of the different environmental and health needs of the countries, to determine where differences mainly reflect specific natural resource, environmental and socio-cultural settings or where they appear likely to distort competitiveness[1]. The essential question is whether input intensity and other aspects of agricultural production practices combined with competing non-agricultural demands for resources has created different levels of intensity of

---

[1] The most common understanding of 'distortion' appears to be that prices for market goods and services do not reflect properly true scarcity values or - in the case of public goods - social costs. Both the scarcity and public goods values depend on preferences and site-specific conditions (vulnerability of sites, differences in pressures). We are referring to those regulatory or support situations where the public agri-environmental action either imposes controls or offers support such that the marginal social costs of environmental protection do not equal marginal social benefits.

environmental problems attributable to agriculture. Without this crucial contextual analysis, which is difficult to provide for individual measures in practice, it is incorrect to draw conclusions about trade distortion from information about standards. Within this perspective the chapter proceeds to a preliminary assessment of the implications of these differences in standards for the relative competitiveness of EU agriculture, vis-à-vis its main competitors on the world market. This is done by relating operational constraints to potential costs, with special reference to selected importing and exporting sectors of agricultural production.

Methodology used and difficulties faced

Comparison of standards versus comparison of on-farm constraints

A broad comparison of environmental and human health standards, as expressed in national and regional regulations or policies, was not sufficient for exploring the economic implications of the policies, since different countries have approached issues of concern through a wide variety of measures. For example, in the EU there are many policy objectives for water quality, which can affect farm-building standards with regard to siting, design, scale, waste management etc. In Canada, farm buildings for livestock are critically affected by spatial planning mechanisms designed mainly to control nuisance, especially odour, which determine minimum separation distances between installations. These rules will certainly have impacts upon water quality, but they are not regarded as part of water quality policy *per se*. Thus when comparing apparent constraints at farm level, the approach needs to be applied with a sensitivity to the fact that environmental measures can have multiple effects beyond the problems that prompted action. So in the Canadian context, farmers meeting the separation distance standards generate joint effects - reduced air quality conflicts, lower stocking rates and less risk of groundwater contamination. It was therefore decided that in principle, the most meaningful scale at which to determine and describe standards in such a way that these could be compared between countries, was to attempt to identify the level of on-farm operational constraints - i.e. the direct effects of standards on producers. By trying to translate standards into the constraints imposed on farmers it was hoped to obtain a common framework for comparison between countries.

Selection of key environmental and health-related issues

A standard set of key issues in environment and human-health policy was selected. While there were differences in the main issues related to agriculture in each country and the weighting given to specific concerns varied considerably, most of the leading issues were similar. The key issues were grouped under seven main themes as follows:

- water quantity and quality (nutrient enrichment by nitrates and phosphates, water born sediments, pesticides, and issues connected with irrigation);

- soil related issues (including soil erosion, contamination and salinisation);
- air quality (including odour, ammonia emissions, pesticide drift, crop burning and noise);
- nature conservation, biodiversity and landscape protection and management (including endangered species, habitat conservation and protected landscapes);
- genetically modified organisms (GMOs);
- farm animal welfare (including housing for livestock, transport conditions and slaughter);
- human-health issues (the use of hormones in livestock rearing, the ingredients used in animal feed, pesticide residues in food, safety of those applying agrochemicals, hygiene rules for dairy farming and veterinary requirements).

Under these seven main headings a standard set of 18 more specific issues was developed and used.

A mix of national, federal, regional and local measures

Environmental and health standards are commonly set through a mix of national, federal, regional and local measures, in each country. The research work revealed that a significant proportion of the standards imposed on farmers in the countries studied derives from local and regional measures, rather than national legislation. Within the EU, there is a range of different levels at which standards are developed, stretching from the Community itself to the municipality level, resulting in significant differences within regions and countries, as well as between Member States. In the United States, there is a body of federal legislation applying to the agricultural sector but a considerable proportion of the standards considered are developed by individual states or authorities at a more local level. The pattern is similar in Canada and local legislation is also important in both Australia and New Zealand.

Furthermore, standards do not always apply to all producers within a particular administrative region. For example, some regulation applies only to larger producers, for instance those with more than a given number of pigs or poultry. Other standards apply solely to sensitive areas, which may be a relatively small proportion of the administrative region. In certain cases, a county level authority applies standards in a discretionary way. It may be the body responsible for issuing consents for new developments.

The form of standards and legislation affecting agriculture

The previous chapters have indicated the wide variation in the nature and status of the policy instruments, which impose constraints of different kinds on individual farmers, in the countries studied. In addition to the various forms of legislation and regulations, there are codes of practice and guidelines, which are frequently not binding but may be widely adopted, as well as entirely voluntary schemes. In the domestic context, the main categories include:

- Legislation and regulations imposing standards *directly* on farms, for example minimum standards for hygiene, animal welfare, the disposal of pesticides etc.
- Legislation and regulations affecting the *availability* of certain products to the producer, such as pesticides, which will have cost implications.
- Legislation and regulations, which impose obligations on farmers by affecting their practices *indirectly* (for example, minimum standards for water quality which can be respected only by adhering to a limited range of farming activities).
- Legislation establishing *procedures* such as controls on land use and the construction of buildings, consent procedures for removing landscape features, etc. The operation of these procedures imposes direct constraints upon producers but to a certain extent, each case is treated individually, making overall assessment of impacts very difficult.
- *Codes of practice*, which may be entirely voluntary (e.g. organic production), quasi legalistic or, in a few cases, binding. Such codes may not be mandatory in themselves but failure to comply with them may expose a producer to prosecution if pollution occurs, for example.
- *Cross-compliance* measures that apply only to those producers receiving benefits under a public programme. Penalties for environmental infringements may be introduced through a reduction of direct payments.
- *Voluntary standards initiated by public agencies* and promoted widely to producers.
- *Voluntary standards developed by processors, retailers or other downstream markets*, which may affect a large proportion of producers and in some cases be 'quasi-obligatory', given market structures.

The existence of legislation and regulations promulgating standards needs to be interpreted in light of enforcement of those standards. Standards can be enforced in various ways, including peer pressure, civil litigation and criminal prosecution.

In addition to the main categories of standards and legislation in the domestic context, in an international context there are also examples of binding standards imposed by governments in importing countries which must be complied with in order to enter their markets, therefore overriding any domestic standards for certain producers. This is a particular consideration for major exporting countries targeting high-value markets, such as Australia and New Zealand. Importing country constraints are commonly binding for specific agricultural commodities. For example, domestic slaughtering and processing standards for sheepmeat in New Zealand might be lower than domestic production standards in the EU. In that case, New Zealand's lamb and mutton for export is already produced to EU standards, which makes it difficult to identify any clear effect on competition. Indeed, this strategy is one of assuring access to the EU market and enabling competition by New Zealand.

The effects of standards on competitiveness

Regulation that internalises external social costs, which may imply creating differences in the cost of primary production between areas that have external cost problems and those that do not, influence trade (relative to a situation where external costs are not internalised) but do not distort it. In theory, environmental and health legislation can influence international competition in agricultural products in three different ways (Bredahl *et al.*, 1998):

- Creating differences in the cost of primary production in different countries (e.g. due to requirements for housing, hygiene, input bans on restrictions and price effects such as taxes, etc). Such differences also include a forced shift in technology and constraints in production levels through upper limits on stocking density.
- Affecting the cost and timeliness of delivery for different categories of producer (e.g., border inspections for phytosanitary purposes, which affect importers).
- Altering the ability of products to meet consumer demand attributes (e.g. enabling producers to provide improved food safety characteristics because of tighter standards on carcinogenic pesticide residues). Meeting the consumer demand attributes also includes altering the ability of products to meet importing country requirements (e.g. banning the use of methyl bromide for fumigation of fruits, which may have been used in order to meet importing country restrictions on alien pests).

This chapter concentrates on the first category, i.e. the economic implications of compliance with on-farm constraints arising from environmental or human-health related standards. However, all three categories may be relevant to trade discussions.

COMPARISON OF ENVIRONMENTAL AND HEALTH-RELATED ISSUES

The development of environmental and human health-related policies for agriculture in the EU and non-EU countries has generally progressed through stages. To begin with, measures tended to concentrate upon waste-treatment processes and end-of-pipe measures, treating individual environmental and health problems where they arose, rather than seeking to prevent them. However, there has been a gradual evolution towards policies setting minimum environmental, health and welfare performance standards (e.g. maximum concentrations of pollutants, often expressed as parts per million) and taking more preventive measures to reduce pollution or enhance positive outcomes at source. Finally, there has been growing interest over the past decade or so, in market-linked mechanisms for promoting positive environmental action, either through public intervention to create markets for environmental services, or through labelling and other commercially-driven schemes linking environmental, health and welfare attributes in sales and marketing promotions.

This section identifies main differences regarding environmental and health-related issues and basic features of legislation in the five countries. A comparison of issues across countries is an essential part of understanding the context of any constraints imposed upon farmers. Unless this is done, it is possible that the findings of the analysis might be misinterpreted. For example, if a country has no policies to restrict nitrogen use this might be completely appropriate, if:

- Current nitrogen application rates are low.
- There is no evidence of water contamination or other pollution by nitrates from agricultural sources.
- The public in that country is willing to accept a certain level of pollution that another will not.

Such contextual information is relevant in trying to answer the key questions of whether environmental and health related standards imposed on agriculture in the EU are higher or lower than in those of the other countries under consideration? Standards may appear to distort competitiveness in case relative case prices do not reflect properly scarcity values, also including social costs. However, higher standards in one country that are imposed to combat more serious externality problems that exist in that country relative to a trading partner do not necessarily distort competition.

Water quality and quantity

Nutrient enrichment by nitrates and phosphates

Compared to non-EU countries, eutrophication in the EU is more of a widespread concern in relation to agriculture. It is an issue at the local level in the non-EU countries. Nutrient contamination of water by agriculture is a high priority issue for the EU and the USA, and both have substantive policy programmes in place. The problem of nitrogen leaching is very much related to the intensity of agriculture, as well as climate, geography and water demand, and supply characteristics. Nitrate levels in drinking water in the EU are still above the EU limit in a significant number of areas, which is partly due to the intensity of the operations (i.e. high rates of fertiliser application and high livestock stocking densities) and the density of population. Similar policy objectives exist in all the countries in relation to nitrate levels in drinking water, based on WHO standards. The EU tends to be more interventionist than elsewhere, in relation to policies specifically for agriculture, affecting drinking water quality, and has established a rather comprehensive Nitrates Directive. In the USA regulation is used selectively by the federal government and most states; mostly for large point source emitters. Although this is now changing, the medium and small size farmers have been largely exempt from water pollution regulation up to now.

Sediments in water

Sediment loading of fresh water is a major issue in several countries, but of more limited regional concern in the EU. The reasons appear to be related at least partly to differences in the environmental context of agriculture, but there are also cultural factors involved as well. Large reservoirs of relatively marginal land in the USA and Australia have historically tended to move in or out of production in response to economic conditions. Many of these soils are relatively prone to erosion, with sediments potentially contaminating watercourses on a large scale. In the EU, the nature of this problem differs greatly between regions. There are undoubtedly serious issues in some areas, particularly in southern Member States and in the East German Länder. However, the issue has received relatively little policy attention in Europe, up to now. By contrast, in the USA, soil erosion, sediment and related problems have been the dominant conservation and environmental issue for agriculture since the 1930s and remain serious problems in many regions.

Pesticides

Water quality concerns related to the use of pesticides are closely connected to other issues involving pesticides, including pesticide drift, affecting air quality and biodiversity, and human health concerns arising from pesticide uses and residues in food. These issues are treated together, because they are all related to the usage of pesticides in agriculture.

   Water quality problems tend to cause some serious concern in each of the countries analysed, although much less in New Zealand than elsewhere. Residues of pesticides in drinking water potentially are a major human health issue in the EU and USA. There are similar but less pronounced concerns in Canada and fewer in New Zealand, while in Australia the primary issues are associated with aerial spraying and surface water contamination. Although they vary in intensity, there are a large number of other pesticides-related concerns in all five countries, including spray drift, and direct impacts on human health and biodiversity. It is therefore unsurprising to find that all five tend to monitor and report on a similar range of issues in relation to the impacts of pesticide use and related standards.

   Most countries have concerns regarding the human health aspects related to the level of pesticide residues and other contaminants in food. Standards are introduced to ensure that harvested and livestock products are safe for human consumption. This is an important issue for all the countries. It has a high priority in the USA, although it is difficult to compare with the EU, where the organic sector is now growing rapidly. Legislation is widespread but it is an area where producer protocols and quality assurance schemes are widely applied in order to reduce the risks of food products exceeding permitted residue levels.

Irrigation

Irrigation has emerged as an environmental issue of growing importance in recent years in a significant number of countries. Excessive withdrawal of groundwater, depleting resources and adversely affecting ecosystems is a widespread concern but there is a range of other issues, including salinisation, increased use of agrochemicals, enhanced pollution risks, losing biodiversity, etc. The issues are clearly linked to the practice of irrigation however diverse the local impacts, as for pesticides.

Irrigation accounts for a major share of water use in most countries and excessive groundwater extraction levels cause concern in several regions in the EU, Australia and the USA. Limitations on the availability of water are a major concern in Australia, and to a lesser extent also in New Zealand, while the issue is not of major concern in Canada. Institutional management systems for irrigation water commonly only consider the costs of access and delivery, but not the opportunity costs of water, which generally include diminished supplies for other users, such as recreation, fishing or urban settlements. This can divert water to agriculture at the expense of other more highly valued water uses, encouraging more use of irrigation water than would otherwise be the case.

Soil quality and soil erosion

Salinisation

Salinisation is mainly an issue in the drier regions of the countries studied, usually associated with irrigation, as discussed in the previous section. It is a major problem with high priority for policy in Australia. The information available indicates more local salinisation problems in the EU, USA and Canada. It is not an issue of concern for agriculture in New Zealand.

Soil contamination and acidification

Soil contamination by heavy metals is at least of some local concern in the five countries. Soil contamination is mainly concerned with heavy metals in the soils. The application of sewage sludge is an important source of heavy metals in the EU and the USA. Cadmium in phosphate fertilisers is an issue in Canada and Australia. Soil acidification is an issue of concern in the EU, Australia and New Zealand, occurring on a large scale in Australia.

Soil erosion

The scale and severity of erosion varies not only with the type of agriculture, but also to local climatic conditions and soil properties. Soil erosion by both wind and water has been a significant policy concern in the USA for many decades, primarily because of off-farm effects, and similar issues are faced in the arable plains of the Canadian wheat belt. Water related erosion problems are of major concern with high priority in policy in Australia and New Zealand. By contrast, these issues are increasingly recognised as a concern in some regions of the EU but as yet controls in most areas are relatively weak.

Air quality

Odour and noise (nuisance)

Air quality problems related to noise and odour from livestock mainly are an issue of concern for production systems near urban settlements and may attract considerable public attention in certain localities. Local governments are usually responsible for the policy response. Less populated regions of the world enjoy a comparative advantage. It is identified as a problem at regional and municipal level in the EU, USA and Canada, but not in Australia and New Zealand.

Ammonia

Emissions of ammonia are more of a concern in the EU than elsewhere, although only in specific regions where intensive livestock production is concentrated.

Crop burning

Air quality problems related to the burning of crop residues have been a general problem in parts of the EU, USA and Canada. Due to legislation in the USA, Canada and several Member States in the EU that controls the damages, it is not of significant widespread concern in any of the countries examined.

Nature conservation, biodiversity and landscape

The loss of biodiversity is a sensitive on-farm issue in the EU, which differs from the other countries because of the small proportion of natural habitat and dependence of many species on land managed for agriculture. In North America, by contrast wilderness preservation is the predominant concern. Farming systems in Europe are important for landscape maintenance and voluntary incentives are offered to fund specific land management practices needed for landscape, biodiversity and general environmental objectives. Farmers often live in cultural landscapes, which require management and protection, thus limiting the acceptability of certain production systems promoting agricultural intensification and priorities to remove landscape features. The combination of biodiversity and landscape is a major issue with high priority for policy in most of the EU, and on a lesser scale to Australia and New Zealand. Biodiversity rather than landscape is of significant concern in the USA, with regional impacts, but not an issue of major concern in Canada.

Genetically modified organisms (GMOs)

This is a new and fast moving issue of major debate. The technology and its application is most advanced in the USA and Canada, with lower uptake in Australia, little in the EU and almost nil in New Zealand. The reasons for differences appear to be related more to consumer and environmental attitudes and awareness of potential risks than to biophysical resource differences between the countries.

Animal welfare

The main animal welfare concerns relate to the treatment of farm animals during housing, transport and slaughter. Legislation and other measures typically aim to ensure that the treatment of animals meets certain standards of animal husbandry, including freedom from unnecessary suffering or abuse.

Animal welfare issues related to housing, transport and slaughter are a major public concern with high priority for policy in the EU. For example, battery cage size for laying hens is a perpetual issue. There is no broad public concern that animals are generally abused in agriculture in the USA and Canada, but certain specific issues are of concern in Australia and New Zealand.

Human-health

Hormones and animal feed ingredients

The use of hormones, antibiotics and various animal feed ingredients are of significant concern to European consumers because of the potential risks to human-health. They are an issue, but not of major concern, to consumers in Canada, Australia and New Zealand and appears not to be an issue of concern in the USA at present.

Hygiene rules for dairy farming

Hygiene rules apply to a wide range of farm production methods and sectors, particularly cattle, pigs, poultry and other livestock. Hygiene rules applying to pigmeat, however are largely off-farm and therefore mainly apply to the processing stage. In the dairy sector, the relationship with the final product is often closer. The rules were chosen for comparison in this chapter, because of the relevance of these rules in the context of on-farm constraints. Wide similarities are observed across countries on the issues related to human-health and hygiene conditions associated with dairy farming. As well as constraints, public policies can give rise to incentives or other assistance for the farm sector, including training and advice.

Veterinary requirements and conditions to control animal diseases

All countries have policies and procedures for disease control as exemplified below. However, the degree of public concern varies greatly. Veterinary requirements and conditions to control animal diseases are issues of concern in all the countries examined, partly because of the high costs involved following an outbreak of animal disease (e.g. the foot-and-mouth disease with more than 1,500 outbreaks in the United Kingdom only). Human health concerns related to farm-level livestock health issues (e.g. disease control) are strongest in the EU, which appears partly the result of recent experience and sustained media attention.

Leading environmental and health issues

Table 9.1 highlights the issues selected across the five countries. It identifies leading environment and health issues, on the basis of the available evidence and the views reflected in the national reports. Some evidence is clearly documented but there is also an important element of judgement. For example, there is widespread data indicating contamination of groundwater by pesticides in certain regions. The scale of public concern, reflected in the media, parliamentary debates etc. is more difficult to compare across countries.

A simple ordinal scoring system is applied in the table. It is not meant to provide an overall judgement regarding the state of the environment of any of the countries involved. In countries where issues are not perceived to be of major public concern, this would not imply that there is not a problem, for example animal welfare is less of a public policy issue in the USA than in Europe and other countries. Mistreatment of farm animals housed and transported may occur in the USA, but broad sustained public interest remains limited. Where an issue appears both to be an issue of significant concern and there has been substantive response in the form of public policies or initiatives in the private domain, three stars are assigned rather than two.

For a variety of historical, geographical and economic reasons, the economic development of agriculture has followed different paths in many of the countries considered in this chapter. These development paths have produced different types and degrees of environmental stress, human-health and animal welfare concerns. In the EU a sizeable proportion of production originates from intensive farming systems but average farm size is small. In the other countries larger farms predominate including sizeable areas of low input arable or pastureland. However, they also contain significant intensive farming sectors, including large-scale livestock. Most industrialised countries have acted to internalise the external effects of agricultural production on natural and environmental resources to some degree, but they have acted in diverse ways.

The EU, the USA and Canada have a general framework of environmental and health legislation at the federal or country level, with principles implemented at the state, regional or even municipal level. In many cases it is complemented with private sector actions and initiatives. The instruments applied and the operational constraints are commonly defined at sub-national level. In Australia, most powers to make environmental legislation rest with the states and decisions mainly occur at local level. A national framework for environmental planning exists in New Zealand. Standards can be imposed at national level, but in practice there are national guidelines and standards are determined at regional and local level. Australia and New Zealand are advanced along the private sector route, with extensive use of voluntary guidelines. Such guidelines put constraints on farming, because failure to comply may limit sales to downstream markets, even though they have a different legal status from regulations.

Table 9.1   Issues of concern in the five countries

Issue	EU	USA	Canada	Australia	New Zealand
Nutrient enrichment by nitrates and phosphates	***	***	*	*	*
Sediments	*	***	*	*	**
Pesticides (including drift and applicator safety)	***	***	*	**	*
Irrigation	**	**	*	***	**
Salinisation	*	*	*	***	-
Soil contamination	*	*	*	*	*
Soil erosion	**	***	*	***	***
Ammonia	**	-	*	-	-
Odour and nuisance	***	**	***	-	-
Crop burning	*	*	-	-	-
Biodiversity, landscape	***	**	*	***	***
GMOs	***	*	*	**	***
Animal welfare	***	*	*	*	*
Hormones (and animal feed ingredients)	**	-	*	*	*
Pesticide residues in food	**	***	**	**	**
Hygiene rules for dairy farming	**	**	**	*	*
Veterinary and animal diseases	**	*	*	*	*

-    No issue;
*   Issue identified as a problem, but not of major concern;
**   Issue identified as a problem, and of significant concern;
***   Major issue with high priority in policy.

The great diversity of policy instruments used within countries implies that the on-farm constraints arising from a web of different measures will likely show a wide range as well. Regulatory approaches, as found more in the EU, USA and Canada, in principle indicate more rigid, binding standards but in practice this will not always apply at the farm level, especially where implementation and enforcement are weak.

Most policies have been reactive in the sense that measures were introduced in response to perceptions of emerging or documented problems. Differences in culture and preferences account for some modest variation in the types or levels of remedial actions undertaken. Most countries employ a combination of technical and information assistance programmes, voluntary programmes, cost sharing and producer compensation as well as criminal and civil sanctions. New Zealand (and to some extent Australia) stand out in that the cost-share programmes are very few, have limited funds and are primarily to encourage voluntary actions.

In all five countries regulation plays an important role in policies for pollution control (pesticides, water pollution, hygiene, and food quality). A far more diverse picture is observed in relation to policies for soil, nature conservation, landscape and issues related to recent public concerns about modern technologies (e.g. GMOs, antibiotics in animal feed, and hormones). With some exceptions, such as soil erosion control, the pattern of intervention appears heavier in the EU than elsewhere.

Legislation and command-and-control measures are relatively widely used in the EU to protect physical resources, as well as to advance human-health and animal welfare objectives. Regulation is currently used selectively in the USA to control water pollution and permissible pesticide compounds, and, to a lesser extent, air pollution by the federal government and in most states. Regulation is mostly for large point source emitters. The medium and small farmers are largely exempt from water pollution regulations, but this is changing. Codes of Practice for protecting resources are rather common in Canada, Australia and New Zealand. However, often a producer can be sued if he or she is not in compliance with the relevant Codes. This makes the Codes 'binding' in a way not dissimilar to regulations.

Public concerns about the potential animal welfare issues surrounding modern production systems are rather widespread in the EU. By contrast, animal welfare is a less prominent issue in the other countries examined, and there is no evidence of broad public concern that animals are being abused in agriculture. Although transport distances for animals are lengthy in Australia, for example, there appears no major public concern about stress or mistreatment of animals during transport and slaughter.

COMPARISON OF ENVIRONMENTAL AND HEALTH-RELATED STANDARDS

This section identifies the major differences regarding environmental and health-related standards imposed on agriculture in the EU and the other four countries. In so far as differences are apparent, an attempt is made to establish whether the EU standards are higher or lower than in those of the other countries under consideration.

The approach is thematic. Where possible, farm level operational constraints are identified. Comparisons can then be made with due caution and in light of existing environmental problems, animal welfare and human health issues in each country.

Several significant differences between countries were described in the previous section. These include differences in the policy instruments adopted, the degree of reliance on regulatory or voluntary measures and the extent to which issues are addressed at national, regional or local scales. Many policies vary significantly between regions and they vary in scope - with some applying only in highly specific areas. Effectively, there is a high degree of spatial variation in on-farm constraints within the countries analysed, such that some policies are implemented by a single national prescription while others are determined through myriad local rules. This is due to the extremely variable pattern of legislation and standard setting, much of which finds expression as a constraint on farming at the local level where the cumulative expression of local, regional, national and higher level standards apply. Therefore, the level of detail presented regarding on-farm constraints unavoidably differs between the issues discussed, and the national level of analysis can be an obstacle to the establishment of clear comparisons between farm level constraints among the countries under consideration.

Water quality and quantity

Nutrient enrichment by nitrates and phosphates

A wide range of on-farm constraints applies to farming in the five countries to meet the requirements of policies to control nutrient enrichment by nitrates and phosphates and associated pollution from livestock wastes and fertilisers. Many of these are multi-purpose.

The application of manure to agricultural land is one of the most important and direct ways of creating potential pollution problems from nutrients. Limits on the application of nutrients are comparable in the EU, USA and New Zealand. Maximum limits on the application of nutrient inputs exist in the EU (170 kg of nitrogen from livestock manure), USA and New Zealand. They are implemented by several Member States (EU) and at state level (USA), and are introduced in light of the high incidence of excess nutrient problems in much of the north-western part of Europe and some states in the USA. New Zealand has limits in place at regional level, which range from 150 to 200 kg N ha^{-1}, and voluntary measures also control the application of mineral fertilisers. No limits are in place on the amount of nutrients applied in Canada and Australia.

Large new livestock production units are controlled through permitting systems in all the countries examined but the control of smaller units is generally determined at the sub-national level or by the Member States in the EU. Controls on units below a certain size threshold appear limited in some countries. In some Member States permits do not allow for an increase of production capacity of intensive livestock production units with high stocking rates. This constraint is not observed in the other countries studied. Most permits are multi-purpose designed to restrict negative impacts on air and water policy, odour, noise etc. Separate building regulations may also apply. There are specific regulations for larger units (CAFOs, confined animal feeding operation) in the USA and IPPC Directive (Integrated Pollution Prevention and Control) in the EU. The IPPC threshold is lower than that for CAFO. In both cases there is considerable scope for authorities at the regional/state level to set specific constraints on production. A less regulatory approach is evident in Canada, Australia and New Zealand where constraints apply through a variety of other mechanisms including codes of practice, regional district plans in New Zealand and guidelines for beef cattle feedlots in Australia.

Nutrient budgets need to be prepared by certain groups of farmers in the EU, USA, Canada and Australia. Nutrient budgets are however, not a uniform requirement in all these countries. They are required by large production units for rearing pigs and poultry in some parts of the EU and the USA, and the efforts required for providing such budgets are rather similar in these countries. In Canada and Australia, the provision of nutrient budgets is only required for starting (or enlarging) a livestock operation; it is not required in New Zealand.

Manure storage requirements exist in all the countries studied, often associated with permits for livestock farms mainly in regions with a concentration of intensive livestock production (to control nuisance from odour and meet restrictions on manure application during part of the year). The constraints they put on agriculture are more stringent in some parts of the EU, relative to the other countries under consideration.

The use of buffer strips to control pollution runoff is rather common in the countries studied. The size of such strips tends to be limited to a few metres in the EU, which appears smaller than equivalents in the other countries studied. Obligatory buffer strips are relatively unused and it is more common for them to be combined with cost-sharing programmes in the EU, USA and Canada.

Land use planning rules are common in the countries studied, but they are mainly applied at sub-national level. Separation distances are applied to livestock production units close to watercourses and residences in the USA, Canada and Australia. Constraints on farming and the cost implications may be largest in regions with intensive livestock production units, with high population densities or close to urban settlements. On-farm constraints of this type tend to be stricter in the EU relative to the other countries studied.

Charges and taxes on the use of fertilisers imposed on agriculture currently do not exist in the EU, although some countries have introduced levies to control excess of manure and there is a fertiliser tax in Sweden. Elsewhere, taxes on fertilisers only exist in the USA. Such charges affect farm resource allocation to a limited extent only. Levies on surplus manure only exist in some areas with intensive livestock production units in the EU. They do not exist outside the EU.

Pesticides

The authorisation procedure applied may reduce the number of active ingredients available for agricultural use. The number of authorised active ingredients in the EU and the USA shows a downward trend, mainly due to reregistration procedures. The constraints that such authorisation procedures put on farming remains uncertain, but the impact is potentially big. The number of active ingredients is increasing in Australia and New Zealand and the compounds are relatively expensive; the number also tends to increase in Canada. While the number may be growing in Australia, many others are also being banned, deregistered or highly restricted.

The usage of pesticides faces a wide range of restrictions across the countries under consideration. They range from applying Good Farming Practices to strict control and prohibition of certain pesticides in specific zones. Rules on safety and on-farm constraints on the use of pesticides read very much the same and there is not very much difference across the countries examined. Dosages are commonly recommended regarding use (when, where, how, application rates). In addition, several Member States of the EU put restrictions on farmers, limiting the use of pesticides in environmentally sensitive areas. Cost-sharing programmes exist in several Member States to farmers in zones where the use of pesticides is banned.

Taxes on use of pesticides exist in the EU, the USA and Canada. No environmental taxes apply to the sales of pesticides in Australia and New Zealand. Where existing, the tax tends to be small, and mainly aimed to generate revenues for supporting legislation, for example by offering compensation to farmers who invest in environmentally-friendly equipment. Taxes on pesticides are a few percent only in most of the EU. The tax is high in a small number of Member States, and amounts to a third of retail price in Denmark.

Aerial spraying of pesticides is mainly controlled to limit spray drift and reduce pollution effects of pesticides use in the air and water. Aerial spraying of pesticides is one of the most restricted pesticide application practices. It is prohibited in parts of the EU and Australia. Rules read all very much the same in terms of conditions, but it still is a permitted use in the USA, Canada and New Zealand. It is controlled in all countries examined, and is mostly applied through hired labour. Special rules apply on teaching of best practices and a licence is commonly needed for carrying out aerial spraying.

The use of pesticides is restricted within a certain distance of watercourses in the countries studied, either through mandatory measures (EU, USA, Canada and Australia) or voluntary measures (New Zealand). Where existing, such buffer zones commonly are wider than in the EU.

Safety rules on the application of pesticides exist in all countries, and mandatory training of applicators of pesticides is commonly used. Controls on storage and disposal of pesticides are mainly to serve occupational health. Also, disposal of containers may pose serious risks for the environment and human exposure. Storage and disposal of pesticides are restricted in the countries examined, including either rules to dispose containers of pesticides via take-back programmes of industry or government, or through rules to rinse containers.

Irrigation

The extraction of water for irrigation purposes competes with other water uses (industry, domestic water supply, tourism, etc.) as well as having potentially significant impacts upon the environment, particularly in arid regions where water supplies are scarce.

Absolute quantitative restrictions to limit the extraction of water for irrigation purposes do not apply at the national level in any country but they may apply in those regions of each country where water is particularly scarce. Irrigation use is increasingly based on time-limited permits. Quantitative restrictions on extraction are usually aquifer-specific and generally appear tighter in New Zealand and Australia than in other countries. Water charges appear to be lower than opportunity costs in most cases, where they apply. The EU does not appear to apply more stringent standards than elsewhere, but the trend in several countries is towards increasing costs to more adequately reflect externalities.

Soil quality and soil erosion

Salinisation of soils occurs in several regions of the EU, USA and Australia, usually where irrigation is applied. It is not an issue in Canada or New Zealand. In the former three countries, voluntary measures have been taken by the agricultural sector to address these problems.

Soil contamination from heavy metals and other toxic substances partly occurs from agriculture itself, since fertilisers and livestock manure contain trace elements including metals and salts. The EU, USA and Canada have limits for heavy metals in soil. Such limits affect the composition of sewage sludge.

Several constraints on sewage sludge will affect on-farm activities in an indirect manner only, because the constraints put to the application of sewage sludge primarily affect those who undertake to apply sludge. This is normally not done by farmers. Constraints on cadmium in inputs apply in Australia and these may increase costs of the inputs used. Controls on salts from feedlot effluents in Australia may also affect farm costs.

Farming practice is restricted to control soil erosion and sediments in the USA, Australia and New Zealand. Proper land management to control soil erosion is required in the USA to be eligible for payments on highly eligible land. There are few standards at Member State level to control soil erosion or sediments in the EU and no on-farm constraints of this kind in Canada. However in all countries, incentives, farm extension services, codes of practice or voluntary industry standards may be in place to improve land management practices.

Air quality

Control of odour, ammonia and noise may involve constraints upon intensive livestock production units. Distance and siting rules constitute the principal on-farm constraints to control air quality, mainly to limit odour and noise, and to a lesser extent also to limit emissions of ammonia, in Canada, Australia and New Zealand. Additional on-farm constraints in the form of conditions on planning consents for livestock installations are quite significant in the EU, partly reflecting its higher population densities and the greater proximity of agricultural and settlement areas. Similar conditions apply only to a relatively small number of very large installations in the USA. In the non-EU countries more emphasis is given to air quality aspects related to odour and nuisance while ammonia has been an important concern in the EU. Constraints also apply to farmers in the EU during the application of manure, to reduce emissions of ammonia and control odour nuisance. Such constraints are less strict or non-existent in the other countries.

Burning of crop residues such as straw can cause significant nuisance to nearby settlements. Crop burning is being phased out mainly in the EU and the USA. When applied in Australia and New Zealand, guidelines exist to prevent nuisance or management of fires. It is not controlled in Canada.

Nature conservation, biodiversity and landscape

The issue of biodiversity and landscape includes endangered species, habitat conservation, cultural heritage, ancient monuments and protected landscapes. In all four of the New World countries there are significant areas of 'natural' or relatively wild habitat in which the integration of production agriculture with traditional landscapes has not applied. It is usually these areas that are seen as a principal refuge for biodiversity, as well as important scenic landscapes that should be preserved. By contrast in the EU the majority of valuable habitats and species, as well as important landscapes, occur on land which has been farmed for many centuries where biodiversity and landscape quality depend upon the maintenance of some kind of active management, usually through appropriate agriculture or forestry. At the micro scale, farmland areas in all five countries include 'interstitial areas' of rough land or semi-natural habitats, which can provide important sites for wildlife. At the macro scale, Europe is relatively unusual in having large areas of farmed land where a great proportion of its important species and valued landscapes are uniquely found. These differences are reflected in the degree of on-farm constraints in the different countries to protect plant and animal species.

There appear to be species protection measures in all five countries although the emphasis in New Zealand and Australia seems to be more on alien species control. In the USA and EU, farm level constraints tend to be limited to small areas, and/or compensation is provided to meet them.

Governments at federal and provincial/state level in all five countries have legislated to protect remaining valuable non-farm habitats such as wetlands from drainage, or bush or forest from clearance, for farming. All countries implement legislation to control protected and alien species, and the evidence collected suggests that this may lead to more on-farm constraints in the USA, Canada and the EU. In addition, many EU measures exist to protect extensive areas of valuable farmland habitat and many threatened species which are dependent upon specific kinds of farming activity, while this phenomenon is mainly confined to more limited examples in the other countries. However, compensation is offered in a substantial proportion of these instances.

Genetically modified organisms (GMOs)

In the five countries examined, various definitions of GMOs are applied, and there are significant differences in the extent to which GMO use on farms is permitted. Genetically modified herbicide tolerant crops will likely decrease costs of weed control and may facilitate the adoption of minimum tillage. Herbicide resistance and new weed species problems that arise as a result of this technology will be dealt with by traditional methods. However it may be too early to assess the competitive effects of these differences, given rapidly changing consumer preferences in each country, and ongoing developments as regards environmental responses to GMO use.

Controls on the authorisation of GMOs are very strict in the EU and New Zealand, whereas the application is most advanced in the USA and Canada. It is slowly gathering uptake in Australia. This is primarily related to consumer attitudes, the political decision making process and awareness of potential risks than to biophysical differences between the countries. However, it is not apparent that the competition effects of these differences in standards is negative for EU and New Zealand producers since there is a growing market for GM-free foods, particularly in the EU.

Animal welfare

The treatment of animal welfare issues focuses on production systems applied. Requirements on housing mainly apply to laying hens, calves and pigs.

Battery cages are the dominant method to maintain laying hens in the EU- and non-EU countries. Housing constraints in the EU and New Zealand are comparable. Space requirements, which apply to battery cages in Australia are similar to those in the EU, but stricter rules are introduced in Australia to growing heavier hens. Recommended guidelines for battery cages in the USA are much smaller than required size in the EU. However, layers used in America are smaller than the European brown layer. White laying hens in the USA on average are about 1.6 kg, whereas in the EU these figures are 1.9 kg (brown laying hens) and 1.7 kg (white laying hens). This implies that the living weight per unit of space in the USA (5.1 g cm^{-2}) exceeds that of the EU (4.2 g cm^{-2} for brown hens and 3.8 g cm^{-2} for white hens). Stocking density limits in the EU (in terms of live weight per unit of space) are substantially below that of the US. Animal welfare constraints to grow laying hens in Canada are between those in the EU and the USA.

Free-range production systems are used in the EU and Australia. No commonly accepted definitions of such production systems exist in the USA and New Zealand. The system is not applied on a commercial basis in Canada. Differences between the EU and Australia reflect the housing conditions. Climatic conditions in Australia allow animals to be kept outdoors throughout the year. Free-range chickens need to be provided sufficient shelter to prevent pain for the animal. Also, in the EU, hens need to have continuous access to the outdoors during the day. Stocking density limits of such runs in the EU ($1,000$ hens ha^{-1}) is lower than that of Australia ($1,500$ hens ha^{-1}).

Buildings with pigs in the EU, Australia and New Zealand must comply with certain minimum space allowances for rearing pigs in a group; in contrast, they are based on recommended space allowances in the USA and Canada. Significant space differences exist for growing pigs. Space allowance for pigs of around 100 kg are more restrictive in New Zealand (0.85 m^2 per animal), Canada (0.76 m^2 per animal) the USA (0.72 m^2 per animal), relative to the EU and Australia (0.65 m^2 per animal).

Rules apply to the transport of farm animals, and may include requirements regarding journey times, resting times, feeding and watering intervals and space allowances during transport. Requirements for the transport of farm animals restrict practice of primary producers to a limited extent only. Restrictions for the transport of animals do mainly rest with other persons at the stage of transport. Constraints on journey time are more restrictive in the EU relative to the USA and Australia. Livestock producers need to provide animals which are fit and healthy for travel. This should also consider the nature of the journey.

Animal welfare issues for the protection of animals at the time of slaughter put down rules such that animals must be spared any avoidable pain or suffering during slaughter. This issue has arisen in the USA lately, with some groups complaining that many animals are not stunned before slaughter, due to lack of enforcement. Rules on slaughter of animals primarily restrict the equipment and facilities of slaughterhouses and put constraints to primary production to a limited extent only. Primary producers commonly pay inspection controls of animals. Rules on slaughter are broadly similar in the EU, Australia and New Zealand, since the large abattoirs and meat slaughtering plants are audited periodically by inspectors from the EU.

Human-health

Hormones and animal feed ingredients

Hormones are used to promote the growth of production of livestock; additives are provided to animal feed to prevent production losses. The use of such compounds are restricted in some countries because of the potential human health effects. The use of Bovine Somatotrophin (BST) is prohibited in the EU, Canada, Australia and New Zealand. This puts constraints to individual producers because the use of BST would enhance production. BST is approved for use in the USA, and it is currently applied on 30% of dairy herds in this country. The ban on using BST is introduced, either because of public concerns about the risks to human health (EU); animal health concerns (Canada) and because of restrictions put by the EU for the import of dairy products (Australia and New Zealand). The use of growth-promoting hormones for beef production is banned in the EU. It is allowed in the USA, Canada, Australia and New Zealand.

Rules on the use of additives in compound feed are rather restrictive in the EU and Australia. Such bans will increase production costs per unit of output marketed. Several of the antibiotics, which are currently banned in the EU, are approved for use in the USA, Canada and New Zealand.

Hygiene rules for dairy farming

The extensive range of legislation on food hygiene in the EU applies mainly to the storage, processing, transport and marketing of food rather than at farm level. The pattern is similar in other countries. However, the dairy hygiene rules are an important example with farm level implications. The on-farm constraints are apparently broadly consistent between the countries studied, covering a similar range of considerations to control hygiene and diseases. Hygiene rules in Australia and New Zealand are in line with those for EU dairy producers. Dairy producers in these countries apply standards similar to EU farmers for their export to the EU. The main incentive approaches used for the agricultural sector tend to focus on the adoption of management practices, the provision of training and advice. The measures are either public funded or primarily market-driven.

Veterinary requirements and conditions to control animal diseases

Several countries have compulsory measures; others have voluntary codes, to control animal diseases. These measures might in practice imply the same constraints to farmers. All five countries have programmes to control diseases in livestock to protect the public from any health hazards that might arise from diseased livestock. Systems of animal health monitoring inspection are in place. Control over the outbreaks of disease, eradication programmes, vaccination policies, slaughter policies etc. may be defined at the regional/state rather than the federal level in some cases but it is not clear that farm level constraints are substantially different. Naturally, constraints arise when outbreaks of disease occur and remedial measures are needed. Such costs do not indicate systematically higher standards in a particular country however.

Main differences in standards

The high level of concern about food safety, farm animal welfare, landscape and nature conservation, nuisance created by more intensive livestock farms, GMOs, growth promoting hormones and certain other technologies creates a significant weight of policy interventions in the EU which goes beyond that experienced in the other countries - although the panorama of other issues covered by legislation is not dissimilar. Different public preferences clearly play a part in this distinctive approach but the reliance on intensive practices in many regions, the close proximity of farm and urban environments and the relative lack of wilder areas play a part too. The policy interventions in the EU clearly generate farm level constraints on a corresponding scale.

The tendency to rely on regulatory and command and control policy measures in the EU may give rise to more formal constraints than in countries relying on codes of practice, voluntary procedures etc. but the level of implementation needs to be taken into account before firmer conclusions are drawn with regard to competitiveness.

Where there are similar concerns, with respect to nitrate pollution of water for example the underlying standards are often broadly comparable between countries, even if the policies associated with the standards are different. This needs to be seen in light of the much higher incidence of excess nutrient problems in much of northwestern Europe compared to the other countries. There are certain areas where EU standards are more restrictive particularly with regard to landscape, GMOs, growth hormones and farm animal welfare (housing systems for rearing laying hens). It is particularly difficult to quantify the significance of constraints related to landscape because many of them arise from local land use planning, zoning and development control procedures. By contrast restrictions on GMOs are tighter in the EU than in North America.

Table 9.2 highlights the policy areas across the five regions, and attempts to identify the main constraints they place upon farming. It identifies leading policy areas, which put major constraints upon farming, on the basis of the evidence as presented in this report. A simple ordinal scaling system has been applied in the table, to indicate relative levels of constraint. We have focused on those policies that internalise an external cost attributable to agricultural production. A qualitative judgement is given on the differences between countries. Differences in on-farm constraints could be major or limited. The intermediate categories have been introduced where the differences in on-farm constraints appear to impact farming to a limited extent only. No reference is made to impacts where producers are subject to the same legal constraints or where no constraints apply to farming.

Table 9.2 Constraints upon farming which internalise external costs to environment and health

Issue	EU	USA	Canada	Australia	New Zealand
Nutrient enrichment by nitrates and phosphates	***	**	*	*	*
Pesticides	***	***	***	**	*
Sediments	-	**	-	**	**
Irrigation	*	*	*	**	**
Salinisation	-	-	-	**	-
Soil contamination	**	**	**	*	-
Soil erosion	*	-	-	**	**
Odour, nuisance, ammonia	***	**	**	*	*
Crop burning	*	*	-	-	-
Biodiversity, landscape	***	*	*	*	*
GMOs	***	*	*	**	***
Housing (laying hens)	**	*	**	**	**
Transport	**	*	*	*	*
Slaughter	*	*	*	*	*
Hormones and animal feed ingredients	***	*	*	**	*
Pesticide residues in food	*	*	*	*	*
Hygiene rules in dairy farming	**	**	**	**	**
Veterinary requirements, control of animal diseases	*	*	*	*	*

- No apparent constraint upon farming and no obvious policy in place in relation to agricultural practices;
* Policy in place, but not considered to imply a real constraint for farming;
** Policy identified as a constraint upon farming but unlikely to be a major cost to the sector as a whole (e.g. limited geographical coverage or low-cost implications);
*** Policy indicated as a major constraint upon farming which implies significant on-farm costs.

There are several areas where differences between the five regions in the constraints imposed at farm level do not appear to be very significant, taking account of the different measures applied. These include pesticide residue standards for foodstuffs, standards applying to slaughter of farm animals, hygiene rules affecting dairy farms, conditions relating to veterinary requirements and the control of the main livestock diseases.

There is a further set of issues where there are clear differences in standards but these appear unlikely to affect costs very significantly in the main agricultural sectors. The relevant issues are standards applying to soil contamination, erosion and sedimentation, sediments in water courses, irrigation, the burning of crop residues and certain animal welfare standards, including the housing of farm animals and transport away from the farm.

More significant cost differences arise in relation to:

- nutrient enrichment of water, including manure management;
- pesticide authorisations and levels of pesticide permitted in water and other environmental elements;
- emissions and nuisance arising from livestock farms, including ammonia and odour;
- biodiversity and landscape constraints on the farm;
- the regulation of GMOs;
- the permitted use of certain substances in livestock production, including growth hormones in beef, BST and certain animal feed constituents including a number of antibiotics.

Considering the influence of the environmental and health contexts in each country, we can see that to a certain extent, differences in on-farm constraints can be rationally explained as similar degrees of policy response, once set in the context of different environmental or health conditions in each country. For example, nitrate contamination in the EU occurs particularly in regions where there are concentrations of intensive livestock production.

In the case of nutrient enrichment of waters, tighter on-farm standards may be necessary in the EU for consumer protection reasons, because its farms are usually closer to, and have a more direct impact upon, other water users and in particular, drinking water supplies. However, the quality of the aquatic environment may suffer in all countries, including those without such constraints, even though their farms are located far from centres of population and the water affected by agriculture is not critical for drinking purposes. However, in a country such as New Zealand where the rate of water flow through the environment is generally claimed to be faster than would be common in much of Europe, the environmental impacts of nutrient enrichment of waters could indeed be lower than it would be elsewhere.

For pesticides, differences in standards between countries are unlikely to be greatly tempered by environmental context. In all five countries, the potential impact of pesticide use upon flora and fauna is likely to be equally significant. It therefore suggests that differences between countries do not simply reflect differences in environmental context, but likely differences in public preferences for risk aversion.

By contrast, issues related to air quality, including odour, nuisance and ammonia discharges, are partially context-related. Odour and nuisance controls are directly related to perceived disadvantages to nearby households, and these are more likely to be a problem in a densely-populated country such as those in the EU than they would be in any of the other countries considered. While ammonia production by intensive agricultural installations may be similar in all five regions, the need to impose standards specifically for farming depends upon the proportional contribution of farming to overall ammonia production and to related concerns about acid rain. These are highly regionally specific and vary according to other factors including the degree of industrial pollution experienced in a particular area. Thus the relative importance of applying ammonia reduction targets to agriculture is likely to vary greatly between the countries examined in this chapter.

It is very difficult to determine, using the information from this analysis, the extent to which differences in farm-level constraints related to biodiversity and landscape conservation are context-dependent or not. Firstly, all countries have a significant number of species that are in decline or threatened, but the degree to which these species are affected by on-farm practices varies significantly between countries. It seems clear that farmland practices can significantly affect biodiversity and landscapes in all five countries but whether the micro-level policies reflect micro-level differences in the agri-biodiversity interface cannot be assessed within the context of this assessment.

Finally, in relation to differences for GMOs, hormones and ingredients in animal foodstuffs it seems clear that these relate to a heightened consumer awareness or concern in the EU as compared to certain other regions in this chapter, most notably the USA. From the US viewpoint, greater EU consumer concern may be largely a reflection of 'scare-mongering' tactics by environmental pressure groups, combined with the particular history of the BSE epidemic in Europe. However many EU stakeholder groups would claim that their caution reflects an appropriate assessment of risk while the response of the USA has given inadequate consideration to the issues of long-term and imperfectly-understood impacts upon human health and the environment. The intermediate position of Canada, Australia and New Zealand in these respects appears to reflect either similar approaches to risk-aversion (e.g. for Canada, in respect of hormones) or greater export-orientation in setting standards (e.g. for Australia and New Zealand). These factors mean that they are more likely to match the more cautious approach of the EU, in their standards.

COSTS OF COMPLIANCE

The maintenance and introduction of standards may, in principle, impose both recurring and non-recurring costs on farmers. Investment costs may arise directly from the introduction of new standards but may also be necessitated by the need to meet existing standards. There may also be opportunity costs for farms required to meet certain standards - such as the loss of the option to expand the capacity of a livestock unit because of environmental controls. These costs need to take into account a wide range of different options for the farm concerned and are more difficult to establish than direct costs. The literature on opportunity costs and shadow prices is particularly sparse, although the impact on competitiveness may be considerable.

The chapters showed that very few detailed studies have been made of the costs for particular sectors of agriculture in the five regions, of meeting different environmental and health-related standards. There is a very broad range of issues involved in determining how these standards actually affect producers' costs. As a result this analysis is necessarily qualitative in nature - although we used quantitative evidence when it was available.

While most new standards are perceived to increase costs, this is not necessarily the case. Standards can result in improved efficiency, for example better use of nutrients on the farm, potentially cutting costs and increasing gross margins. There is evidence from the USA of precisely this effect on some livestock farms.

Most of the literature on the impact of environmental, animal welfare and health related constraints on agriculture suggest that the overall impact on production costs in the sector concerned is relatively modest. This is true of the relatively few studies available in the EU as well as elsewhere. Nonetheless, it is clear that certain groups of producers can be severely affected and may even be forced to abandon production. It should be noted that standards have been tightened recently in several countries, including the EU and these changes have not usually been analysed in the literature.

In general the literature suggests that compliance costs for meeting environmental requirements would normally be less than three to four percent of gross revenue, provided that producers have sufficient flexibility in selecting the means by which they meet their obligations. However costs are significantly higher in the livestock sector than for crop production. Compliance costs may result in a significant loss of competitiveness in the most heavily regulated regions as a result of the cumulative impact of different regulatory regimes. This may in turn affect the location of livestock production, particularly for intensively managed stock. Crop farms seem less likely to face compliance costs on a scale, which will affect future location.

One of the main difficulties faced was the very limited detailed number of studies that have been made on the relative costs for particular sectors of agriculture in the five countries. The main problem faced is that the coverage of the available literature does not include all the relevant commodities and some countries have more than others. In addition, international comparison of cost estimate is problematic because of things like exchange rates and differences in production practices unrelated to environmental regulation.

A general conclusion that we can draw from the available assessments is that the costs of environmental compliance in primary agriculture, at least as indicated by the available evidence, are not particularly high. When producers are allowed flexibility in selecting the means by which they can achieve various environmental objective targets, the available analyses indicates that compliance costs are generally less than 3-4% of gross revenue. Moreover, the cost estimates are generally for the short run, and would be expected to decline over time as operators become accustomed to the constraints, and find new practices and technologies to lower the added expense.

CONCLUSIONS

The chapter disclosed the extent to which the impact of standards varied between farms, even within a particular region or target group, e.g. dairy farms. Many studies indicated that variations were accounted for by factors such as:

- the biophysical conditions of the farm and the surrounding area;
- the technical options available to the farmer;
- the production systems in use and the capacity to adapt the enterprise to changing conditions;
- farm size (as well as the scale effects, it is notable that some standards apply only to larger units);
- the investment cycle on the farm, including the age of buildings and equipment affected the timescale for renewal, discount rates etc.
- when new standards are introduced, the extent to which farms may decide to abandon a particular management practice or form of production.

Detailed empirical studies based on farm interviews rather than solely modelling exercises are required in order to establish the range of incidence of cost effects accurately. The high cost of such studies is clearly one of the principal reasons for their scarcity. Published work provides useful cost estimates, most often expressed in terms of aggregate costs for a particular group, sector or region or as an estimated range. However, these studies are too variable in their basic assumptions, target groups studied etc. to provide more than a rather general comparison.

Competitiveness

One general conclusion is that overall, the magnitude of costs applying to agriculture as a result of environmental and health-related standards is relatively small in all but a few sectors and cases, usually representing no more than 5% of total costs. If this is so, then we could expect the competitive effects of differences in standards to be relatively small, also. However, the material presented also highlights the uneven impacts of costs at sub-national level. This is often overlooked in cost studies, which tend to aggregate producers and make calculations based upon regional or national averages.

Costs in agriculture may be somewhat higher than the available studies for other sectors (e.g. manufacturing, services, and chemical), but they do not appear to be large enough to drive location of production decisions generally. In crop production, we did not find evidence to suggest that compliance with environmental regulations has been or will be a driving force determining the location of production.

With respect to livestock production, it may be a different story. The costs of compliance with nutrient regulation and measures to control odour and nuisance from intensive livestock production units are increasing in several parts of the world (e.g. EU, USA and Canada). It is primarily a question of finding a location for a facility that reaps the available size economies and at the same time is far enough away from adjacent land uses. And here, there are significant differences within and among countries. The compliance costs of producing pigs and poultry have increased during the past 10 years in the EU, USA and Canada, and this may have a significant effect on the location of production in the future.

A body of cost studies in relation to particular standards and sectors are available. However, it is not appropriate to use these studies to draw firm insights into competition effects because:

- Costs data is highly dependent upon farm structures and management practices in each country or region and they may vary considerably. For example, most dairy production in New Zealand involves cows living outsides all year, while this practice would be comparatively rare in other countries. Average herd sizes also differ considerably. There are thus difficulties in attempting to use the data for comparative purposes between similar sectors in different countries.
- Variation within countries, in terms of both farm structures and environmental constraints, may be significant. Exporting producers may be largely concentrated in particular parts or sectors within each country (e.g. pig producers in Denmark, CAFOs in the USA) and these areas may face specific and different standards that are not be reflected in studies, which are often based upon industry averages.

Key issues of concern in the EU

The most important environmental and health-related issues, which are of major concern in the EU and with high priority in policy are:

- Nutrient enrichment by nitrates and phosphates, which also is a major issue in the USA.
- Pesticides (including pesticide drift and applicator safety), which also is a major issue with high priority in policy in the USA and Canada.
- Odour and nuisance, which also is a major issue with high priority in policy in Canada.
- Biodiversity and landscape, which is important in Australia and New Zealand but which is tackled very differently in each case.
- GMOs, which are also a major issue with high priority in policy in New Zealand.
- Animal welfare, which has a lower ranking in the other countries.

Other issues also are identified as a problem in some countries:
- Irrigation, which is of similar importance in the USA and New Zealand, and has higher priority in policy in Australia.

- Soil erosion, which has higher priority in policy in the USA, Australia and New Zealand.
- Ammonia, which has more priority in the EU. Ammonia also is a problem in Canada, but for reasons of odour.
- Hormones and animal feed ingredients, which has much more priority in the EU.
- Pesticide residues in food, which has higher priority in the USA and has similar relevance in all the other countries studied.
- Hygiene in dairy farming has similar relevance in all countries.
- Veterinary requirements are less of a problem in the other countries than the EU.

Large similarities in the use of standards to control public concerns

Most policies have been reactive in the sense that measures were introduced in response to perceptions of emerging or documented problems. There are many more similarities than differences in approaches to control environment and health concerns across the countries examined, e.g. many voluntary approaches and relatively few binding constraints compared to the level of environmental constraints upon non-agricultural sectors. However, there are very real differences in approach to resolving the problems, such as the mostly voluntary approaches with federal payments in the USA and Canada, the reliance on industry voluntary compliance in Australia and New Zealand, and a greater mix of regulatory and voluntary approaches in the EU. Moreover, there are dynamic changes occurring in the approaches developed within countries, such as the rise of state and local programmes in the USA and the trend toward more regulation as policy devolves to those levels.

At this point, it is hard to substantiate a claim that the costs of environmental regulation and standards in agriculture are significant for most issues, and in total, in most countries. As evidence, there are no apparent large shifts of industrial location from one country to another, i.e. land bases have stayed fairly constant. There have been shifts within countries, such as for confined animal operations in the USA. Intensive livestock production also tends towards a further concentration within the EU.

The trend is toward more regulation and binding constraints, but not wholesale across the industry, such as for large animal units in the USA but not small ones. This trend will continue as the industry is transformed into larger farm units and the demand for EQ rises further from income growth. Hence, the need to search for low cost approaches.

Some of the most striking differences, such as EU controls on GMOs, animal welfare, and animal feed are reflections of country preferences and attitudes towards risk, for different types and levels of environmental protection. Hence, it is debatable whether they should be seen as trade barriers or distortions.

Empirical limitations faced

- Estimation of the environmental compliance costs in agriculture is in its infancy. The assessment made represents the first systematic effort to characterise the structure of the environmental and health regulatory regimes within which the agricultural sector operates in the EU, the USA, Canada, Australia and New Zealand.
- This analysis is the first attempt to compare constraints to farming of environmental and health-related standards. The relative competitiveness of EU agriculture vis-à-vis those of some key competitors is affected by the accumulation of on-farm constraints. This is not examined in detail here because of the limited detailed studies that allow for assessing cumulative impacts and direct comparisons between countries.
- Competitiveness of EU agriculture is examined by comparison of on-farm constraints by policy field. Competitiveness also needs to consider innovation strategies of farmers to meeting constraints resulting from legislation and emerging market developments, such as 'green' product innovations.

Resource limitations faced

- The analysis revealed that the variety of environmental and health related standards at sub-national level within all the countries studied, placed severe constraints upon our ability to achieve a thorough comparative analysis at country level.
- Our approach has therefore been illustrative and qualitative, and we have been unable to conclude with a concrete assessment of cost differentials and competitiveness effects, however, indications of broad areas where standards seem likely to have the most significant impacts upon sectors, have been provided.
- A much larger and more exhaustive assessment would be necessary in order to provide thorough analysis of these issues, but we hope that by highlighting the most promising avenues for further investigation we can assist in the targeting of future resources to this area.
- Future studies should, in our view, be focused on a narrow range of key sectors and might also be most usefully confined to the production sectors, which are most export-oriented, within each country. Issues of environmental context, financial compensation and voluntary initiatives, and other non-environmental constraints upon production costs will continue to be essential considerations for these studies.

General lessons about the relationship between regulation and trade

- The available assessments indicate that the share of environmental compliance costs per unit of output is not particularly high, which implies it likely has only a modest effect on trade.

- Environmental regulations (by any level of government, or by civil liability) that legitimately internalises an external costs (e.g. nuisance, riparian rights) does not distort trade unless such external costs occur in the economy of some other trading partner and those costs are not being internalised, or unless the regulations are applied to such an extent that the marginal social benefits remain below the marginal social costs.

- Different countries, by virtue of differences in standards of living, consumer and political preferences, technology or geography, can legitimately apply different standards to determine whether something is a non-internalised external costs.

REFERENCES

Bredahl, M.E., Northen, J. and Boecker, A. (1998) Trade Impacts of Food Quality and Safety Standards. University of Reading, Center for Food Economics Research, UK. Working Paper.

Brouwer, F., Baldock, D., Carpentier, C., Dwyer, J., Ervin, D., Fox, G., Meister, A. and R. Stringer (2000) Comparison of Environmental and Health-related Standards influencing the relative Competitiveness of EU Agriculture vis-à-vis main Competitors in the World Market. The Hague, Agricultural Economics Research Institute (LEI), Report 5.00.07.

Environmental Standards in Developing Countries

10

Ulrike Grote

INTRODUCTION

International trade in agricultural products has rapidly increased in the last decades. But so did the importance of mandatory and voluntary environmental standards for products and production processes which may affect either public health or the environment. When environmental standards differ between countries, they have the potential of affecting trade, and differing costs of compliance for meeting environmental standards may impact on the competitiveness of countries. In the agricultural sector, little empirical evidence on the cost implications of standards has been given so far. However, given the importance of agricultural products for developing countries as their major export goods, this question is of utmost importance to them.

This chapter offers an overview of the political and legal debate on environmental standards, including an analysis of the treatment of standards under the World Trade Organization (WTO). I will then analyse and compare environmental legislation on vegetable oils (soybean, rapeseed and palm oil), grain (maize, barley and wheat) and broiler chickens in Brazil, Germany and Indonesia. All three traded products and countries play a significant role on the international market, and future trade liberalisation may lead to an increasing interrelation and competition between these products and countries. Of special interest is the question to what extent environmental standards affect total production and processing costs, and thus international competitiveness. I will then examine to what extent standards are transferable from one country to other countries, as well as implications for WTO actions.

THE DEBATE ON ENVIRONMENTAL STANDARDS

The WTO aims at helping trade flow smoothly, freely, fairly and predictably. It came into force on 1 January 1995, and was established by the Marrakech

Agreement which marked the end of the Uruguay Round of the General Agreement on Tariffs and Trade (GATT) in 1994. The WTO replaced the Secretariat of the GATT (the institution) which dates back to 1947 and incorporated the amended GATT (the agreement) into its new WTO Agreements. While the GATT only deals with trade in goods, the other WTO Agreements also cover services and intellectual property. The membership increased from 128 GATT signatories as at the end of 1994 to 140 WTO members in the year 2000. Over three-quarters of the WTO members are developing countries.

The GATT established the two basic directions aiming to reduce tariffs and other trade barriers, including environmental standards. The heart of the GATT is based on the following two principles which have been also adopted by the other WTO agreements: The *most favoured nation requirement* (Article I GATT) and the *national treatment requirement* (Article III GATT). Both articles affect the treatment of standards under GATT.

The most favoured nation rule (Article I) requires that if special treatment is given to one country, then it must be also given to all other WTO member countries. This rule has two exceptions. First, it does not apply to regional trade agreements, and second, GATT allows member countries to apply preferential tariff rates, or zero tariff rates to products coming from developing countries, while still having higher rates for other member countries.

According to Article III, imported products ('like products') must be subject to the same regulations and requirements as similar domestic products. Discriminating against imported 'like products' offends GATT principle. But the GATT does not define the term 'like product'. Since two products produced in different countries will never be completely identical, the decision according to which criteria two products are alike is made case by case. The relevant criteria include the final use of the products, consumer tastes and habits, the products' properties, nature and qualities as well as commercial substitutability (UNEP and IISD, 2000).

Products are also considered as like products, even if they have been produced based on different Process and Production Methods (PPMs). The major argument of the panels in the famous tuna dolphin dispute between the USA and Mexico was that no discrimination in terms of import bans between like products (tuna) based on different non-product-related PPMs (catching methods for tuna) is allowed, as long as the characteristics of the final product is not affected. Thus, the WTO does not envisage to incorporate any requirements on non-product-related PPMs into its rules so that domestic environmental degradation caused by the production of a tradable good in the producing country cannot be addressed by the use of trade measures through another country. The use of trade measures to enforce such non-product-related PPM standards risks an extraterritorial imposition of standards of the importing country on its trading partners, and would interfere with the sovereign right of countries to exploit their own resources and set their own standards and rules for activities within their borders.

Exceptions from the prohibition to use trade measures are included in Article XX of GATT. Article XX (b) allows trade restrictions necessary to protect human, animal and plant life or health, while Article XX (g) allows restrictions relating to the conservation of exhaustible natural resources if such measures are made effective in conjunction with restrictions on domestic production or consumption. However, such restrictions are not allowed to be applied in a manner which would constitute a means of arbitrary or unjustifiable discrimination between the same conditions, or a disguised restriction on international trade. In general, the use of GATT Article XX for exceptions to the general trade agreement on the grounds of environmental justifications will be subject to continued substantial, multilateral debates.

Based on Article XX, the Technical Barriers to Trade (TBT) Agreement and the Sanitary and Phytosanitary (SPS) Agreement have been developed under the WTO system. Both agreements have been adopted to ensure that the standards do not create unnecessary barriers to trade. The TBT Agreement lays down the rules for preparing, adopting and applying technical regulations, standards and conformity assessment procedures, for example, for human, animal or plant life or health, for the protection of the environment or to meet other consumer interests. Technical food standards, for example, refer to characteristics such as size of a product (e.g. fruit or fish), its quality provisions, or labelling. Conformity assessment procedures include technical procedures like testing, verification, inspection, laboratory accreditation, certification or independent audit. The SPS Agreement specifically concerns the application of food safety and animal and plant health regulations, for example to protect human life from risks arising from additives, contaminants, toxins or plant- or animal-carried diseases; or to protect animals and plant life from pests or diseases. SPS measures can take many forms, such as quarantine regulations, certification or inspection of products (e.g. in slaughterhouses), specific treatment or processing of products, setting of allowable maximum levels of pesticide residues, permitted use of only certain additives in food, or health-related labelling measures.

Both agreements allow individual countries to set their own standards. However, countries are obliged to inform other countries about the introduction and use of new technical regulations or standards. In the case of SPS standards, other WTO member countries can request scientific proof and a risk assessment justifying the necessity of a standard if they feel they are being discriminated against. Both agreements encourage countries to base their domestic standards on internationally-agreed standards.

For developing countries, both agreements include an Article on Special and Differential Treatment. The TBT Agreement acknowledges that although 'international standards, guides or recommendations may exist, in their particular technological and socio-economic conditions, developing country members adopt certain technical regulations, standards or conformity assessment procedures aimed at preserving indigenous technology and production methods and processes compatible with their development needs. Members therefore recognise that developing country members should not be expected to use international standards as a basis for their technical regulations or standards, including test methods, which are not appropriate to their development, financial and trade needs.' It further recognises 'that the special development and trade needs of developing country members, as well as their stage of technological development, may hinder their ability to discharge fully their obligations under this Agreement. Members, therefore, shall take this fact fully into account.' The SPS Agreement allows scope for the phased introduction and longer time periods for complying with SPS measures. To take into account the financial, trade and development needs of developing countries, further exceptions can be granted. In addition, an Article on technical assistance for developing country members is included in the TBT and SPS Agreements, e.g. to provide training and improve infrastructure in developing countries.

ENVIRONMENTAL STANDARDS IN THE DEVELOPMENT CONTEXT

Given the increased public awareness and concern about environmental problems, especially in higher-income countries, developing countries are experiencing increased pressure with respect to environmental issues. Thus, there is increasing concern in developing countries about green protectionism restricting their access to export markets in developed countries. This relates closely to the use of standards, especially non-product-related PPMs, and conformity assessment procedures under the TBT and SPS Agreements, as well as the related and controversial debate on eco-labelling. Many developing countries face difficulties in complying with PPM-related standards. In addition, they are concerned about the compliance cost to meet environmental standards and the impact on the international competitiveness.

The use of PPMs

Non-product-related PPMs are not covered under the WTO. From an environmental point of view, it appears reasonable to discriminate at the border between products that were produced in environmentally-friendly and -unfriendly ways. However, there is the fear that the rapidly increasing development and use of PPMs as non-tariff barriers may be motivated by economic rather than environmental considerations. Another fear relates to the suitability of standards for different environments. While environmental standards might be appropriate in the country where they have been developed, they might not be easily transferable to the environment in other countries. The environmental priorities set by governments in low- and high-income countries are also often different. Given the limited financial resources and high levels of poverty, many governments in developing countries give lower priority to environmental problems than to social issues, for example.

The use of PPMs also plays a role within eco-labelling programmes which are now operating in most developed countries and also increasingly in developing countries (Vossenaar, 1997). Eco-labels inform the consumer, that a product and/or its production process are - compared with traditional products and processes - especially environmentally-friendly. While some programmes promote product attributes like 'recyclable', 'degradable' or 'ozone-friendly', others include whether production methods have been environmentally friendly. For example 'dolphin-safe tuna' is caught with methods which avoid unnecessary killing of dolphins in fishing nets, or tropical wood is certified when it comes from sustainably managed forests. The price premium which the consumer pays for the labelled product functions as an incentive for the producer to pay attention to environmental protection by internalising environmental costs during production or processing.

Theoretical analysis has shown that certification is an attractive solution for the problem of eco-unfriendly methods of production (Grote *et al.*, 1999). It is considered as a voluntary and market-driven approach. Positive environmental effects have been found in the textile and leather tanning industry in India (UNCTAD, 1995; OECD, 1997a). For example, the exports of jute from Bangladesh - for the production of eco-bags as an environmentally-friendly alternative to plastic packages - increased (Wyatt, 1997). And finally, due to eco-labelling, the fertiliser- and pesticide-intensive flower production has become not only more sustainable in developing countries like Ecuador, Kenya or Tanzania, but also in industrial countries like the Netherlands (Grote, 2000).

Adverse trade impacts have been identified as a consequence of eco-labelling programmes for foreign producers and suppliers of input materials, especially in the pulp and paper, footwear, textile and timber markets in Brazil, Bangladesh, Maldives and Laos (UNCTAD, 1995). Some Colombian textile companies have stopped exporting to certain developed countries due to the high costs of obtaining the label and the recognition that without the eco-label, the products could no longer compete.

Eco-labelling is also acknowledged by the WTO as an effective instrument of environmental policy. However, this trend is likely to result in further conflicts under the WTO, as there is no agreement on the extent to which eco-labelling may include non-product-related PPMs. Eco-labelling is covered by the TBT Agreement, but the WTO generally considers non-product related PPMs as not conforming with its rules (Chang, 1997).

Another concern of developing countries refers to the use of trade measures as enforcement mechanisms for PPMs in Multilateral Environmental Agreements (MEAs). PPMs with transboundary and global environmental impacts are harmonised within MEAs. There are hundreds of agreements by now that address shared environmental problems like climate change, management of migratory and endangered species or loss of biodiversity. The MEAs are based on co-operative action between countries, and enforced by the threat of trade restrictions imposed on countries that neglect their environment. Because there is inconsistency of measures applied between certain MEAs and WTO rules, an amendment of WTO Agreements has been proposed. So far, there have been no disputes but there is concern about potential conflicts resulting from discordant MEAs and WTO rules and obligations (Sampson, 1999).

Two international environmental agreements in which longer-term PPM implications are identified are the Basel Convention on the Transboundary Movement of Hazardous Waste and the Convention on Biodiversity. A key goal of the Basel convention is to alter upstream production methods and reduce wastes before they occur. In the Biodiversity Convention, there are three very general PPM requirements, like the conservation of biological diversity, the obligation of parties to manage biological resources in a sustainable manner, and the obligation of parties to build an equitable distribution of the benefits arising from the economic use of genetic resources (Vaughan, 1994). The later is also closely related to the Cartagena Protocol on Biosafety which regulates the trade with genetically modified organisms (GMOs).

Both conventions mention certain PPMs which are to be followed by countries in order to achieve environmental goals. Unlike environmental agreements with elaborate strict controls, however, the gap between stated objectives and current implementation remains wide in the conventions. It is unclear to what extent PPMs which are only broadly and vaguely defined will be enforced.

To date, no GATT contracting party has formally objected to the use of trade policy to address environmental problems in MEAs. Nor have there been objections to the bans on trade in ivory, rhino horn or tiger products that are part of the Convention on International Trade in Endangered Species (CITES), or to the trade provisions in the Basel Convention on trade in hazardous wastes (Anderson, 1995).

The impact of standards on competitiveness

Developed countries have voiced concerns about losing their competitiveness due to the costs of meeting relatively high environmental standards.

Past studies on competitiveness and increasing environmental standards were based on very different methods, time periods and countries (Nordström and Vaughan, 1999; Helm, 1995). Therefore, results of these studies differ significantly. In some studies, it was found that environmental standards resulted in a decreasing export share of polluting products from industrialised countries - whereas developing countries exported a greater proportion of polluting products (Low and Yeats, 1992; Sorsa, 1994; UNCTAD, 1994). Other studies, however, showed that trade of environmentally-intensive products from the USA and Japan increased (Kalt, 1988; Sorsa, 1994). For Europe, Jenkins (1999) found little evidence of a general loss of competitiveness for environmentally-intensive industries. Also a comprehensive empirical study referring to different sectors in Germany did not find a systematic relation between environmental costs and international competitiveness (Felke, 1998).

Tobey (1990, 1993) tested whether world trade suffers from the imposition of environmental policy, but found little empirical evidence for it. According to him, the primary reason seems to be that the costs of pollution control have not been very large in pollution-intensive industries and countries with stringent pollution control policies. In the USA, estimates suggest that control costs amount to about 2-2.5% of total costs in most heavily polluting industries. Based on water-pollution data from China, Dean (1999) analysed the impact of trade liberalisation on emission growth. She found that while increased trade openness in China directly aggravates environmental damage by inducing an expansion of polluting sectors, income growth indirectly decreases emissions. Mani and Wheeler (1999) found that 'pollution-haven' effects meaning that lower trade barriers will not result in developing countries specialising in pollution-intensive industries, are insignificant because production is primarily for the domestic market, not for export. The increase in the developing countries' share of dirty-sector production is attributable to a highly income-elastic demand for basic industrial products. As income levels have increased, this elasticity has declined, and the stringency of environmental regulations has been raised.

According to the OECD (1997b), direct environmental costs are estimated to make up only 1-5% of the total production costs in the industrial sector. Similarly, Dean (1992) and Jaffe *et al.* (1995) pointed out that for most industrial producers environmental costs make up only a small part of the total costs. For the agricultural sector, there is little empirical evidence on the costs of compliance with environmental standards.

In spite of some empirically shown negative effects of environmental standards on competitiveness, authors are very careful about their interpretation. Other factors like wage level, education level, political and economic stability or the distance and size of markets as well as the infrastructure are considered to have a substantially greater effect on international competitiveness than environmental costs.

The impact of standards on competitiveness may also be analysed with a more dynamic innovative approach. This is based on the assumption that increasing competition and relatively higher environmental standards may positively affect companies' innovative power (Porter-Hypothesis). These companies search for new ways of increasing their productivity and thus competitiveness by reducing environmental pollution, substituting input factors with cheaper material or reducing losses (Porter, 1991). Eco-labelling which informs consumers about the production of environmentally-friendlier products tends to fill a gap in the market and thus allows price advantages (Grote *et al.*, 1999). Alternatively, companies may transform their waste into marketable goods to earn an additional income (Porter and Van der Linde, 1995).

It can be summarised that existing studies show ambiguous results about the effects of environmental standards on the competitiveness of companies, especially in the industrial sector. Given the lack of data and empirical evidence in the agricultural sector, an empirical study has been conducted for selected agricultural products in two developing and one developed countries. The results will be presented in the following.

THREE EXAMPLES IN AGRICULTURE[1]

This comparative study provides some insights into the discussion on environmental standards by analysing the production and processing of vegetable oils (soybean, rapeseed and palm oil), grain (maize, barley and wheat) and broiler in Brazil, Germany and Indonesia. All three traded product groups and countries play a significant role on the international market. After the identification of environmental problems and the legal environmental framework in the respective

---

[1] The case studies were conducted in cooperation with the Center for Advanced Studies and Applied Economics (CEPEA) of the University São Paulo, Piracicaba, Brazil and the Center for Agricultural Policy Studies (CAPS), Jakarta, Indonesia. For detailed results of the case studies, see Grote *et al.* (2001).

countries, a cross-country comparison of the cost implications and the transferability of standards to other countries will be conducted.

Brazil

Brazil is one of the biggest producers and exporters of soybeans and broilers worldwide. The production of soybeans increased steadily from a volume of almost 10 million tonnes in 1978 to more than 30 million tonnes in 1998. Thus, with about 20% of world production, Brazil is the second biggest producer worldwide. Vast areas of savannah (approximately 200 million ha), known as Cerrado, in Brazil's central states are available for further expansion of the production of soybeans and maize which at the same time form the basis for an expanding poultry industry. The production of broilers amounted to about 10% of world production in 2000. Both soybeans and broilers from Brazil have been increasingly exported to European countries. The only important arable product imported by Brazil is wheat (OECD, 1997c).

Environmental issues

The production of soybeans has rapidly expanded from the South and Southeast states in Brazil to the central states and now moves further to the North and Northeast. A concern of environmental groups, parts of the Brazilian civil society and other countries is the expanding agricultural use of the ecologically sensitive dry savannah, the Cerrado. While some field studies have shown that the production of soybeans, if carefully introduced into this area, can be beneficial to the fertility of the soil (Bayer, 1994), experience has shown that a severe degradation of the soils and a loss of biodiversity resulted after a few years of growing soybeans.

At the federal and state level, there is also an increasing concern about deforestation and the impact of the use of fertilisers and pesticides in agriculture on the quality of river water; the use of fire as land clearing method; and water resource management in general.

Apart from these federal and regional concerns, some Brazilian farmers from the more traditional South and Southeastern regions of Brazil (São Paulo, Minais Gerais and Goias) have identified the following environmental needs and problems from their perspective: disposal of chemical packages; high erosion risks in conventional farming unless no-tillage systems are used which reduce erosion; and disposal of dead broiler chickens.

Environmental legislation

Many environmental laws applying to the agricultural sector, are developed at the federal level, but also increasingly at state and local levels. The following more general federal laws set the legal framework for the agricultural sector in Brazil:

• The National Environmental Legislation (No. 6.938/81) defines targets, planning and application mechanisms of the national environmental policy.

- The Agricultural Policy Law (No. 8.171/91) is the base for setting agricultural policy instruments. It solely requires a 'reasonable' use of soil, water, animal and plant and other natural resources.
- The Law on Environmental Crime (No. 9.605/98) defines environmental crimes and their sanctions, including fines, and in some cases imprisonment. It also sets obligations for the removal of the environmental damage which includes offences like cutting down forests, polluting waters or air.
- In addition, the recent CONAMA (National Environmental Council) Resolution no. 237/97 had an impact on the industrial but also on the agricultural sector. It requires approval of agricultural activities like livestock production if a certain size of production unit is exceeded, the economic exploitation of wood and forestry by-products, and introduction of exotic and/or genetically modified species.

The most important, but controversially debated environmental regulation imposed on Brazilian farms is the Forestry Code (Law No. 4.771/65 from the year 1965). It has two main parts that affect land use and land cost: one is called the *Legal Reserve*, and the other one *Permanent Preservation Areas (PPA)*. Within the legal reserve, each farmer is obliged to permanently set aside at least 20% of his land for the purpose of maintaining or replanting local species, mainly trees. Within the PPA, farmers have to permanently set aside strips along rivers, lakes and waterways for nature conservation. There is no compensation or support to the farmers for implementing these measures. While currently, these laws are insufficiently enforced, it is expected that enforcement of the Forestry Code will be stricter in the future, due to stronger control from the government.

At federal and state levels, laws and standards, like the Federal Law on the Planning of the Water Resources (No. 9.034/94) are being established. It has been announced by the government that there will be a charge for the use of groundwater, and existing water prices will be increased. However, there is no final agreement about the implementation of increasing water user fees.

For the state of São Paulo, there is a Law on the Rules of Orientation for Water Policies and Integrated Systems of Water Resource Management (No. 7.663/91) which:

- proposes a decentralised model of water management;
- designates hydrographic basins as planning and management basis; and
- recognises water as a public good with economic value and demands therefore pay for use.

However, water is still drawn by farmers from their own wells in Brazil. Also waste water disposal is presently still free of charge, but the introduction of fees for waste water disposal has been announced by the Brazilian government.

In general, it is noticeable that in the state of São Paulo with greater population density, more and stricter laws and decrees have been developed than in the less densely populated state of Minais Gerais for example. The enforcement of the jurisdictional regulations in São Paulo is also more effective than in Minais Gerais.

With respect to plant protection, there is the Law on Chemicals (No. 7.802/89) which determines the use, transport, disposal and final removal of packages and the use of chemicals to secure the safety and health of people. As the case study shows, the disposal of packages for pesticides is problematic for many farmers in Brazil (Grote *et al.*, 2001). No law on fertiliser use exists.

In Brazil, there are also regulations on the use of fire as a land clearing instrument. Decree No. 2.661/98 defines rules for the conservation of the environment by setting standards for precautionary measures for the use of fire as an instrument of land clearing in agriculture and forestry. It also prohibits burning in some regions and situations but concedes the possibility of a license to burn. In São Paulo, Decree No. 42.056/97 sets regulations on the use, conservation and protection of agricultural soils. In detail, it determines that fire generally must be avoided, and it defines cases in which fire is tolerated or prohibited. It also determines the procedure and alternative methods of sugar-cane burning before harvesting.

With respect to soybean production in the Cerrado, no environmental legislation exists to decrease potential negative effects on the environment. There is no law on animal welfare in animal production, including broiler production.

At processing level, there is a Law on the Establishment of Industrial Zones in Regions with Critical Pollution (No. 6.803/80). This law names conditions under which the creation, operation and other details of new industries in regions with critical pollution are licensed, and it specifies environmental standards for emissions of gases, noise, residues and sewage etc.

Indonesia

Indonesia is one of the main producers and exporters of palm oil, which competes with other vegetable oils. The production of palm oil increased from 1.5 million tonnes in 1987 to around 4.5 million tonnes in 1996, amounting to about 20% of world production. The exports increased steadily reaching around 3 million tonnes in terms of volume and almost US$1.5 billion in terms of value in 1997. Since the mid-1980s, the plantation area more than tripled, with further areas being still available for expansion.

Environmental issues

The conversion of forest and other lands to oil palm plantations is still partly done by rapid and insufficiently controlled fire clearance. Forest fires of catastrophic dimensions have been the result. Closely related to the burning issue is a general concern about the protection of primary forests. Specific to palm oil processing, the management of waste water is of environmental concern.

Environmental legislation

With respect to the use of forestland and palm oil production, the following laws have been developed, especially in the last 5 years: The recently renewed Forest Law No. 44/99 and Decree No. 41/99 (which is a Renewal of Decree No. 5/67) prohibit the use of primary forests, and thus aim at the protection of primary rainforests and the conservation of biodiversity. Most other environmental legislation in Indonesia primarily refers to decrees on the management of natural resources and land use; in addition, there are environmental administration and pollution control laws.

With respect to oil palm plantations, the following decrees exist: The Decree of the Ministry for Forestry and Estate No. 376/Kpts-II/1998 sets criteria for forestry areas that are suitable for plantation use. The criteria include slope, elevation, rain fall, average temperature, soil depth and minimum area (10,000 ha). The Decree No. 22/99 on Regional Autonomy enables regional governments to issue licenses for land use. This is even the case for plantations of more than 50,000 ha which was previously only granted by the central government.

For all of the above mentioned laws and decrees, enforcement has been judged good by the Center for Agricultural Policy Studies (CAPS), Jakarta. However, for the following two decrees referring to the banning of fire as a method for land clearing, there has been no enforcement in the past: The first one is the Decree of the Minister of Forestry and Estate No. 728/Kpts-II/1998 setting standards on the use of forests and the clearing of forestry areas. The second one is the Decree No. 38/KB.110/SK/DJ.BUN/05.95 of the Director General of Estate Crop which regulates the development of land by including a ban on the use of fire as a method of land clearing. There is also evidence that land had been cleared without considering other legal guidelines, such as on slopes over 8% or on anti-erosion measures along contours. In addition, trees had been moved into waterways or commercial timber had been left partly burned in the field (Barber and Schweithelm, 2000).

There exists a federal regulation on the conduct of environmental impact assessments. Furthermore, the Decree of the Minister of Demography and Natural Environment No. Kep.48/MENLH/II/1996 sets a safe noise level of 70 dBA within 10 meters distance from the source, while in some locations of the case study, the noise level reached 88 dBA. Similarly, dust content in the air reached 17 to 50 $\mu g/m^3$, which is much higher than the safe level of 0.26 $\mu g/m^3$ defined in the Decree of the Minister of Demography and Natural Environment No. 02/MENKLH/I/1988. These two laws are relevant during the land clearing period when establishing a plantation. The Law Lampiran B. KepMen LH No. Kep-512/MENLH/10/1995 and Decree No. Kep. 04-05/MENLH/I/1998 of the Ministry of Environment set very specific quality requirements on sewage and waste water from processing which is applied on fields or redirected into surrounding rivers. Control and enforcement of these decrees, however, is in most cases judged as weak, mostly due to the remoteness of the production areas. As in Brazil, there is no law specifically referring to the use of fertilisers or pesticides. However, for oil palms in Indonesia for example, the fertiliser requirements are often tested voluntarily by means of a leaf analysis.

Germany

In Germany, the production of wheat, rapeseed and broiler chickens plays an important role in agriculture. Within Europe, Germany is the second biggest producer of rapeseed. The area for rapeseed production expanded rapidly in the 1980s; however, it stagnated or even decreased in the 1990s. In 2000, the production of rapeseed amounted to close to 10% of world production. The export volume reached close to 1.9 million tonnes being dominated by processed products. At the same time, Germany was also a big importer of around 1.6 million tonnes of mainly unprocessed rapeseed. Within Europe, Germany is also one of the biggest oilseed processors. In 1998, close to 4 million tonnes of both, rapeseed and soybeans have been processed in Germany. While two third of the rapeseed is supplied domestically and one third is imported from the EU, soybeans for processing are all imported, mainly from the USA (1.8 million tonnes in 1998) and Brazil (1.3 million tonnes). About 20-25% of total agricultural land in Germany is used for wheat production. However, the share in world production amounted to only 3.6% in 2000. Exports of wheat amounted to almost 1.8 million tonnes in 1996/97, while imports reached some 1.2 million tonnes during the same year. The broiler production increased steadily in Germany but with imports still playing a major role for meeting the demand. The major supplier of the German market is still the Netherlands.

Environmental problems

Typical environmental problems occurring in German agriculture include decreasing soil fertility due to erosion and compaction caused by intensive mechanical cultivation and long periods of bare soils without protective plant cover, particularly in the cultivation of row crops like maize and rapeseed (Brink and Baumgärtner, 1989). In addition, acidification of soils and contamination of waters occur due to intensive use of mineral fertiliser (nitrogen and phosphate) and inappropriate treatment of organic manure (OECD, 1998). Biodiversity is endangered via the destruction of small-sized biotopes in the agricultural landscape such as pools and hedges due to consolidation of farmland; furthermore, there is a concentration on the growing and cultivation of few high yielding varieties and breeds (Brink and Baumgärtner, 1989).

In broiler production, environmental problems are mainly caused by the large volumes of waste and manure as well as odours from the poultry production units. A study of Thomann (1997) shows that in the so-called OBE-region in Lower Saxony in the northern part of Germany (covering the counties of Bentheim, Emsland, Cloppenburg, Osnabrück and Vechta), there is a nutrient surplus of phosphorus that cannot be compensated by plant growth anymore. Farmers operating in the area need to prove that they have sufficient own or leased land available for spreading the manure according to the limits presently set by the fertiliser regulation. In case sufficient land is not available, they may also conclude contracts over a longer time period with cooperatives which again sell the manure to other farmers.

Environmental legislation

In addition to several general environmental protection regulations that indirectly affect agriculture, there are environmental regulations that are especially aimed at agricultural production. Standards that have been considered relevant in this context relate to the protection of soil, water and air, the conservation of biodiversity and landscape as well as animal welfare. Most standards include relatively precise requirements referring for example to maximum permissible fertiliser rates or to stocking rates in broiler production.

The Forest Protocol or the Federal Law on Soil Conservation and the Law on Forest Soil establish a Code of Good Agricultural Practice to prevent soil erosion/compaction and the conversion of biodiversity areas. It also prohibits the burning of forests or deforestation. In addition, there is a Federal Law on the Conservation of Nature which determines requirements with respect to landscape protection (hedges, harvesting time) and preservation of species.

Given the relatively high livestock density and relatively high intensity of production, very strict regulations for organic fertilisation apply in Germany. To prevent soil degradation, the Federal Law on Soil Protection which has been passed in 1998, requires that farmers comply with the Code of Good Agricultural Practice in mineral and organic fertilisation. Furthermore, Directive (EC) 91/676 regulates the use of mineral and organic nitrogen fertilisers for the protection of water and soil, and is translated into German law through the so-called Fertiliser Regulation. It specifies for example, that annual nitrogen balance statements have to be produced for farms with more than 10 ha. The application of nutrients has to be based on plant demand considering the conditions of location and production on the individual farms. Violations can be penalised as an offence.

Certain regulations and restrictions or prohibitions of the use of pesticides result from the Pesticide Act supplemented by for example, the Ordinance for the Protection of Bees from Risks associated with Pesticides or the Pesticides Maximum Quantities Ordinance, and the Sewage Charge Act. They set limits or ban the application of pesticides depending on the type, amount, and time of application. In addition, the laws determine the regular control of sprayers. Compared to other countries of the EU, Germany has set relatively high standards to protect human health, animals, and ground water. For example, the use of pesticides such as atrazin, dicofol or captan has been banned in Germany.

Machines and equipment have to be controlled regularly at the German Technical Supervision Agency (TÜV) to ensure user safety and to prevent technical failures that may lead to pollution. There are also requirements for a special facility area designated for cleaning machines and equipment. Farmers who decide to build these facilities have to construct a concrete area with oil separator and sewage treatment plant. Also for the establishment of a gas station, the farmers have to cover the cost for technical supervision, for a building permission and for a retaining basin.

For the German oilseed processing mills, strict standards have been established for explosion protection, air pollution control and the treatment of waste water. These differences refer not only to investments needed to meet environmental standards but also to operating cost.

With respect to broiler production, there are a number of laws referring to the reduction of emissions and the protection of the welfare of animals in Germany.

- Federal air pollution law and law on environmental impact assessment: compulsory licensing of intensive livestock farming (for stables with more than 14,000 chickens); minimum distance to communities and running/stagnant waters; filter facilities; additional environmental impact assessment is needed for stables with more than 85,000 chickens.
- Decree on fertilisers; law on emission of sewage; VDI guidelines (reduction of emission with respect to livestock farming); techniques of manure storage, manure and ventilation techniques; sewage disposal.
- Law on the assessment of environmental sustainability: building permits and assessment of environmental sustainability; regulations on buildings and other facilities; impact assessments when intruding on nature and landscapes.
- Decree on poultry meat hygiene. Voluntary agreement of the poultry finishers of Lower Saxony on population density: inspection by veterinarians (health of the animals in general; population density when keeping in free-range conditions).
- Law on building and construction: building codes on insulation, ventilation, lighting, structural engineering, protection against fire and appearance.

There is also a decree on animal protection during transport. This decree sets guidelines for the transport of broiler chickens to the slaughterhouse, including requirements on the transport time period, size of the transport cages etc.

The enforcement of German regulations are generally judged as being good. Policy on agricultural environment includes control and incentive elements. Legal regulations for the protection of air and water have mostly control elements, while incentive elements like grants are included in measures aiming at the protection of resources, nature and landscapes (Neander and Grosskopf, 1996).

Cost implications of environmental standards

The methodological base for the detailed cost analysis at production level was the International Farm Comparison Network (IFCN) which defines typical farms in different regions of the three countries. For each farm, a comprehensive physical and economic data set was obtained in close consultation with farmers to calculate the total production costs. In Brazil, three typical farms for plant production and two broiler producers were chosen. In Germany, a total of five typical farms were defined, while in Indonesia, the analysis was based on two palm oil plantations. These typical farms and plantations also provided the information about the environmental standards and laws. Given the standardised procedure and identical definition of cost positions, IFCN allows an international comparison of data. At processing level, the cost data is based on balance sheets and expert interviews. Those figures have to be interpreted as guidelines only.

The cost analysis has shown that the impacts of environmental standards on total production costs of the typical farms are low (Grote *et al.*, 2001). In Germany, environmental standards are responsible for 0.3-4.4% of the total production costs for rapeseed and grains on the typical farms. These additional costs derive from higher technical safety standards for regular control of pesticide devices (e.g. < US$0.002 per 100 kg grain) and special facility area (e.g.

US$0.006-0.03 per 100 kg grain) as well as from environmental standards in plant protection and fertiliser use (soil sampling, e.g. US$0.03 per 100 kg grain). In plant protection, simazin is banned from use in grain production in Germany. To quantify the cost of this ban, additional costs for German farmers for using more expensive substitutes were calculated and amounted to US$0.4 per 100 kg grain. Neglecting and not enforcing current environmental laws in Brazil (Forestry Code) and Indonesia (Zero-Burning law) resulted in large cost savings, with possibly significant adverse environmental externalities (such as overextended land use and forest burning). In the case of Brazil, e.g. they were calculated to amount to 15-23% of the total production costs for the Brazilian soybean farmers.

The typical German broiler producer incurs additional costs of 2.7% of the total production costs. Costs caused by German regulations which do not exist in Brazil, were calculated and then deducted from the total production cost of the German broiler chicken farm. They include for example regulations on the number of windows needed in the stable, number of feed and drinking troughs or a platform for dung.

At processing level, the results are ambiguous. Environmental standards in the investigated German oil mills cause 5% of the total processing cost. For soybean in Brazil, the corresponding values range between 0.5 and 1% of the processing cost and for the two Indonesian palm oil producers between 0.4 and 1.1%. Based on expert interviews, costs of environmental standards for the processing of broiler chickens in Germany are estimated, however, to amount to 17% of the total processing costs compared with 4% in Brazil. The largest share of additional cost in the considered enterprises results from the voluntary use of the air-cooling system in Germany compared with the less expensive water-cooling system used in Brazil. The shift to the air-cooling system in Germany was based on the expectation that the farmers would receive a higher price for better quality of the final product through the carcass classification scheme. However, this change of the system has not been introduced yet.

The international comparison also revealed some significant differences in total cost between the typical farms. For example, with respect to rapeseed and soybean production, the costs of the Brazilian farms amount to 46-73% of the costs of the German farms. For the oil palm production in Indonesia, even lower costs were calculated. Compared with these values, the costs deriving from environmental standards as a percentage of total cost, are insignificant for the international competitiveness of German farms. Other factors like the wage level, prices for land, machines, buildings and equipment are the decisive determinants of the total international cost differences and thus competitiveness. But even if the environmental laws in Brazil and Indonesia were strictly enforced, a significant comparative advantage in agriculture would remain for the typical producers of oilseeds and fruits as well as broiler chickens.

Comparing environmental standards and assessing their transferability

A comparison of the environmental legislation in the three countries reveals the following differences:

- In Germany, there are more and relatively higher environmental standards with respect to the selected agricultural products than in Brazil and Indonesia.
- The enforcement of the laws is better in Germany. However, it must be mentioned that in Brazil, some environmental legislation has been developed by individual states at the more local level with better enforcement than the federal laws.

However, an international comparison of environmental standards and their relevance cannot be given without regard to the different production systems and their environments:

- With respect to plant production, the following features have been identified: First, Brazil and Indonesia use pesticides (atrazin, simazin in Brazil and sevin [carbaryl] in Indonesia) which are no longer approved in Germany. This has to be judged against the following background information: First, all three pesticides are approved and very important in other EU countries and the USA; second, economic reasons (small market) have prevented the chemical industry from extending the permit of a pesticide; and third, the pesticide is not suitable for local environmental conditions (e.g. hazardous for the European honey bee which does not exist in Asia).
- The intensity of production and also the level of yields on typical Brazilian farms in oilseed and grain production are lower than the German ones. The nutrient efficiency differs depending on the products and local conditions. To draw conclusions from any of these figures about the necessity or availability of internationally harmonised environmental standards for fertiliser, e.g. for the protection of ground water, is therefore not possible and inappropriate.
- At processing level, the need for and the setting of the standards is mainly determined by processing methods and natural climatic conditions. For example the regulations for buildings with respect to explosion and emission protection for rapeseed processing are very strict in Germany. For the processing of soybeans in Brazil, these standards are of no relevance because the processing takes place in open buildings. There is no danger of explosion due to the accumulating hexane. Palm oil in Indonesia is gained through pressing, rather than extraction.
- In broiler production, the environmental and animal welfare protection standards in Germany are higher than in Brazil. However, it cannot be concluded that the animal welfare conditions in Germany are better than in Brazil. Many of these regulations refer to buildings which are not relevant for Brazil where the broiler chickens are normally kept in open Louisiana-type stables. Lower stocking rates have been found on the typical farms in Brazil. However, it also cannot be concluded that animal welfare conditions are therefore better in Brazil, because good health of the broilers also depends, e.g. on the temperature. In addition, experience has shown in Brazil that higher stocking rates resulted in increased competition for food and water of the animals, leading to skin damage and rejection by consumers.
- The German regulations regarding odour emissions and landscape protection are relatively strict. The introduction and implementation of these regulations

is mainly based on the high population density and given preferences of a society for a high environmental quality compared with the other two countries where the per capita income level is significantly lower. In Brazil, these regulations play only a minor role since the broiler producers are mostly located far from settled areas with relatively low population density.

IMPLICATIONS FOR WTO ACTIONS

The review and comparison of the environmental legislation in Brazil, Indonesia and Germany show adjustments to local conditions and development levels, and indicate that a transferability of environmental standards to other environments is often inappropriate. Existing procedures in the TBT and SPS Agreements take this into account and allow countries to establish their own standards. The TBT Agreement even proposes the preservation of indigenous technologies and production methods. Nevertheless, the WTO is bound to be invoked more often in the future as the number of process-based rather than product-based standards and regulations increases.

Several non-product-related PPMs have been identified in the case studies like the burning of forestland which results in the production of dangerous and toxic emissions like smoke; the use of herbicides which affect drinking water but are not detectable in the final agricultural product; or the potential degradation of natural ecosystems converted to crop production. Since these PPMs are not covered under WTO, trade limitations cannot be applied. The danger of green protectionism is considered high if the use of non-product-related PPMs was permitted under WTO. In addition, an enforcement of non product-related PPMs with the help of trade measures has no impact on the actual root environmental problems. Instead, the national enforcement of already existing environmental regulation needs to be strengthened - with the help of technical assistance granted to developing countries under the WTO Agreements, and only where global environmental goods (like rainforest) with transboundary impacts (e.g. severe smoke and haze in neighbouring countries, caused from land clearing with fire in Indonesia) are concerned, they should be defined and controlled within MEAs (Grote et al., 2001).

It is noted in the Article on Special and Differential Treatment of the TBT Agreement, that developing country members should not be expected to use international standards which are not suitable for their development levels or financial capacity. Nevertheless, more conflicts arise in the light of the problems which developing countries face in meeting international standards as set by the TBT and SPS Agreements. The WTO does not set international minimum standards on its own, but makes reference to the work of existing international organisations as the Codex Alimentarius Commission. Through such reference, these fixed standards gain much more importance, since they have to be applied under the obligatory international law of WTO and its sanction mechanisms. There is a need to strengthen the capacity of developing countries to ensure their participation in the development of environmental standards. The TBT and SPS Agreements both include provisions for developing countries - either in terms of technical assistance or in terms of extending deadlines for the introduction of standards; however, these endeavours are often not sufficient to achieve a long-term improvement.

There is uncertainty, to what extent existing national environmental legislation is implemented. In Brazil and Indonesia, the enforcement is often judged poor. For better enforcement, environmental agencies have to be strengthened at national, regional and local levels. An efficient control and sanction system would help to ensure a more efficient implementation of the existing laws. For example, negative consequences of burning could be fought with better controls and incentives at the local level. While bans on burning have not proved effective, temporal restrictions of burning during the years of El Niño or during other critical periods may be enforceable. In addition, the adoption of other forms of land clearing should be promoted. Also a change in forest policy is needed to make the sale of timber products in the native country more attractive, so that the products would be sold rather than burned. Limiting licenses for the establishment of tree plantations, or the introduction of a license for burning would reduce smoke and other problems of uncontrolled burning in Indonesia (ICRAF and ASB Indonesia, 1998).

According to the innovation-oriented approach, competitiveness should be discussed on the basis of product quality. Securing high product quality through eco-labelling could represent a comparative advantage at international level. Higher product quality justifies higher prices for labelled products giving an incentive to the producer to pay more attention to the environment during production or processing (Grote et al., 1999). Thus, the higher price margins can compensate for or exceed the relatively low cost margins which result from the consideration of environmental standards. This is equally true for PPMs and for product standards.

In the case studies, eco-labelling for palm oil that has been produced without burning as a method of land clearing is an option to develop incentives for sustainable management of oil palm plantations (Glover and Jessup, 1999). Individual enterprises like Unilever use the practice of sustainable palm oil production in Malaysia or Kenya as a marketing instrument in order to strengthen their competitiveness[2]. The Indonesian Ecolabeling Institute (LEI) was established in 1998 to certify timber. The first official certification process started in mid-1998. Major challenges in implementing the certification process arise from the need to build capacity for assessing and monitoring logging operations and to create an efficient administration. In the longer term, LEI also considers certification of plantations. In Brazil, an eco-label was introduced for sustainable soybean production[3]. Control and international certification take place through the Biodynamic Institute in Brazil. However, to reduce the risk of conflicts, the transparency of eco-labelling schemes must be increased. The criteria selected for eco-labelling should not only be based on domestic environmental priorities and technologies, without reflecting the environmental conditions in developing countries. In addition, an independent monitoring of the schemes is needed so that consumers do not lose their trust in eco-labels because of inappropriate use.

---

[2] cf.: http://www.unilever.com/public/env/review/environ/public/05food/05foodfr.htm>
[3] cf. homepage: <http://www.laser.com.br/ibd/proyeder.htm>

CONCLUSIONS

The results of the case studies have shown that the fear about the cost implications of environmental standards for some typical agricultural producers in Brazil, Indonesia and Germany and their impact on the international competitiveness is not justified. In Germany, environmental standards cause additional costs of only 0.3-4.4 % of the total production costs for rapeseed, grain or broiler. In Brazil and Indonesia, no actual costs to meet environmental standards have been identified. However, current environmental laws do exist such as the Forestry Code in Brazil and the Zero-Burning law in Indonesia. Neglecting and not enforcing these laws result in large cost savings for the typical producers - with possibly significant adverse environmental externalities (such as overextended land use and forest burning). In the case of Brazil, for example, they were calculated to amount to 15-23% of the total production costs for the Brazilian soybean farmers. However, even if the typical Brazilian and Indonesian producers would meet the environmental standards and cover the compliance cost, their total production costs would be still significantly below the total costs of the typical German producers. At processing level, the results were more ambiguous.

The number and complexity of environmental standards increases with the level of development. And so does the enforcement of environmental laws. This has been shown for comparable products, e.g. vegetable oils, from Brazil, Indonesia and Germany. However, most of the stricter standards which are to be found in Germany are of no relevance for the typical farms and plantations in Brazil and Indonesia (e.g. standards for buildings).

Environmental regulation seems to be well adjusted to the country-specific environments and conditions, like climate, population density and the differences in the availability of natural resources. To a large extent it is also due to national economic and social conditions, including the state of development of the country. The existing procedures in the TBT and SPS Agreements consider this need by allowing individual countries to set their own environmental standards, with some special and differential treatment for developing countries. However, enforcement of the existing national legislation in Brazil and Indonesia needs to be strengthened.

The often significant costs to a country or enterprise which originate from neglecting environmental protection, need to be considered more fully when evaluating environmental standards. Repetto (1995) estimates, that the economic costs annually arising in industrialised countries through environmental damage account for 1-2% of the GDP and even up to 4% in newly industrialising countries. In the case of Indonesia, it has been calculated that the total economic costs of the 3-months haze and fire period in 1997 amounted to more than US$4 billion. This estimate includes among others the medical costs, productivity loss, impacts on timber, agriculture and biodiversity, fire-fighting costs and tourism impacts due to fires and smoke (Glover and Jessup, 1999). At processing level, the case studies showed ambiguous results with low compliance costs in oilseed processing but relatively high costs in broiler processing in Germany.

Further research is needed regarding the assessment of environmental damage and the sustainability of production systems in individual countries. There is debate in Brazil, whether and to what extent increasing soybean production affects the Cerrado ecosystem. In Indonesia, there is uncertainty about the effects of the conversion of forest into oil palm plantations. In Germany, environmental problems evolve from the relatively high intensity of production whose environmental costs need to be more closely assessed.

Eco-labelling offers the opportunity to give incentives to farmers to use environmentally-friendly PPMs. Some initiatives already exist in developing countries, as has been seen for palm oil and soybeans. However, financial and technical assistance for the development of standards, their enforcement and for the establishment of own certification and monitoring systems are needed in developing countries. Transparency of and trust in eco-labelling programmes need to be increased to avoid that the programmes are used as non-tariff trade barriers.

REFERENCES

Anderson, K. (1995) The entwining of trade policy with environmental and labour standards. In: Martin, W. and Winters, L.A. (eds) *The Uruguay Round and the Developing Economies.* World Bank, Washington D.C., pp. 435-456.

Barber, C.V. and Schweithelm, J. (2000) *Trial by Fire. Forest Fires and Forestry Policy in Indonesia's Era of Crisis and Reform.* World Resources Institute, Washington D.C.

Bayer (1994) *What does El Niño have to do with growing soya beans in Brazil?* In: Courier Agrochem 1/94, Bayer AG, Leverkusen.

Brink, A. and Baumgärtner, M. (1989) *Wachstumslandwirtschaft und Umweltzerstörung.* Vol.1, Rheda-Wiedenbrück.

Chang, Seung Wha (1997) GATTing a Green Trade Barrier. *Journal of World Trade,* 31, 137-159.

Dean, J.M. (1992) Trade and environment: A survey of the literature. In: Low, P. (ed.) *International Trade and the Environment.* World Bank, Washington D.C., pp. 15-28.

Dean, J.M. (1999) Testing the impact of trade liberalization on the environment: theory and evidence. In: Fredriksson, P.G. (ed.) *Trade, Global Policy, and the Environment.* World Bank Discussion Paper No. 402. World Bank, Washington D.C., pp. 55-64.

Felke, R. (1998) *European Environmental Regulations and International Competitiveness.* Nomos Publisher, Baden-Baden.

Glover, D. and Jessup, T. (1999) *Indonesia's Fires and Haze. The Cost of Catastrophe.* IEAS and IDRC, Singapore and Ottawa.

Grote, U., Basu, A.K. and Chau, N.H. (1999) *The International Debate and Economic Consequences of Eco-Labeling.* Center for Development Research, ZEF Discussion Papers on Development Policy No.18, Bonn.

Grote, U. (2000) Sustainable development in the flower sector with eco-labels? In: Humboldt-University Berlin (publ.), *ATSAF-proceedings of Deutschen Tropentag 1999 in Berlin.*

Grote, U., Deblitz, C., Reichert, T. and Stegmann, S. (2001) Umweltstandards und internationale Wettbewerbsfähigkeit: Analyse und Bedeutung - insbesondere im Rahmen der WTO. Vauk publisher, Kiel.

Helm, C. (1995) *Sind Freihandel und Umweltschutz vereinbar?* Wissenschaftszentrum, Berlin.

ICRAF and ASB Indonesia (1998) *Alternatives to Slash-and-Burn in Indonesia.* Summary Report, International Centre for Research in Agroforestry, Bogor.

Jaffe, A.B., Peterson, S.R., Portney, P.R. and Stavins, R.N. (1995) Environmental regulation and the competitiveness of U.S. manufacturing: What does the evidence tell us? *Journal of Economic Literature,* 33, 132-163.

Jenkins, R. (1999) *Environmental Regulation and International Competitiveness - A Framework for Analysis.* Conference paper, University of East Anglia.

Kalt, J. (1988) The Impact of Domestic Environmental Regulatory Policies on US International Competitiveness. In: Spence, M. and Hazard, H. (eds) *International Competitiveness.* Cambridge, MA., Harper and Row, Ballinger, pp. 221-262.

Low, P. and Yeats, A. (1992) Do 'dirty' industries migrate? In: Low, P. (ed.) *International Trade and the Environment.* World Bank Discussion Paper 159. World Bank, Washington, D.C., pp. 89-104.

Mani, M. and Wheeler, D. (1999) In search of pollution havens? Dirty industry in the world economy, 1960-1995. In: Fredriksson, P.G. (ed.) *Trade, Global Policy, and the Environment.* World Bank Discussion Paper No. 402, World Bank, Washington D.C., pp. 115-128.

Neander, E. and Grosskopf, W. (1996) Agrarpolitik für eine 'nachhaltige' Landwirtschaft. In: Linck, G., Sprich, H., Flaig, H. and Mohr, H. (eds) *Nachhaltige Land- und Forstwirtschaft.* Springer-Verlag, Berlin, pp. 543-564.

Nordström, H. and Vaughan, S. (1999) *Trade and Environment.* WTO, Geneva.

Organisation for Economic Co-operation and Development (OECD) (1997a) *Eco-Labelling: Actual Effects of Selected Programmes,* OECD, Paris.

Organisation for Economic Co-operation and Development (OECD) (1997b) *The Effects of Government Environmental Policy on Costs and Competitiveness: Iron and Steel Sector,* DSTI/SI/SC(97)46, OECD, Paris.

Organisation for Economic Co-operation and Development (OECD) (1997c) *Brazilian Agriculture: Recent policy changes and trade prospects.* OECD, OECD Working Papers, Vol.V, No. 55, Paris.

Organisation for Economic Co-operation and Development (OECD) (1998) *Agriculture and the Environment. Issues and Policies.* OECD, Paris.

Porter, M. (1991) America's Green strategy. *Scientific American,* April, p. 68.

Porter, M. and Van der Linde, C. (1995) Green and competitive. *Harvard Business Review,* September-October, 120-134.

Repetto, R. (1995) *Jobs, Competitiveness, and Environmental Regulation.* World Resources Institute (WRI), Washington D.C.

Sampson, G.P. (1999) *Trade, Environment, and the WTO: A Framework for Moving Forward.* Overseas Development Council, ODC Policy Paper, Washington D.C., February.

Sorsa, P. (1994) *Competitiveness and Environmental Standards: Some Explanatory Results.* World Bank, World Bank Policy Research Paper 1249. Washington D.C.

Thomann, B. (1997) *Mengen- und Nährstoffvergleich organischer Rest- und Abfallstoffe für die OBE-Region.* ISPA-Mitteilungen Heft 30, Oktober 1999, Institut für Strukturforschung und Planung in agrarischen Intensivgebieten, Hochschule Vechta.

Tobey, J. (1990) The effects of domestic environmental policies on patterns of world trade: An empirical test. *Kyklos,* 43, 191-209.

Tobey, J. (1993) The impact of domestic environmental policies on international trade. In: OECD, *Environmental Policies and Industrial Competitiveness.* OECD, Paris, pp. 48-54.

United Nations Conference on Trade and Development (UNCTAD) (1994) *Sustainable Development. Trade and Environment - The Impact of Environment-related Policies on Export Competitiveness and Market Access.* UNCTAD, TD/B/41 (1)/4, Geneva.

United Nations Conference on Trade and Development (UNCTAD) (1995) Trade, Environment and Development Aspects of Establishing and Operating Eco-Labelling Programmes, UNCTAD, TD/B/WG.6/5, Geneva.

United Nations Environmental Programme (UNEP) and International Institute for Sustainable Development (IISD) (2000) *Environment and Trade.* IISD, Manitoba.

Vaughan, S. (1994) PPMs and International Environmental Agreements. In: OECD, *Trade and Environment: Processes and Production Methods.* OECD, Paris, pp. 127-136.

Vossenaar, R. (1997) Eco-Labelling and International Trade: The Main Issues. In: Zarilli, S., Jha, V. and Vossenaar, R. (eds) *Eco-Labelling and International Trade.* Macmillan Press and St. Martin's Press, Geneva, pp. 21-36.

Wyatt, C. (1997) Environmental Policy Making, Eco-Labelling and Eco-Pakaging in Germany and its Impact on Developing Countries. In: Jha, V., Hewison, G. and Underhill, M. (eds) *Trade, Environment and Sustainable Development. A South Asia Perspective.* St. Martin's Press, New York, pp. 51-68.

Public Concerns and Consumer Behaviour in Japan

Theo H. Jonker and Ikuo Takahashi

INTRODUCTION

Farmers and the agri-food industry (including food processors and retailers) increasingly need to respond to the wide range of consumer demands, both domestically and globally. The current chapter identifies the role of public concerns from the perspective of a major importer of food. Japan is highly dependent on the imports of food products for its calorie supply. The self-sufficiency rate of food production is only 40%, if measured in calories. The country mainly depends, among other products, on the imports of beef, pork and poultry. Government policy aims to increase the self-sufficiency rate in the next 30 years to reach a level of some 50%. Increasing the efficiency of farming is one of the means to achieve that. In total there are around 3 million farmers, and only 5-10% are considered to be efficient producers. Only 4% of the farmers aged below 65 years are employed in agriculture on a full-time basis.

This chapter examines the role of public concerns in Japan. We offer insight into how Japanese consumers perceive issues, such as human health, environment and animal welfare in relation to food products. Furthermore, the chapter identifies strategies chosen by traders and food-industry in response to these perceptions.

IMPORTS AND DEGREE OF SELF-SUFFICIENCY

Japan is the largest net importer of agricultural and food products in the world. In 1996-1997, imports of agricultural and food products (excluding marine products and wood) amounted to US$40 billion and exports were US$1.6 billion (Silvis and Van Bruchem, 1999, p. 44). In 1997, meat and meat preparations had a share of 19% of total import value of agricultural and food products (Jonker, 1999, p.

30). In that year, the main countries exporting agricultural products to Japan are the USA (a share of 38%), followed by China and Australia, each with 9%, and Canada with 5% (Toda, 2000, p. 4). According to this author most of the population is well aware of the fact that their country is the largest net importer of agricultural and food products in the world and many of them are concerned about it. He indicates that resolving this situation is one of the main reasons behind drafting new legislation, namely the Basic Law on Food, Agriculture and Rural Areas, enacted in July 1999. The new law 'aims at raising a food self-sufficiency ratio, recognising it as one of the major political targets of the country to be attained' (Toda, 2000, p. 1). During the period from 1965 to 1998, the food self-sufficiency ratio in Japan showed a sharp decrease. It decreased from 73% to 40% on a calorie supply basis and from 62% to 27% on a grain basis (MAFF, 2000, p. 8).

Concerns by Japanese consumers about the dependency of the country on imported food are related to a lack of information about the production methods applied in agriculture and the human-safety aspects of agricultural products. According to Toda (2000, p. 31), although the Japanese government stresses the safety of agricultural products in the market place, strong objections are voiced by some scientists. Since the consumers may have difficulty in judging such scientific results, they have doubts about the safety of agricultural products. This applies especially to imported food, as well as production processes applied domestically[1]. Consumers are mainly concerned about food safety issues related to imported food. These concerns are partly induced by the mass media. Information in the press often is presented in a sensational manner, and insufficient to present factual information in an unbiased manner.

In the past decades, imports of agricultural and food products have increased. Imports increased in particular for specific commodities, including fresh vegetables, fruits, meat and milk and dairy products. Food imports increased due to liberalisation of agricultural trade, a decline in domestic production, changing consumption patterns and the high exchange rate of the Japanese yen. Structural features of the agricultural sector also influenced the increase of imports of agricultural and food products. Japanese agriculture is characterised by, among other things, small farm size and higher production costs of agricultural products as compared to the major exporting countries in the developed world. The latter is at least in part due to the high economic growth rates. Domestic farmers face major difficulty to compete on the global market and liberalisation of agricultural trade was therefore synonymous to abandonment of domestic production. The types of agriculture that are least competitive are the ones which depend mainly on agricultural land. Efforts have been made to compete against imported products through improving the quality of agricultural products. In order to achieve this, the financial resources needed are mobilised and improved technologies are adopted. However, domestic agricultural products, which could

---

[1] Conclusions are drawn from the Food Consumption Monitoring Survey from April 1998. Some 99.1% of a total sample of 1,021 housewives did respond and results are presented in 'Food Safety' by MAFF. According to this survey, almost 90% of the respondents felt anxious about imported raw materials of food, and slightly more than 80% did feel concerned about the production methods applied in agriculture.

not secure the highest quality demands, or those, which were less adapted to climatic and soil conditions, were replaced by imported products. Table 11.1 illustrates the trends in agricultural production since the mid-1960s. Gross agricultural production shows a declining trend during the 1990s. Furthermore, the share of food expenses in total living expenses gradually decreased from 40% during the 1960s to reach a level of around 26% since the mid-1990s.

Table 11.1 Agricultural production and food consumption in Japan

Year	Gross agricultural production[a] (100 million yen)[c]	Living expenses[b] (per household) (yen)	Expenditures on food[b] (per household) (yen)	Share of food in total living expenses (%)
1965	31,769	580,753	232,305	40
1970	46,643	954,369	346,145	36
1975	90,514	1,895,786	649,887	34
1980	102,625	2,766,812	867,393	31
1985	116,295	3,277,373	957,528	29
1990	114,927	3,734,084	1,030,125	28
1991	114,869	3,994,772	1,093,797	27
1992	112,418	4,070,693	1,100,050	27
1993	104,472	4,073,758	1,088,181	27
1994	113,103	4,062,079	1,074,313	26
1995	104,498	4,008,832	1,041,108	26
1996	103,166	4,005,152	1,037,955	26
1997	99,113	4,071,303	1,053,715	26
1998	99,441	4,009,765	1,049,988	26

Sources: a) MAFF (1995, 1999) and MAFF-website; b) Annual Report on the Family Income and Expenditure Survey, Statistics Bureau, Management and Co-ordination Agency; data for 1991 to 1998 from MAFF (1995, 1999). Data for 1965 to 1990 from Research and Information Service for Food and Life (2000); c) 1 US$ equals 113.7 Yen (31 December 1998).

The new Basic Law on Food, Agriculture and Rural Areas intends to limit the dependence of the country on imported products, and the legislation strives for an increase of the food self-sufficiency ratio. Legislation was at least in part a response to the increasing imports of agricultural and food products. This new legislation may affect consumption patterns and, consequently, affect food imports. The Basic Plan - which is an integral part of the new Basic Law - for Food, Agriculture and Rural Areas, is to set up specific targets of food self-sufficiency ratio. The Annual Report on Food, Agriculture and Rural Areas in Japan (MAFF, 2000, p. 8) explains the following: '[in] the Basic Plan developed in March 2000, major target food self-sufficiency ratios (2010 as a target year) were set up; 45% on a calorie supply basis, 30% on a grain basis, and 62% on a grain-for-staple foods basis respectively. Although it is essential that more than 50% of all calories supplied by foods should be covered by domestic production, these ratios were determined in light of the attainability by 2010.' Furthermore, MAFF (2000, p. 8) also indicates that domestic agricultural production will be promoted: 'The national government assumes great responsibility in assuring the availability of the food supply to its people. Since there are certain limitations on stockpiling and importing, it is important to increase domestic agricultural production as much as possible in order to secure a stable food supply.'

In March 2000, the Japanese Ministry of Agriculture, Forestry and Fisheries (MAFF) proclaimed new guiding principles for dietary life (i.e. eating habits). They are directed to the general public and replace the previous guiding principles dating from 1985. The new guiding principles call on, among other things, paying attention to one's health, but also - indirectly - promote the consumption of Japanese-style food. The mainstream consumer considers Japanese-style food as more wholesome than non-Japanese food. This thought is also promoted by different publications. Makuuchi (1995) has been a best seller, promoting a drastic change in eating habits. The author argues that the Japanese should only consume traditional Japanese food, mainly because it is better for their health. Traditional Japanese ingredients are rice, vegetables, seaweed, fish and soybeans. Products like oil, fat, sugar, milk, pork, eggs, flour, bread and pasta, mainly originate from abroad. Makuuchi (1995, p. 87) also indicates, that alternatives are available for Western products, such as, rice for bread, soyamilk for cow milk, and sake for whiskey. These types of dietary recommendations may contribute to diminishing imports of agricultural and food products.

CONSUMER PREFERENCES AND PUBLIC CONCERNS

This section identifies some main views from trade and industry regarding the perceptions by consumers on issues related to human health, environment and animal welfare. We build on consumer perceptions developed by Blandford and Fulponi (1999). The objective is to indicate which perceptions most significantly influence the decisions by Japanese consumers to buy food products. Before doing so, we will present some key features of Japanese consumer behaviour.

Trends and conformism

Japanese consumers are highly susceptible to information from the media and other sources. As a consequence, booms - sudden interests for certain products - occur frequently, as the mainstream Japanese consumers follow trends. A boom can start very fast, but consumers can lose their interest in the particular product just as fast. Often, emotional aspects play a role in connection to these trends and publicity on health issues has a major impact. Well-known examples are the cocoa (for chocolate milk) boom in 1995 and the banana boom in 1999, following television programmes featuring these products. Mass communication and articles in the press largely affect consumer behaviour. Saison Research Institute (1998) concludes that the influence of mass communication is very strong, especially in the case of health and food.

The presence of these strong trends is sometimes used as an argument to show that Japanese consumers become more individualistic and wish to express their individuality. However, one should judge on it in a more balanced way. The importance of individuals conforming to group preferences is emphasised by the following statement.

> Yet someone who simply wears or acquires the latest fashion - which is overwhelmingly the case in fad-obsessed Japan - is simply following the crowd, and is surely demonstrating the inadequacy of their individuality rather than the strength of it. (...) By identifying with a fad they are really stating that their 'individuality' conforms to a group preference. It is just another form of group identity (Henshall, 1999, p. 157).

Although the observation of Japanese following trends might be an argument in favour of non-conformism, in fact, it is just another argument for the conformism of the mainstream consumers. Henshall (1999) puts it as follows:

> Displays of deliberately outrageous non-conformist behaviour are also quite common. However, such behaviour is usually selectively outrageous, sufficient to send a protest message of defiance to teachers, parents, and other elders and figures of authority but not sufficient to incur peer disdain or serious and permanent marginalisation by mainstream society. Typically it involves such statements as orange-dyed hair, body piercing, or 'way out' clothes. Paradoxically, some of this 'non-conformist' behaviour is patently conformist, but to a different set of norms. (...) In other words, in this and similar cases, the youth who appears to eschew the 'system' is not a true free-wheeling rebel, but one who has simply - and almost always temporarily - chosen an alternative set of norms to follow (Henshall, 1999, p. 116).

Schütte and Ciarlante (1998) also argue that the supposed Americanisation of the Japanese society is doubtful. Although the country faced a pervasive American influence over the past 50 years, the inhabitants remain Japanese in thought, behaviour and lifestyle. A transition of Japanese society from *hitonami* ('alignment with society') towards *seikatsusha* ('designing one's own personal lifestyle to reflect one's values') is widely debated. However, the country remains a highly conformist society.

Consumer perceptions in relation to food products

Good taste and high quality (which also includes safety for human health) are the main priorities - or even prerequisites - when consumers select a food product. Main arguments for Japanese consumers to choose commodities are freshness of the product, its appearance and place of origin. Freshness is by far the most important issue. It even plays a role in the case of long-life products. Japanese consumers also connect freshness to health.

Packaging and wrapping relate to a product's appearance and, thus, they are important. For example, in the case of confectionery, items are wrapped individually. Other reasons for the individual wrapping are safety and freshness, but it may also be connected to Japan's cleanliness culture.

Consumers also attach importance to the region or country of origin. Knowledge on the place of origin gives consumers more confidence about the safety of a product. In the case of many food products, the fact that it is domestically produced gives the consumer a positive feeling about its quality and safety. Several serious food-poisoning scandals occurred in 2000, and various products were involved, including contaminated milk, sour desserts, mouldy buns and rotten tofu. The incidents, caused by negligence of high-profile food companies, received much attention in the media. Such food-poisoning scandals by domestic food manufacturers may affect the consumer's attitude, but there is no evidence whether it has diminished the consumers' confidence in domestically produced food products.

Various consumer surveys are available about the perception of domestic versus imported food products, including a 1999 survey by the Japanese Consumers' Co-operative Union among 295 Co-op members. The mainstream of consumers perceives quality (freshness, safety, taste and wholesomeness) of domestic food products to be higher than foreign products. Several arguments could be given in favour of the perceived quality difference between foreign and domestic food products:

- Japanese farms are small-scale, and farmers can therefore pay more attention to their products and take better care of it.
- Domestic products generally are more expensive than imported products, and consumers tend to link a higher price with higher quality.
- Transportation time required for the import of fresh products tends to be long, and Japanese consumers consider this requires treatment with chemicals or preservatives and, consequently, think that these products are not natural.

Despite this perception of quality differences between domestic and imported products, imports of agricultural and food products have risen. As indicated before, this development is mainly due to liberalisation of agricultural trade, a decline in domestic production, changing consumption patterns, high exchange rate of national currency and structural features of Japanese agriculture.

Japanese Consumers' Co-operative Union (1999) indicates that consumers do not feel reluctant to buy foreign food products if they have experience purchasing the particular product from abroad. A large share of the consumers in this survey mentioned they do not mind about the foreign origin of garlic (37.6%), asparagus (37.6%), paprika (45.4%), oranges (50.2%) and cherries (31.2%). Consumers tend to prefer domestic products for many other fresh foods. It generally also applies to processed food, given that consumers have experienced eating it before. Consumers, for example, do not mind about the foreign origin of cheese (60.7%), wine (62.4%) and canned tomatoes (58.0%).

Public concerns regarding health issues

Japanese consumers are far more concerned about health issues than about the environment and animal welfare. It is supported by different publications, including Saison Research Institute (1998) on reassurance and safety of food. This report indicates that consumers are most concerned about their health. We will now elaborate on this health concern and provide an insight into consumers' motives. Health issues are the main concerns in relation to food products and they play a significant role in the consumers' purchasing behaviour. However, it is difficult to pinpoint what health concern actually means and what the specific concerns are. One can wonder whether the current consumer attention on health is nothing more than a fad, especially in the case of young women. However, there is more to it than that. It is the image of health in its broadest meaning that plays a role. Furthermore, presently consumers are highly influenced by the media and opinion leaders. For example, several television programmes almost exclusively focus on health issues.

Attempting to pinpoint the specific anxieties of consumers in connection with health leads to the following. Japanese consumers suffer from stress and fatigue. They worry about the insufficient amounts of vegetables and wholesome food they consume. Furthermore, the fact that consumers suffer from allergies for artificial ingredients is often heard. Amemiya (2000) highlights that 72% of the respondents to a consumer survey mention the balance of nutrition to be important. The key words in the perceptions of Japanese consumers regarding health issues are nutritional balance and natural products.

There is ambivalence in the attitude of consumers concerning the importance of nutritional balance. Japan has a large number of fast-food restaurants, the consumption of instant and convenience food products is high, and the available space to cook at home is limited because kitchens are small. Therefore, consumers - especially in urban areas - realise that their intake of fat and calories is higher than it was some decades ago. They become more sensitive to wholesome food and the balance of nutrition, but consumers continue their eating pattern, although they are aware of the fact that it is not optimal for their health. Nonetheless, the consumption of salty snacks ('sunakku-gashi') has decreased, because it negatively affects one's health.

The Japanese consumer has a preference for 'natural' products and is concerned about artificial ingredients. This might be due to the Japanese culture that puts priority on purity. Aesthetics, particularly of purity, is more important than in most other parts of the developed world. Purity is not identical to undiluted or clean, it can include such concepts as perfection and normalcy (Henshall, 1999).

One significant motivation for this attitude of Japanese consumers is their anxiety about the future. Life expectancy of the population is among the highest in the world, and people are concerned to live their life in a healthy condition. The population also wants to maintain their good health for as long as possible. Many Japanese worry about the consequences of the ageing society and they feel uncertain about their financial position after retirement. They are also concerned about whether their future pensions will suffice and they wish to avoid personal medical expenses (in addition to the health insurance fee). Furthermore, the economic depression - and the attention about it in the media - contribute to their feeling of uncertainty. Japanese society also changes rapidly. The lifetime employment system is being replaced by an employment system that is driven by skills and competence. The concerns are present in all age groups, since the younger generation perceives their parents' worries. On the one hand, the young part of the population faces better opportunities to achieve their ambitions due to the changing employment system. On the other hand, however, the education system does not fully encourage individual ambition yet. A survey - undertaken by the Economic Planning Agency - indicates the significant changes that society has faced between 1986 and 1998 (Saison Research Institute, 1999). In 1998, consumers are concerned most about the economic situation (such as living expenses) (52%), their health conditions (50.2%) and their nursing care when they become elderly (29.5%). These figures showed a major increase since 1986. At that time, the shares of interviewees who had worries about these issues were 26.5%, 33.6% and 8.6%.

Public concerns regarding environmental issues

Environmental issues are wide ranging and include nature protection as well as pollution of the physical environment. The perception on the environment can be viewed in a number of ways. We will first focus on environmental concerns in general, and more specifically also look at packaging. This is considered relevant, because packaging is an important aspect in Japanese business culture.

The environmental awareness of mainstream Japanese consumers is increasing, but is significantly lower than in Western Europe. In Japan, the government is the driving force behind the consumers' environmental concerns. Although certain actions of consumers may in the first instance seem to stem from environmental concerns, they are in fact due to human health concerns. Consumers' personal health, and not the environment, is their motive to buy organic food, grown without use of pesticides and fertilisers, or vegetables and fruits with limited pesticide use. That can be deduced from, for example, the retailers' communication to the consumers, since their messages refer to health issues. Furthermore, the reason behind the discussions on waste incinerators and their dioxin emission is the fact that consumers increasingly suffer from allergies and bronchial and pulmonary symptoms. The announcement of the 'NO! PVC (polyvinylchloride) Campaign' by the Consumers Union of Japan (1999, p. 10) clearly mentions that it is directed against 'dioxins (hormone disrupting chemicals)'.

The environmental awareness also varies among the different consumer groups. Children learn about the environment at their schools. Both middle-aged and elderly people, as well as housewives raising children, care about the environment. It is mainly people in their late teens, twenties and - partly - thirties, who are not concerned at all. They are occupied with their starting careers and do not yet have the responsibility of raising children.

Japan is known as a country where the consumer considers the appearance of a product essential, and, as a consequence, packaging and wrapping as well.

Surplus packaging tends to be decreasing, and several supermarket chains promote the re-use of shopping bags. However, gift packaging remains to be important. A distinction should be made between ordinary products for personal use, and gifts. Packaging may decrease for the former. For the latter, the appearance will continue to be very important. Consumers are primarily concerned about how the present will be perceived by the recipient of the gift. One could say that social rules or virtues exceed environmental concerns. This notion also applies to products that one consumes in public. Then the appearance (i.e. packaging and brand name) plays a role.

Public concerns regarding animal welfare

Consumers of food products are hardly concerned about animal welfare issues (housing and raising of animals). One reason might be that animal welfare concerns are hardly observed in the final product. It does not directly affect product quality, but only relates to the production method applied. The main argument of consumers to buy meat from free-range chicken or pigs is a (purported) better taste, rather than the welfare of the animal.

Moving to a vegetarian diet for animal welfare reasons hardly exists in Japan. A limited share of the population only eat the traditional Japanese-style products (Makuuchi, 1995), and they are mainly connected to the Buddhist religion. Furthermore, there are special Buddhist-style restaurants where vegetarian meals are served. However, the mainstream of Japanese population sees such restaurants as an alternative to other restaurants, serving a specific type of food. They typically do not visit such restaurants for reason of animal welfare concerns.

In case the media would give more attention to the issue, animal welfare concerns may become more important in the nearby future. Furthermore, the saturated Japanese market requires manufacturers and retailers to look for new products and for new features of existing products. Since attention to animal welfare is something new in the Japanese marketplace, it can attract consumers. In case animal welfare becomes a new issue, it might be the main argument for consumers to buy these products instead of the manufacturer's corporate policy.

PUBLIC CONCERNS AND THE RESPONSE BY THE AGRI-FOOD SECTOR

This section examines the main response by domestic companies in the agri-food supply chain to consumer perceptions on human health, environment and animal welfare. Emphasis is given to any adjustments made in the corporate strategies of companies, including the organisation of their supply chain and their marketing mix. We will first make some remarks related to the corporate strategy. Then, some observations are made related to the marketing mix variables, and classified according to the four Ps of marketing, viz. product, price, place (i.e. distribution) and promotion (Kotler, 2000, p. 15). Some concluding remarks are made regarding the way government responds to the consumer perceptions. We will focus on issues related to human health, since they are most significant to the consumers.

Corporate strategy

One might argue that the Japanese consumer, especially in the major urban areas, shows many characteristics that are similar to those in other parts of the developed world. However, in reality traditional Japanese values remain and the consumer's behaviour and mentality are still rooted in the Japanese tradition. Schütte and Ciarlante (1998) argue that consumers may differ widely across the globe, whereas companies expand their efforts and competition becomes global. Cultural aspects in Asia differ largely from western cultures. Such differences are reflected in consumer behaviour, in terms of taste, preference, perceptions and motivations. The authors also argue that, although similar products might be available in different countries, this would not imply similar patterns of consumer behaviour. Asian countries experience modernisation of society, which however is dissimilar to westernisation. Although McDonalds and Domino pizza are popular in Asia, 98% of all restaurants remain to serve the food indigenous to the local region. In order to support this, Henshall (1999) argues that much of society in Japan has an American flavour. However, it would be a mistake to ignore the specific Japanese situation, and companies like McDonalds and Domino offer special types of menus to adapt to the demands of Japanese consumers. This again requires a product-marketing mix, which needs to be adjusted to the specific Japanese situation.

Product

In order to be successful on the highly saturated and competitive Japanese market, a product needs to have something extra. Thus, manufacturers and retailers permanently search for additional or new features to make their products stand out and catch the eyes of the consumers. High quality, good taste, perfect appearance and good service are prerequisites, and a product needs something in addition. Food producers look for or develop features in line with the public concerns related to human health. Innovative products (i.e. functional foods) are introduced, having a function or making health claims. In addition, manufacturers and traders highlight features of certain products. Some examples of the latter are presented in the following section, since it relates to the promotion variable of the marketing mix.

Eurofood (2000) indicates a world-wide shift away from less harmful products - with lower fat, less cholesterol or reduced salt - towards products, which actively enhance human health. Products that are less harmful to health are taken over, because they do not promote health in the way that functional foods can lower cholesterol or provide added vitamins.

Current marketing approaches of Japanese food manufacturers continue to focus on less harmful products, such as soy sauce (with reduced-salt content), confectionery (with lower-sugar content), and oil (with less 'bad' content). Novel foods also are developed in Japan, and a legal term was introduced to classify certain foods making health claims, namely FOSHU (Food for Specified Health Use). In accordance with the pharmaceutical legislation, producers cannot claim that a product is good for one's health, but in accordance with the Nutrition Improvement Law, the government authorises certain food products to get the FOSHU-certificate. According to Eurofood (2000), the leading sector of FOSHU-products in Japan is probiotic[2] dairy: sales increased from US$22.1 million in 1995 to reach a level of US$823.3 million in 1999. The market for both types of foods is foreseen to increase in Japan, which might have massive potential during the next ten years (AgraFood Asia, 2000b). There are surprising examples of product innovations, which respond to consumers' health concerns. AgraFood Asia (2000b) mentions Japanese consumers who 'are eating soup with collagen[3] to moisturise and help heal damaged skin, and they are consuming soft drinks and chewing gum made with blueberry extract to relieve eye strain'. There are also breweries marketing wholesome beer.

Genetically modified organisms (GMO) were publicly debated in 1999. Labelling was also introduced by that time and several manufacturers selected GMO-free ingredients. This did allow these companies to communicate that their products are 100% GMO-free and sell them at a premium. Although labelling for some of these products would not be required according to the new legislation, manufacturers chose alternatives and indicated their products were GMO-free. It is remarkable that items not requiring labelling also were adjusted to become completely GMO-free. Nowadays, GMO-free products have an added value and consequently are more attractive to the consumers. According to the Secretary General of the NO! GMO Campaign manufacturers select GMO-free ingredients even for products that do not require mandatory labelling, primarily because of general public concerns (Consumers Union of Japan, 1999). Producers of soy sauce and beer (non-mandatory labelled products) have decided to switch to non-GMO ingredients. Although the price of GMO-free cornstarch is 15% higher than ordinary cornstarch, three major beer breweries (Sapporo, Asahi and Kirin) changed to GMO-free ingredients (AgraFood Asia, 2000c).

Two examples of the development of a new product instigated by consumer concerns on health issues are vegetable juice and yoghurt containing fruit. Food manufacturers currently market different types of vegetable juice. One type is especially targeted for women. The present popularity is connected to the fact that Japanese consumers realise that they do not eat enough wholesome food and that they do not cook proper meals (including vegetables). Hence, vegetable juice is perceived as a good alternative. Furthermore, a mix of vegetable and fruit juices is an alternative for those who do not eat enough fruit as well. According to AgraFood Asia (2000a), the dairy companies Morinaga, Meiji Milk, Kyodo Milk and Glico Dairy have all added new lines to their yoghurt products. Emphasis is given to yoghurt products containing pieces of fruit, which particularly appeal to

---

[2] Probiotics refer to live microbial feed supplements which beneficially affect mankind by improving intestinal microbial balance. Lactobacilli are commonly used as probiotics, either as single species or in mixed culture with other bacteria.
[3] Collagen is the protein substance of the white fibres of skin, tendon, bone, cartilage and all other connective tissue.

health-and-beauty conscious young women[4]. Several companies, including Morinaga and Meiji have launched products containing blueberries, which are purported to be beneficial to improving eyesight.

Price

The most significant connection between public concerns and the marketing variable 'price' is through the preference by Japanese consumers for high quality products. Japanese consumers have limited knowledge on the price of supermarket items (Takahashi, 1999, pp. 124-128). This is especially typical for fresh meat and vegetables, and suggests that the information cue to choose this type of merchandise is not price but freshness and quality. The association of a higher price with better quality results in a price setting at a rather high level. Consumers generally prefer a reasonable price rather than the lowest possible price. However, when a product becomes less exclusive (i.e. its quality becomes lower), consumers expect a lower price as well. At the time that organic products were rarely available in supermarkets, consumers were prepared to pay a high premium for organic vegetables. Now, organic vegetables have become less exclusive, and consumers also expect lower prices.

Distribution

Major efforts are required on the distribution strategy in the agri-food supply chain to respond to the consumers' 'obsession with freshness' (Food and Agriculture Policy Research Center, 1997, p. 8). Transportation time needs to be minimised and frequent deliveries to retail outlets are vital. Furthermore, informing consumers about the origin of the product also influences the distribution strategy. For example, in the case of fresh vegetables, consumers feel more confident about the safety of a product when they are informed about its origin. The 'sanchoku' system achieves this and has become rather popular. This system allows consumers to buy products directly from the farmer. Consumers who intend to purchase organic (or low-chemical or low-pesticide) products especially apply it. It can be organised in different ways, including mail order. Organic farms (or farms using limited amounts of agro-chemicals) also supplied their products to Co-op supermarkets, since these supermarkets promote such production methods. Presently, other supermarkets also offer organic produce. In order to increase confidence to the consumer, the retailer communicates that the products are produced with great care. This should allow making the consumer feel more familiar with the farmer. In order to achieve that, photographs of farming practices are shown in supermarkets.

Promotion

Promotion includes tools and content. Promotion tools facilitate communication to consumers, and the tools applied depend on the product sold. Four types are

---

[4] Yoghurt is associated with the beauty of a lady's face.

available, including advertising, publicity, personal selling and sales promotion. Advertising and publicity are related to a pull strategy. In contrast, personal selling and trade promotion (which falls within the sales promotion category) are related to a push strategy.

Publicity in the media has a major influence on consumers' behaviour. Advertising through commercials or advertisements is less effective than informative television programmes and articles in the written press. Informative television programmes (cooking or health) are influential. Authorised information in particular, like statements by medical doctors, has an effect on consumers. Certificates often appear at the point-of-purchase, but messages that refer to certain television programmes seem to be more effective. Food manufacturers and retailers normally make sure that sufficient supplies are available of the products featuring in that day's television programmes. Otherwise, their stocks may not suffice. Many of the booms in the previous years, like the red wine and cocoa booms, were the result of informative television programmes or articles in the press.

The content of a message is another main factor promoting sales. Japanese consumers desire abundant information about a product and also decide on it. Detailed background information and easy-to-understand explanations are demanded by consumers (Amemiya, 2000). Important topics relate to features highlighting the exclusivity of a product (e.g. brand, special production method or source of origin) or its beneficial health effects. Manufacturers or retailers may not have the scientific evidence, but they might emphasise that the product contains certain ingredients of which most consumers know they have a beneficial health effect. Through informative television programmes, articles in magazines and highly-popular non-specialist literature on food and health, consumers are relatively well-aware of the (purported) effects of certain ingredients to their health. Even manufacturers of chocolates, chewing gum, and snacks have promoted sales of these products by putting emphasis on health and other product-related benefits (Dentsu Inc., 2000).

In addition to the exclusivity and its beneficial health effect, freshness is also vital to the promotion of a product. The importance of freshness even applies to products that are long lasting, including beer, canned food and chocolates.

Focus Japan (2000) gives an account of the strategy applied by the importers of bananas in Japan. A good marketing strategy is very important to reach healthy product sales. The Japan Banana Importers Association (JBIA), for example, identified bananas as a miracle fruit, with a range of human health aspects. Eating bananas may enhance immunity and make people less susceptible to arteriosclerosis and cancer, hold down blood pressure, stabilise blood-sugar levels, and catabolise cholesterol. Furthermore, they have high nutritional value, are rich in sugar (which quickly turns into energy), potassium (which lowers blood pressure), and magnesium (which steadies nerves). Consumers were informed by the media about the positive effects. Bananas have then gained wide recognition as a food that is good for health. The importers have also successfully influenced the image of bananas for people engaged in sport activities. Eating bananas might also be a very attractive handy substitute breakfast for those who often start the day without a meal.

The marketing efforts of the banana importers have resulted in double-digit rates of import growth, namely 13.7% in 1999, and have created another boom. Stores ran out of inventories despite efforts by importers to increase supplies. The importers and manufacturers were highly demand-oriented, also considering preferences of Japanese consumers. Manufacturers breed varieties that suit the tastes of consumers, who prefer products with higher sugar content. So, new varieties of bananas appeared and they were marketed as Super Sweet and Premium One. These brands were introduced in Japan some 5 years ago. The popularity of bananas also resulted in the development of other products and new processed products were developed, like banana chips, banana drinks, and banana jelly and jam. Imports of these products are steadily growing.

Short-term commercial interests by Japanese companies may dominate over the provision of factually based information to consumers. In spring 2000, manufacturers communicated that certain processed food products in which they used pomegranates, contained oestrogen. The company released this message. In addition, reference was made to the fact that the products would be helpful in the case of menopausal problems and in the case of menstrual irregularities. The Japan Consumer Information Center[5], tested ten different products. Results were published by the Center in 2000, concluding that oestrogen was not detected. Several newspapers addressed the issue as well. For example, the Nikkei Shimbun newspaper published an article on 7 April 2000 entitled 'Zakuro, kônenkishôgai ni kikanai?' (Pomegranate, is it not effective against menopausal problems?).

Legislation from the Ministry of Health applies in case communication from companies refers to medical topics. Otherwise it is the responsibility of the Fair Trade Commission, since this organisation looks after correct labelling and the provision of information.

Response from the Japanese government

National government is well aware of the perceptions and attitudes of consumers regarding food safety, and government also facilitates the consumer driven processes described in this chapter. The Japan Agricultural Standards (JAS) Law has been revised. The main expectations of legislation can be summarised as follows:

> In response to growing demands for appropriate and easier labelling and a standardization system, the revised Japanese Agricultural Standards (JAS) Law was enacted in July 1999, which covers the improvement of the food labelling system such as mandatory description of country of origin for any perishable foods as well as the establishment of a system

---

[5] The Japan Consumer Information Center, which is renamed the National Consumer Affairs Center of Japan, is the main organisation to look after the interests of consumers and solve consumer complaints. It was established in 1970 by the Japanese government in accordance with the Japan Consumer Information Center Law. The objective of the Center is to offer information to consumers, to resolve any complaints from consumers, and product testing related to public health.

for inspection, certification and labelling of organic foods (MAFF, 2000, p. 14).

Labelling is a major point in the revised JAS Law, and is a response to the increasing demand for organic food and confusion of consumers by labels such as 'chemical-free' or 'chemical-reduced' (Japan International Agricultural Council, 1999). National government also initiates environmental legislation, since consumers tend to be not very much concerned about the environment. Recent legislation, for example, was enforced in April 2000, including the Law for Promotion of Sorted Collection and Recycling of Containers and Packaging (Clean Japan Center, 1995). This law covers small- and medium-sized enterprises thus far exempt from the mandatory recycling and it covers containers and packaging in both paper and plastic (MAFF, 2000, p. 12). From 1997, the Law for Promotion of Utilisation of Recyclable Resources (Clean Japan Center, 1991) took effect. This legislation applies to a wide range of products, including packaging (i.e. glass, paper and cans), automobiles and electric appliances. Basically, enterprises, consumers and local public bodies are strongly demanded to accomplish their individual responsibilities to promote recycling. In 2001, new legislation was implemented regarding waste products from food.

CONCLUSIONS

In Japan, public health issues play a more significant role in the decisions taken by consumers to buy products, relative to environmental and animal welfare issues. Different motives are behind this consumer behaviour. They include concerns about the future and cultural aspects. We can conclude that the key words in the consumer perceptions regarding health issues are 'nutritional balance' and 'natural products'.

As part of such public concerns, there are specific features of Japanese consumer behaviour. Good taste and high quality (also including safety) are first priorities when consumers choose food products. Issues of great significance to Japanese consumers in their purchasing behaviour are the product's freshness, appearance and place of origin.

Japanese trade and industry are well aware of these preferences. They use the commercial opportunities such concerns also offer and respond in various ways. Since the Japanese market is highly saturated and competitive, Japanese manufacturers and retailers are continuously looking for additional or new features to make their products stand out and catch the eyes of the consumers. Product development leads to innovative products. This applies to functional foods. On the one hand, Japanese food manufacturers develop products that are less harmful from the perspective of public health. In addition, products are developed which actively improve public health. When scientific evidence allows, Japanese government can certify them as FOSHU products.

Communication through the media to inform consumers has a major impact on consumer behaviour. Booms - sudden interests for certain products - regularly occur in Japan. The agri-food industry is well aware of using this in their promotion of food products.

Government aims to increase the self-sufficiency ratio and increase domestic agricultural production. However, Japan will remain to depend largely on the import of agricultural and food products. The significant role of public health issues poses many opportunities for exporters of food products. It also leads to the development of innovative products that are not yet present anywhere else in the world. Furthermore, it may also affect perceptions of consumers elsewhere in the world.

Market surveys and clear business strategies are vital to successfully seize these opportunities. In-depth research will yield information on the particular issues that motivate present-day Japanese consumers. Furthermore, detailed investigations on product development would identify promising strategies for the Japanese market. The challenge is to touch the right cord with Japanese consumers. Since they are sensitive to information, it is important to find and to communicate the right message.

REFERENCES

AgraFood Asia (2000a) Yoghurt sales increase. In: *AgraFood Asia*, Agra Europe, London, January, p. 9.

AgraFood Asia (2000b) US 'functional foods' find ready market in Japan. In: *AgraFood Asia,* Agra Europe, London, February, p.7.

AgraFood Asia (2000c) GM beer ingredients rejected. In: *AgraFood Asia*, Agra Europe, London, February, p. 9.

Amemiya, H. (2000) Shoku wo meguru torendo; tsuyomaru shôhisha no 'bôei' ishiki (Trends concerning food: consumers' 'defensive' consciousness is becoming stronger). In: *Nikkei Shôhi Keizai Fôramu Kaihô: Shôhi & Mâketingu (Nikkei Consumption Economics Forum: Consumption & Marketing)*. Nikkei Sangyô Shôhi Kenkyûjo (Nikkei Research Institute of Industry and Markets) / Nihon Keizai Shimbun Inc., 186, pp. 8-11.

Blandford, D. and Fulponi, L. (1999) Emerging public concerns in agriculture. *European Review of Agricultural Economics*, 26 (3), 409-424.

Clean Japan Center (1991) *Law for Promotion of Utilisation of Recyclable Resources; Ministerial Ordinances*. Clean Japan Center, Tokyo.

Clean Japan Center (1995) *Law for Promotion of Sorted Collection and Recycling of Containers and Packaging*. Clean Japan Center, Tokyo.

Consumers Union of Japan (1999) *Japan Resources; Consumers Union of Japan Newsletter*. Consumers Union of Japan, Tokyo, Number 110.

Dentsu Inc. (2000) *Japan 2000 Marketing and Advertising Yearbook*. Dentsu Inc., Tokyo.

Eurofood (2000) Functional foods turn from 'less bad' to 'good'. In: *Eurofood*, Agra Europe, London, 3 February, p. 16.

Focus Japan (2000) The health-food boom lifts banana imports. In: *Focus Japan*, JETRO, Tokyo, May, pp. 10-11.

Food and Agriculture Policy Research Center (1997) *Structural Changes in Japan's Food System*. FAPRC, Tokyo.

Henshall, K.G. (1999) *Dimensions of Japanese Society: Gender, Margins and Mainstream*. MacMillan Press, Basingstoke.

Japanese Consumers' Co-operative Union (1999) *Survey about the perception of domestic versus imported products*, October (in Japanese).

Japan Consumer Information Center (1999) *Japan Consumer Information Center* (brochure).

Japan Consumer Information Center (2000) *Jishu chôsa tesuto No. H12-1; Esutorogen wa hontô ni fukumarete iru no? zakuro wo tsukatta kenkôshikôshokuhin (Independent research test No. H12-1 Do they really contain oestrogen? Health aimed food products with pomegranates as ingredients)*. Japan Consumer Information Center, Tokyo, 6 April.

Japan International Agricultural Council (1999) Toward the promotion of organic farming. *Japan Agrinfo Newsletter*, 16 (6), p. 7.

Jonker, T.H. (1999) *Agri-Food Supply Chains and Consumers in Japan; An inquiry into the current situation and the opportunities of five Dutch product groups on the Japanese market*. Agricultural Economics Research Institute (LEI), The Hague, Report 3.99.15.

Kotler, P. (2000) *Marketing Management International Edition*. Upper Saddle River, New Jersey, Prentice-Hall.

Makuuchi, H. (1995) *Soshoku no susume (Plain food recommended)*. Toyo Keizai, Tokyo.

Ministry of Agriculture, Forestry and Fisheries (MAFF); Statistics and Information Department (ed.) (1995) *The 70th Statistical Yearbook of Ministry of Agriculture, Forestry and Fisheries 1993 - 94*. Association of Agriculture and Forestry Statistics, Tokyo.

Ministry of Agriculture, Forestry and Fisheries (MAFF); Statistics and Information Department (ed.) (1999) *The 74th Statistical Yearbook of Ministry of Agriculture, Forestry and Fisheries 1997 - 98*. Association of Agriculture and Forestry Statistics, Tokyo.

Ministry of Agriculture, Forestry and Fisheries (MAFF) (2000) *Annual Report on Food, Agriculture and Rural Areas in Japan; FY 1999 (Summary)*. MAFF, Tokyo (provisional translation).

Nikkei Shimbun (2000) Zakuro, kônenkishôgai ni kikanai? (Pomegranate, is it not effective against menopause problems). In: *Nikkei Shimbun*, Tokyo, 7 April, p. 38.

Research and Information Service for Food and Life (2000) *Shoku seikatsu deeta bukku 2000 (Food and Life Data Book 2000)*. Association of Agriculture and Forestry Statistics, Tokyo.

Saison Research Institute (1998) *'Shoku no anshin & anzen' ni kansuru seikatsuishikichôsa; seikatsusha no ishiki, kigyô no taiô, yûshikisha no ninshiki kara (Life consciousness' research on 'reassurance & safety of food'; the consciousness of human beings, the response of companies, from the perception of scholars)*. Saison Research Institute, Tokyo.

Saison Research Institute (1999) *Shôhi & seikatsu sutairu repooto (Consumption & lifestyle report)*. Saison Research Institute, Tokyo.

Schütte, H. and Ciarlante, D. (1998) *Consumer Behaviour in Asia*. MacMillan Press, Basingstoke.

Silvis, H.J. and Van Bruchem, C. (eds) (1999) *Landbouw-Economisch Bericht 1999 (Agricultural Economic Report)*, Agricultural Economics Research Institute (LEI), The Hague.

Takahashi, I. (1999) *Shôhisha Kôbai Kôdô & Kouri Mâketingu eno Shazô (Consumer Shopping Behaviour and Retail Marketing)*. Chikura Shobô, Tokyo.

Toda, H. (2000) General Introduction and Summary - Changes in Agriculture and Food Consumption in Japan Created by Increased Imports of Agricultural Products. In: Food and Agriculture Policy Research Center, *Effects upon Japan's Food Market by Increasing Agricultural Imports*. Food and Agriculture Policy Research Center, Tokyo, pp. 1-40.

Evaluating Environmental Trade Disputes in the Post-Seattle World

Glenn Fox

INTRODUCTION

Trade liberalisation and environmental protection appear to be on a collision course. Events in the streets of Seattle in November of 1999 made the general public acutely aware of a social conflict that has been emerging for some time. A controversial set of environmentally related trade disputes over the previous 10 years provide compelling evidence that the trade liberalisation agenda and the environmental protection agenda are no longer separable. Most environmental organisations have concluded that the General Agreement on Tariffs and Trade (GATT), the World Trade Organization (WTO) and the North American Free Trade Agreement (NAFTA), when settling trade disputes involving environmental issues, have gotten it wrong virtually every time and that nothing short of fundamental reorganisation of these institutions will correct this situation. The instrumental purpose of this chapter is to examine these calls for fundamental structural change in the institutions that regulate international trade. This instrumental purpose is, however, merely a means to a more general end. The current state of acrimony between environmental activists and the largely governmental agencies engaged in the ongoing process of liberalising trade is hindering progress on both agendas. This chapter is an attempt to find some common ground and to identify a way forward out of this looming gridlock.

DIVERGENT PERSPECTIVES BETWEEN TRADE LIBERALISATION AND ENVIRONMENTAL PROTECTION

Critical differences in perceptions about the relationship between trade and environmental values are now evident. Four of the most important are:

1. Some environmental critics of trade policy argue that if trade liberalisation increases standards of living, this will lead to more natural resource consumption and increased threats to environmental values. Advocates of trade liberalisation and some environmental groups (Van Putten, 1999, p. 2) argue that reduction of protectionism improves resource use efficiency and thereby contributes to a better matching of resource employment with individual preferences and it has *even* been suggested that, in the cases of agriculture, forestry and fisheries, existing protectionist measures actually encourage excessive natural resource use and that trade liberalisation can improve standards of living while at the same time reducing environmental degradation. Furthermore, some writers have argued that there is a relationship between the level of economic development of a society and measures of environmental quality such that, at least at moderate to high levels of development, increases in standards of living seem to be correlated with improvements in environmental quality. This relationship is often described as an environmental Kuznets' Curve.

2. Environmental critics of trade liberalisation argue that reducing trade barriers will lead to more transportation of goods and people, which puts more pressure on energy resources and increases risks of air and water pollution. Trade advocates counter that increased international transportation is offset by reduced intra-national transportation and that the net effect may be small.

3. Critics of trade liberalisation fear the relocation of dirty industries to less demanding regulatory regimes, creating a 'race to the bottom' as governments compete with one another to attract investment and stimulate employment (Van Putten, 1999, p. 3). The race to the bottom results in pressure to harmonise national regulations and the fear is that harmonisation will occur around the lowest existing standards. Advocates of trade liberalisation counter that evidence of races to the bottom is hard to find and that environmental compliance costs, where they have been estimated, seem to be small.

4. Finally, environmental critics of trade liberalisation charge that the track record of the WTO and other trade bodies, in resolving disputes over environmental trade measures, reveals a fundamental conflict between trade liberalisation and environmental protection. According to this view, it is inevitable that environmental measures will be found incompatible with member obligations under the WTO, especially when the WTO itself conducts the assessment, since panellists who adjudicate disputes are appointed from the ranks of trade policy experts, not environmental experts. Defenders of these institutions maintain that dispute settlement processes have given due regard to environmental issues and that the best way forward is to develop WTO compatible environmental policies, not to dismantle the multilateral trade liberalisation process.

This chapter addresses the fourth of these perceptions. A substantial literature already exists, some of it represented in other chapters in this book, on the first three controversies. The evidence regarding an environmental Kuznets' Curve has been examined by Goklany (1999), De Bruyn *et al.* (1998), Selden and Song

(1994) and Grossman and Krueger (1991). Anderson and Strutt (1996) have characterised the environmental impact of trade liberalisation in agriculture. Runge and Fox (1999) concluded that there is no evidence to suggest that the environmental effects of trade liberalisation under the NAFTA with respect to the North American cattle feeding industry are large. The question of policy harmonisation has been investigated by Bhagwati and Hudec (1996), Leebron (1996), Casella (1996) and in a series of conferences convened by the International Agricultural Trade Research Consortium (Loyns *et al.*, 1997, Loyns *et al.*, 1998 and Loyns *et al.*, 2000). Specific examinations of agricultural environmental policies were undertaken by Lindsey and Bohman (1997), and by Ervin and Fox (1998). Harmonisation of technical regulations related to food safety were analysed by Bredahl and Holleran (1997) and Josling (1997) addressed conceptual and definitional issues. Evidence on the existence of a race to the bottom has been examined by Levinson (1996), Klevorick (1996) and by Ervin and Fox (1998). Other chapters in this volume present some new empirical findings on the environmental compliance costs for agriculture.

Much less interpretive literature has been written on the lessons that might be learned from the short list of recent environmental trade disputes in the GATT and the WTO that could help guide future environmental policy development. Much of the literature that is available was produced by environmental groups in anticipation of the 1999 Seattle WTO meeting. Sampson's (2000) recent book is one of the first attempts to respond to the charges levelled against the WTO on its environmental record.

ENVIRONMENTAL PROPOSALS FOR REFORM OF THE WTO

Many leading environmental organisations have produced policy papers on the relationship between trade liberalisation and environmental stewardship and on the way that environmental trade disputes have been treated in the GATT and the WTO. Most of this literature was made available in the few months immediately preceding the WTO Seattle meeting in November 1999. Taken together, the findings of this literature are unequivocal. The WTO, before it the GATT, and, when it is included in the analysis, the NAFTA, constitute an institutional impediment to the achievement of sustainable development globally. The aborted Multilateral Agreement on Investment (MAI) is generally lumped in with these existing trade agreements as an example of something that the world is better off without, environmentally speaking.

The Policy and Research Units of Friends of the Earth England, Wales and Northern Ireland as well as Friends of the Earth International have produced a series of reports calling for environmental reform of international trade relations, including the proposed MAI and the WTO (Friends of the Earth England, Wales and Northern Ireland, 1998, 1999a, 1999b, Friends of the Earth International, 1999a, 1999b, 1999c, 1999d, 1999e). The Friends of the Earth proposals are wide ranging and, if implemented, would represent a fundamental change in international trade and environmental policy making relationships. Their assessment is that:

> As we approach the 21st century, the world needs trade rules that reflect society's current values and needs. Our existing trade rules and institutions and indeed the current global economic system are out of date and do not do this. Instead, they undermine biological and cultural diversity. They are still based on the pursuit of profit regardless of social and environmental costs; and inequitable access to, and the overuse of, limited natural resources. Critically, current rules also prevent the maintenance and development of locally-appropriate and sustainable systems of commerce (Friends of the Earth England, Wales and Northern Ireland, 1999a, p. 35).

These proposals are expressed (Friends of the Earth England, Wales and Northern Ireland, 1999a, pp. 33-35) as six recommendations:

- To cease negotiations to initiate a new comprehensive round of negotiations that would bring new issues into the WTO.
- To implement a moratorium on WTO challenges to laws designed to promote and protect development, environment and health.
- To protect local, national and international environmental and social laws from unfair challenges and weakening by trade rules.
- To address issues related to an international agreement on investment through the United Nations, not the WTO.
- To open up the process of negotiating future trade agreements to more stakeholders and to release minutes, papers, governmental position papers and draft texts to the public.
- To forgive some or all Less Developed Country indebtedness and reject the 'liberal free market' principles that underlie the strategy of export-led development.

Earlier (Friends of the Earth England, Wales and Northern Ireland, 1999a, pp. 6-7) in the same report in which these recommendations appear, two additional recommendations, that an independent review of the environmental, social and developmental effects of past WTO (and presumably GATT) decisions be undertaken and that the WTO acknowledge that multilateral treaties on the environment, development, health, labour and human rights take precedence over the WTO.

The Earthjustice Environmental Defense Fund (Wagner and Goldman, 1999, p. 20), formerly the Sierra Club Legal Defense Fund, has offered its own list of proposals for the reform of the WTO. The rationale for these proposals is developed in a series of policy research documents (Goldman and Fawcett-Long (1999), Goldman and Scott (1999), Goldman and Wagner (1999), and Wagner and Goldman (1999)). Earthjustice argues that the WTO should be reformed to protect 5 rights:

- The right to have strong environmental standards that use the precautionary principle and protect citizen health and the environment.
- The government's right to limit the harmful effects of production, such as pesticide poisoning of workers, and toxic air and water pollution from factories.
- The consumers' right to know which products are environmentally friendly.
- The right to use the government's purchasing power to protect the environment.
- The right to access information about and to participate in disputes, negotiations, and other proceedings that affect public health and the environment.

Like Friends of the Earth, Earthjustice also calls for a review of the environmental and health impacts of the WTO before any further multilateral negotiations are undertaken, and that, in the interim, a moratorium be imposed on WTO challenges to food safety, health and environmental measures.

The World Wildlife Fund (WWF) also produced a series of trade related policy briefs in the months leading up to the Seattle WTO meeting (Stillwell *et al.* (undated); Perrin, 1998a, 1998b; Mabey and McNally, 1999). The Stillwell *et al.* discussion paper, representing the views of the US Centre for Environmental Law, OXFAM of the UK and the US Community Nutrition Institute as well as the WWF, cites the WTO findings in the EU beef hormone case, the Caribbean banana dispute and the shrimp and sea turtle case as indicative of the failure of the WTO to adequately protect human health and the environment. The shrimp and sea turtle case is described as the most important environmental case ever to come before the WTO (Stillwell *et al.*, Section 2). The WTO panels treatment of the scientific and legal issues in the shrimp and sea turtle case is described as 'unsophisticated' (Stillwell *et al.*, Section 2.2) and 'superficial' (Stillwell *et al.*, Section 2.3). The WTO beef hormone panel decision is equally unfavourably received (Stillwell *et al.*, Section 4). Among other things, the Panel is described as rejecting the precautionary principle as a legitimate basis for public health and environmental policy, a claim that is contradicted by even a casual reading of the panel's reports.

The National Wildlife Federation has offered five proposals for the reform of the WTO (Van Putten, 1999, pp. 7-11):

- The WTO agreements must incorporate explicit rule changes that allow each WTO member country to retain the right to develop and enforce high conservation measures through trade restrictions - even if they exceed the international norm - without running foul of WTO rules.
- WTO rules should explicitly provide for deference to multilateral environmental agreements in instances of conflict between trade rules and trade-related provisions of those environmental agreements.
- The environmental ramifications of any trade agreement must be carefully evaluated before the agreement is concluded and put into effect.

- Trade rules should explicitly allow countries to label products or restrict the importation of products that are produced or brought to market in a way that harms endangered species or the global commons.
- The WTO must adopt more open procedures and increase public access to documents. A permanent role for non-governmental organisations must be established within the WTO structure.

The Earth Island Institute, which was the lead plaintiff in the US Court of International Trade case that ultimately resulted in the WTO dispute over the United States (USA) shrimp embargo, has called for the abolition of the WTO (Fugazzotto and Behera, 1999). The Institute's criticisms of the WTO (Fugazzotto and Steiner, 1998) include the charge that the WTO is not democratic, that panel members who have heard environmental trade disputes have had conflicts of interest and that decisions of panels on environmental cases have ignored scientific evidence in favour of economic considerations.

General themes in environmental NGOs criticisms of the GATT's and more recently the WTO's treatment of environmental trade disputes have tended to focus on some process and some substantive issues. The process issues have to do with the generally confidential way in which trade negotiations are undertaken, the accessibility of panels and documents by civil society organisations and ultimately with the nature of national membership obligations in the GATT or the WTO. As I will argue below, the WTO has improved access to documents dramatically in recent years and it is profoundly easier for members of the general public to gain access to WTO documents than it ever was with the GATT. Nevertheless, there continue to be good reasons for some aspects of trade negotiations being conducted in confidence. In many cases, environmental criticisms of the WTO's handling of environmental trade disputes should really be directed at the conduct of national governments that are party to the relevant dispute. Fugazzotto and Steiner (1998) acknowledge that their quarrel in the shrimp and turtle case is, at least in part, with the Clinton administration.

A final procedural recommendation that has been frequently offered by environmental NGOs is that environmental assessments of trade agreements be conducted before they are implemented. On its face, this is not an unreasonable requirement, but it does reflect an unacknowledged asymmetry. Protectionist measures, which are the things that we are trying to dismantle with trade agreements, have not been subjected to environmental assessments before they were put in place. It is imperative, therefore, that an appropriate reference point be set in the environmental assessment of any trade agreement. It is unwise to identify only the threats of trade liberalisation to environmental protection. These effects need to be considered in the context of threats to environmental values that should be attributed to protectionism and that would be ongoing if the trade agreement is not implemented. Similarly, Anderson and Strutt's (1996) admonition that trade liberalisation can only be welfare enhancing if optimal pollution taxes are in place should be applied equally to protectionism.

The substantive NGO criticisms of the GATT and the WTO have tended to focus on the reluctance of the trade community to grant process standards comparable legitimacy to product standards, on the unwillingness of the WTO to allow one country to impose its own domestic environmental policy instruments on another country and on the perception that the grounds on which article XX exemptions to WTO obligations are allowed are too narrow. I will return to these criticisms later in this chapter.

ACADEMIC ASSESSMENTS

The academic literature and the literature produced by researchers at NGOs without an exclusive environmental focus (for example, Johnson and Beaulieu, 1996; Runge, 1994; Audley, 1997 and Esty, 1994) is only slightly less pessimistic about the prospects for peaceful coexistence between trade and environmental policies than the literature produced by environmental NGOs. Johnson and Beaulieu (1996) and Esty (1994), writing in anticipation of the implementation of the NAFTA agreement and prior to the formation of the WTO raised concerns about the prospects of domestic trade related environmental measures being found inconsistent with treaty obligations under either agreements. Their specific concern was whether environmental policies would be able to meet the GATT or the NAFTA tests for necessity and for least trade distorting means. Johnson and Beaulieu (1996) offered two interpretations of this test; one in which a challenged measure would be compared to all alternative measures, including hypothetical measures, to determine if the actual measure was less disruptive of trade than any conceivable alternative. The second interpretation was that 'necessity' would require that the challenged measure only had to be defended against practical available alternatives. My reading of environmental trade dispute panel reports from the WTO is that this second and less demanding standard seems to have been adopted, allaying these early fears about threats to environmental policy.

Esty (1994), like some of the environmental groups, calls for a comprehensive environmental assessment at the outset of any trade negotiation and for a more transparent process for reaching new agreements and for resolving disputes. However, Esty's recommendation that 'National and international environmental standards should be presumed to be valid and legitimate until shown otherwise' (p. 212) in the settlement of trade disputes suffers from an internal contradiction - what happens when national and international standards differ?- and is also at variance with environmental critiques of the WTOs resolution of the shrimp and sea turtle case. India, Malaysia, Pakistan and Thailand maintained that they had implemented effective and appropriate measures to protect sea turtles in their territorial waters and that Turtle Excluder Devices (TEDs) were not essential or even necessary to those efforts, given the ecological and social conditions that they faced locally. Their charge was that the US embargo did not presume the validity of their national measures. On the other hand, Esty's recommendation that Panels solicit independent outside technical advice on environmental matters was in fact followed by the WTO shrimp embargo Panel and in the EU beef hormone case.

Kerr (2000), writing from a post-Seattle perspective and with the advantage of observing the actual operation of the WTO in the first 5 years of its existence, offers a different interpretation of the relationship between trade liberalisation and environmental protection. His analysis is based on the capture theory of regulation. He suggests that while environmental groups have been successful at the national and even international levels at getting agreement in principle on the need for environmental protection policies, they have been less successful at securing practical compliance with those principles, particularly when compliance carries significant political or economic costs. The WTO, with its ability to impose trade sanctions on nations whose conduct is judged to be unacceptable to, say, members of a WTO panel, is a tempting prize. According to Kerr, the agenda of environmental groups to gain greater access to WTO decision making goes beyond a desire to insulate environmental policies from trade challenges. An additional motivation is to be able to use the considerable incentive of WTO trade sanctions in support of compliance with international environmental agreements. An organisation that was developed primarily to counteract the perverse economic and political incentives generated by domestic firms seeking political protection from competition from foreign firms, in Kerr's assessment, may not be well positioned to resist capture by other domestic interest groups motivated by less directly self-serving goals.

GENERAL THEMES

There are at least two separate themes in criticisms of trade liberalisation that have been forwarded by leading environmental groups. One of these themes is a perception or conclusion that the operation of free markets can either never or only under rare circumstances be compatible with acceptable stewardship of natural and environmental resources. This perception apparently applies equally to domestic and international markets, especially when transactions involve transnational corporations. To use the traditional lexicon of natural resource economics, this view holds that market failure is pervasive when it comes to transactions involving fish, trees, water, air, flora, fauna, minerals and other components of the natural world that are impacted by human activity. This is the longstanding conventional wisdom of natural resource economics textbooks.

More recently, a new perspective has emerged in the natural resource economics literature. This perspective emphasises the role of policy failure, regulatory failure or government failure as an additional contributing cause of pathologies in natural resource stewardship. According to this view, actual examples of natural resource degradation represent complex interactions of the traditional categories of market failure with the often less well-appreciated policy failures. Wolf (1979) identified four main categories of policy failure:

- weak incentives for cost control in the provision of government services;
- 'internalities', which refers to the capture of an agency's mission by its staff resulting in inconsistencies between legislatively intended goals and actual programmes;
- derived externalities, which are analogues to the market failure category of uncompensated external costs but that, in the case of policy failure, arise from the actions of government agencies, not households or firms; and

- distributional inequities, where policy measures unfairly benefit politically influential groups at the expense of other less powerful interests.

Wolf's categories are not necessarily exhaustive. The taxonomy of market failures continues to grow and evolve and the same process is likely to take place with policy failure.

Anderson and Leal's (1991) widely discussed book applied concepts related to policy failure to several important US natural resource management problems. Brubaker (1995) provided a similar treatment for some Canadian issues. The evolution in thinking taking place in natural resource economics has important implications for the diagnosis and treatment of environmental problems generally, but is particularly germane for the apparently widely held perspective in environmental organisations that free markets will generally never get it right when it comes to natural resources. First, it implies that we need to re-examine the historical data that is often cited to bolster the claim that free markets and environmental stewardship are incompatible. As Anderson and Leal's and Brubaker's work often reveal, what appears to be a market failure on the surface often turns out to have policy failure causes when the record is investigated more carefully. Second, if real world regulators are subject to the pitfalls of policy failure, calls for intervention in markets must be more circumspect. As Wolf points out, if the remedy that we invoke has undesirable side effects and the cure can be worse than the disease, then it might be preferable to decide to live with the problem. In any case, the analytical framework of the economic analysis of environmental policy is richer and more complex today than it was a generation ago. We now understand that market and policy failures coexist and interact in ways that had not previously been well understood and that environmental problem solving is at once more complex and more pragmatic.

The controversies around the relationship between trade liberalisation and environmental protection are, in many ways ultimately empirical questions. The emerging literature on these subjects, in my judgement, seems to confirm that there is a relationship between the level of economic development of a society and the priority that members of that society place on environmental protection so that at the currently highest levels of development observed in the world today, economic growth is inversely related to environmental degradation. What is less clear is whether either cross sectional analyses or longitudinal analyses of the historical experiences of currently highly developed economies are a good basis for making predictions about future conditions in the currently less developed countries of the world. Furthermore, the data used to calibrate Kuznets' Curve relationships are often highly aggregate and may mask more complex relationships between types of development and individual indicators of environmental quality. Work in this area has been slow to provide an explanation for the relationships that have been estimated, but recent work by Goklany (1999) is a step in the right direction. In any case, most work in this area has focused on the relationship between measures of domestic environmental quality and income. To my knowledge, the relationship between income and actions to reduce transboundary pollution issues has not been investigated empirically.

Evidence on the race to the bottom seems to suggest that environmental compliance costs, in general, are not sufficiently large to play a critical role in determining the national location of production (see Ervin and Fox, 1998). Work to determine the net environmental effects of trade liberalisation is in its infancy (Kirton *et al.,* 1999). Until more definitive data are available, however, it would seem prudent to at least acknowledge that there are two potentially offsetting effects here. Reduction in protectionist measures, especially in agriculture and forestry in the developed countries, could both improve efficiency of resource use in the narrow conventional sense but also in the broader environmental sense, if artificially high domestic prices fall and subsequently discourage input intensive production systems currently employed in these sectors. On the other hand, overall global transportation of goods may increase[1]. A more balanced perspective on these issues, including a broader appreciation of what the available evidence does and does not say would go a long way to reduce the tensions in the polarised rhetoric that lead up to Seattle.

This chapter, however, is specifically concerned with criticisms from leading environmental groups of the process of trade liberalisation, especially as that process is embodied formerly in the GATT and more recently in the WTO. Calls for a comprehensive environmental assessment of the effects of the WTO are made with the expectation that such an assessment would reveal widespread negative impacts. The track record of the GATT, and more recently the WTO, in resolving trade disputes over environmental issues makes up the bulk of the evidence offered to sustain the conclusion that these institutions have never gotten environmental cases right and that they never will. The most frequently cited cases are, for the GATT, the US tuna dolphin case, for the WTO, the EU beef hormone ban, the US sea turtle and shrimp case and the US Venezuelan gasoline case.

THE NATURE AND FUNCTION OF WTO MEMBERSHIP OBLIGATIONS

Before proceeding with an examination of the various environmental proposals for WTO reform, it is important to understand the implications of membership in the organisation. Sampson (2000) has recently offered a thoughtful and constructive response to environmental critics of the WTO and to the various proposals for reform that they have advanced. He offers the perspective of an insider to the GATT and the WTO as institutions. He reminds his readers that the WTO itself is not a global government. It is an intergovernmental organisation.

---

[1] It seems to be generally accepted that there is no uncertainty here. It is frequently alledged that freer trade will increase global transportation. A Canadian counterexample, however, illustrates that this is not necessarily the case. Long standing transportation subsidies to encourage grain exports particularly from ports on the Pacific coast of Canada have been dismantled, at least in part to comply with more liberalised trade relations in agriculture under the GATT and the WTO. These subsidies encouraged energy intensive east-west grain transportation in Canada at the expense of north-south transportation. The net effect of agricultural trade liberalisation in this context may indeed be to reduce the overall level of resources employed in transportation. On the other hand, Gabel (1994) has concluded that trade liberalisation will lead to an overall increase in energy use in transportation.

Governments must choose to join or to maintain their membership in this organisation. Membership consists of a set of agreements that create obligations and opportunities for signatories. Individual governments weigh the benefits that they perceive in these opportunities against the obligations and decide if it is in their interests to join. The implication of this reminder, in the present context, is that if the operation of the WTO is at cross purposes to the cause of global or national environmental protection, it is the action of the collection of governments that are signatories to the WTO that determines this, not the autonomous action of the WTO itself. The importance of this point resurfaces in his discussion of transparency and the role of NGOs in WTO disputes. Parties to a WTO dispute are nation states, but nation states have wide discretion in determining whom to include in their delegations. It is completely permissible for nation states involved in a WTO dispute to include representatives from environmental NGOs in their delegation. That they typically do not do this can hardly be blamed on the WTO. Sampson's general advice (Sampson, 2000, p. 29) to potential reformers is that proposals for reform need to take into account the institutional nature of the WTO if they are to have any prospect of success.

On the question of transparency and openness, Sampson argues that the WTO is considerably more accommodating on this issue than was the GATT. Access to documents through the WTO document dissemination web site is possible, if not, at least in the experience of this writer, awkward and time consuming. Sampson argues that more can and should be done regarding the timeliness of making documents available, but institutional evolution from the GATT to the WTO has progressed enormously in this area.

In addition to the technical legal implications of WTO membership, environmental proposals for reform should reflect sensitivity to some general principles have guided the post-war process of trade liberalisation if those proposals are to have any prospect of success. I would describe some of these principles as overt. They are acknowledged in policy statements, Panel and Appellate Body reports and in public announcements. There are also, however, some covert principles that may be less frequently proclaimed but that are nevertheless important. These principles have been described in various ways both in official GATT and WTO documents and in analytical and interpretative research exercises. The overt principles are:

- *National treatment and non-discrimination.* This principle requires that firms located in other countries who are signatories to the WTO not be treated worse by national regulatory or taxation authorities than domestic firms. Non-discrimination also applies to so-called 'like products'. Like products, defined in terms of objective physical characteristics of products as those products are presented for import, should be treated in the same manner by trade measures. The resistance to redefinition of like products to include the manner in which products are produced is arguably the most contentious and difficult difference between GATT/WTO principles and their environmental critics. There are examples of process related standards under the WTO SPS agreement, however.

- *The necessity of means.* This principle embodies several related ideas. First, while the GATT and the WTO agreements explicitly recognise the legitimacy of actions by national governments to protect natural resources and environmental values, both reflect a fear that such actions could be subverted by national protectionist pressures. Consequently, when challenged in a dispute, the nation defending such actions is expected to be able to demonstrate to an impartial panel that the measures in question are necessary to the attainment of the environmental outcome that is sought. It is also presumed that convincing technical and scientific evidence can be marshalled to document the relevant hazards. WTO panel decisions have reflected a preference for actions based on multilateral environmental agreements rather than unilateral actions. Criticism that GATT and WTO dispute resolution procedures have failed to give due regard to the so-called precautionary principle typically focus on this necessity principle. Johnson and Beaulieu (1996) take a different tack. They argue that the implicit standard of necessity has been 'Can anyone imagine an alternative means of achieving the stated environmental goal that is less obstructive of trade?' and that that standard is impossible for any actual measure to ever meet.
- *Deference to national sovereignty.* Consistent with the nature of the GATT and the WTO as associations of nation states, with the limited authority to do only what member countries agree to, both agreements reflect a reluctance to affirm the extra-territorial and unilateral application of one nation's policies, regulations or standards on another nation.

The most important covert principle guiding trade liberalisation is that the whole institutional process, under the GATT as well as the WTO, is a charade. It sounds like negotiations proceed in terms of offers, counteroffers, compromises and concessions as nation states grudgingly open their societies to the ravages of competition with foreigners. The appearance, however, is really for the benefit of currently protected domestic industries operating behind protective tariff or non-tariff barriers. Figure 12.1 shows, for the case of a negotiated reduction in an import tariff, that for a country which currently is an importer of the goods in question, the gains to consumers from a reduction in a tariff must always outweigh the losses to the domestic industry plus the reduction in tariff revenue. In the Figure, a country which had formerly applied a tariff of t_0 unilaterally reduces that tariff to t_1. Assuming that the importing country faces a perfectly elastic supply of the imported good at the world price P_w, the price in the domestic market falls from $P_w + t_0$ to $P_{w + t_1}$. Imports of the good in question rise to M_1 from M_0. Domestic producers lose producers' surplus equal to the area bounded by the points $P_w + t_0$ and $P_w + t_1$ on the vertical axis and the points A and F on the Domestic Supply function. Tariff revenues fall by the area ABDE minus $(M_1 - M_0)$ times t_1. However, consumers gain through and increase in consumers' surplus bounded by the points $P_w + t_0$ and $P_{w + t_1}$ on the vertical axis and by points B and C on the Domestic Demand function. The sum of the net gains from this unilateral reduction in a tariff is the areas of the triangles FAE and BCD plus $(M_1 - M_0)$ times t_1.

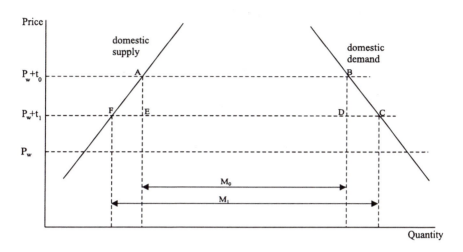

Figure 12.1   Welfare gains from unilateral trade liberalisation

So unilateral trade liberalisation makes a country better off in the overall welfare measures represented in the figure. Nevertheless, in the language of GATT and WTO trade liberalisation, the country agreeing to reduce its tariff is making a 'concession'. The reason for this rhetorical dance is that while it may be true that economists measures of who gains and who loses from this action indicate that the gainers gain more than the losers lose, there are typically many more gainers than there are losers. So individual potential gainers have little incentive to organise themselves to lobby their own government to reduce its import tariffs. On the other hand, the relatively small number of owners and employees of firms protected by the tariff, as well as of owners and employees at those firms that supply inputs to the protected industry often face substantial losses if the protection that they enjoy is eroded. They have plenty of incentive to lobby their government to maintain the status quo. The rhetoric of concessions allows governments to tell their citizens that, yes we have 'given up' access to our domestic market, but in exchange we have 'gained' access to some other country's market because they too have reduced a tariff. Sampson (2000, p. 20) explains that the main implication of this principle is not all of the negotiations in a multilateral round can or even should take place in an open forum. Governments who pursue liberalisation that stands to benefit consumers generally but that will expose currently protected industries to foreign competition might be reluctant to do so if every aspect of the negotiations were open.

THE WTO SHRIMP AND SEA TURTLE CASE

The reputation of the WTO among its environmental critics, in large measure, hinges on the ultimate findings of first a Panel (World Trade Organization,

1998b) and later an Appellate Body (World Trade Organization, 1998c) on a US embargo on imports of shrimp that had been harvested without the benefit of a TED. Like the earlier GATT tuna and dolphin case (General Agreement on Tariffs and Trade (GATT) (1991)), and the later EU beef hormone case (World Trade Organization, 1997a, 1997b, 1998a), the shrimp and sea turtle case is seen as *prima facie* evidence that legitimate environmental measures have a bleak future when challenged in an international trade tribunal. Upon closer examination of the principles that undergird WTO membership obligations and of the actual process followed by the Panel and by the Appellate body, it is less clear that this is an example of good policy derailed by a bad institution.

A timeline of key events

In 1987, the USA issued regulations under the 1973 Endangered Species Act requiring all shrimp trawling vessels either to use TEDs or to restrict the length of time that trawl nets are towed so as to reduce sea turtle mortality from incidental catch. These regulations were implemented in 1990. In 1989, under Section 609 of PL 101-102 required the US Secretary of State to initiate negotiations to establish international environmental agreements on the protection of sea turtles. The same statute imposed an embargo on imports of shrimp harvested with methods that might adversely affect sea turtles. That embargo was to be applied in 1991 to any nation in the Caribbean and western Atlantic regions, subject to a 3 year phase in period. In 1995, the US Court of International Trade found that the limited regional scope of the embargo was not legal and directed the Secretary of State to extend the import ban to all countries harvesting wild shrimp. This extension was required to be in place by 1 May 1996. In a series of communications with the WTO beginning in the autumn of 1996, India, Pakistan, Malaysia and Thailand challenged the 1996 extension of the US embargo as inconsistent with the obligations assumed by the USA as a member of the WTO. A WTO panel was appointed and began its work in 1997, issuing its report in April of 1998. The USA appealed against the findings of this panel and the WTO Appellate Body reported its findings in October of 1998.

Findings of the panel and of the appellate body

The Panel found that the US embargo violated Article XI:1, which prohibits quantitative import restrictions (World Trade Organization, 1998b, p. 272), Article I:1, which requires like products be treated similarly and Article XIII:1 with requires that competing exporters be treated similarly. But these findings are hardly controversial. The USA claimed an exemption to these articles under Article XX. The Panel concluded that the introduction, or 'chapeau', of Article XX indicates that a claim of exemption under that article is subject to the requirement that the measure in question not be applied in an arbitrary or unjustifiable fashion across countries and that the measure not be a disguised effort to insulate domestic firms from foreign competition. After a painstaking examination of the proper interpretation of the conditions that must be satisfied for a WTO member to invoke an exemption under Article XX, the Panel concluded that a qualifying exemption must not 'undermine the WTO multilateral trading system' (World Trade Organization, 1998b, p. 281). The Panel expressed concern not just over the effects of this particular case but also over the effects of the precedent that would be set by the manner in which this case was resolved. Paragraph 7.45 of the Panel report is worth quoting in its entirety.

> In our view, if an interpretation of the chapeau of Article XX were to be followed which would allow a Member to adopt measures conditioning access to its market for a given product upon adoption by exporting Members of certain policies, GATT 1994 and the WTO Agreement could no longer serve as a multilateral framework for trade among Members as security and predictability of trade relations under those agreements would be threatened. This follows because, if one WTO Member were allowed to adopt such measures, then other Members would also have the right to adopt similar measures on the same subject but with different or even conflicting requirements. If that happened, it would be impossible for exporting Members to comply at the same time with multiple conflicting policy requirements. Indeed, as each of these requirements would necessitate that adoption of a policy applicable not only to export production (such as specific standards applicable only to goods exported to the country requiring them) but also to domestic production, it would be impossible for a country to adopt one of those policies without running the risk of breaching other Members' conflicting policy requirements for the same product and being refused access to these other markets. We note that, in the present case, there would not even be the possibility of adapting one's export production to the respective requirements of the different Members. Market access for goods could become subject to an increasing number of conflicting policy requirements for that same product and this would rapidly lead to the end of the WTO multilateral trading system (World Trade Organization, 1998b).

Given this finding, the Panel concluded that it was not necessary to examine the question of whether the US embargo fit into any of the categories enumerated for exemptions in Article XX.

Some aspects of the reasoning of the Panel were challenged as overly broad by the Appellate Body (World Trade Organization, 1998c), although it rejected the US appeal, finding that the shrimp import embargo had been implemented by the USA in an unjustifiably discriminatory manner. In the eyes of the Appellate Body, even though the US policy did qualify as an acceptable case of a measure applied to the conservation of exhaustible natural resources (Article XX (g)), implementation of the embargo required WTO members exporting shrimp to the USA to adopt 'essentially the same policy' (World Trade Organization, 1998b, p. 63, emphasis in the original) regarding the conservation of sea turtles as the USA, without regard to the effectiveness of ongoing efforts to protect sea turtles that might have similar effects but that use different methods than those employed in the USA or to other differences in local conditions in exporting countries. In addition, the Appellate body found that the imposition of an embargo on shrimp harvested in a manner conforming to US conservation policy but by a firm operating in a country which had not been certified as in compliance with US standards was unjustifiably discriminatory. The finding of unjustifiable discrimination also was based on the fact that different countries were allowed different lengths of time to achieve compliance with the US standard. The Appellate body found the US measure to be arbitary given that the USA did not make a good faith effort through diplomatic means to develop a cooperative international agreement to protect sea turtles before unilaterally imposing the embargo, that it failed to develop a transparent and expeditious process for certification and that it did not institute a procedure for the appeal of denial of certification.

COMMENTS ON THE SHRIMP AND SEA TURTLE CASE

The Panel report in this case runs to 424 single spaced pages. The Appellate Body report is a mercifully compact 77 pages, again single spaced. It is clearly not possible to recapitulate in a single book chapter all of the substantive and procedural issues touched on by the Panel or in the appeal[2]. I have elected to highlight a few of the key issues in light of the widespread criticism of both the findings of the Panel and of the Appellate body from environmental organisations. Given the rhetoric surrounding this case, there are several aspects of the panel report that are surprising. First, the report reflects a considerable effort to explore the current ecological knowledge about sea turtle populations and the effectiveness of different approaches to their protection taken in different countries. The USA and India, Malaysia, Pakistan and Thailand submitted scientific evidence on the status of endangered sea turtles as well as on the nature of programmes initiated to protect sea turtles within their jurisdictions. This evidence was subjected to rebuttal arguments. Critics of the WTO's conduct in this case generally never acknowledge the initiative taken by the Panel to consult with technical scientific experts. It seems clear from the biological evidence presented by the USA and by India, Pakistan, Malaysia and Thailand and by rebuttal evidence offered by both sides that scientific controversies exist on several matters related to this dispute. The Panel, on its own initiative, appointed an independent scientific panel of experts. These experts were given access to submissions by both sides and were asked to offer answers to a list of technical questions posed by the Panel in its efforts to reconcile differing interpretations of the available biological evidence.

The Panel report reflects the considerable extent of disagreement about the proper interpretation of existing turtle population data and of scientific assessments of the various means employed to protect sea turtles in India, Malaysia, Pakistan and Thailand. The US defence of its import ban took the position that the national governments of India, Malaysia, Pakistan and Thailand,

---

[2] The steadfast reader can find both the Panel report and the Appellate Body report in the WTO Document Dissemination web site.

despite their protestations to the contrary, had not been able to effectively protect sea turtle populations in their territorial waters or in Southeast Asian waters generally. Not surprisingly, India, Pakistan, Malaysia and Thailand took exception to this interpretation and offered data and scientific studies of their own to rebut the US claim. Much of this rebuttal argued that the species of sea turtles, the methods of shrimp harvest and other local environmental and social factors in the four countries were different from the conditions in the USA, so that while the measure used by the USA on its domestic shrimp fleet might be appropriate in that context, it was not appropriate in the four countries who initiated the dispute. A second surprising aspect of the case is the extent of disagreement over the US characterisation of sea turtle populations as global migratory species. This characterisation is pivotal to justifying unilateral US action. India, Pakistan, Thailand and Malaysia argued that sea turtles should be characterised as regionally migratory or migratory within their territorial waters. A third surprise might be the extent to which the panel report refers to research and technical documents produced by independent scientific research and by non-governmental environmental organisations.

A second surprising aspect of the Panel report is the extensive involvement of countries which are members of the WTO but who were not parties to the dispute. Australia, Equador, El Salvador, the EU, Guatemala, Hong Kong, Japan, Nigeria, the Phillipines, Singapore and Venezuela all made submissions to the Panel. What is striking about this list is that it extends to countries who have no interest or involvement in exporting shrimp to the USA. For that matter, the countries that brought the case to the WTO, according to evidence submitted to the Panel, shipped shrimp valued at something less than $15 million (US) per year prior to the embargo. All of this suggests that, from a WTO membership point of view, this case was seen by many countries as having much broader international implications than the loss of revenue from sales of shrimp in the US market.

A third surprise has to do with the treatment of submissions from NGOs to the WTO. Both the Panel and the Appellate Body engaged in protracted deliberations on the treatment of such submissions. Their findings, while not identical, have important implications for future disputes involving environmental issues in the WTO. The Panel, largely on procedural grounds and concerned about setting a precedent that might overwhelm future Panels with unsolicited documentation, chose to not consider material submitted to it by non-members to the WTO. It did, however, remind the parties to this dispute that they were at liberty to include any material that they deemed appropriate in their own submissions and the USA revised its submission to the Panel to include part of a report prepared by the Center for Marine Conservation and the Center for International Environmental Law as an annex. The Appellate Body overturned the Panel's procedural ruling by observing that any Panel, upon receipt of unsolicited information from a non-member can, *ex post*, make a request for this information.

GASOLINE, TUNA AND BEEF HORMONES

Three other environmentally related trade disputes have figured prominently in especially environmental non-governmental organisation criticism of trade treaties. The first of these cases involved the GATT dispute between the USA and Mexico regarding restrictions on imports of tuna into the US market on the basis of whether that tuna was caught in a manner that adequately protected dolphins. The second case was a WTO dispute between the USA and Venezuela regarding standards on gasoline (World Trade Organization, 1996). The most recent case to be adjudicated was the dispute between the EU and both the USA and Canada regarding imports of beef that had been treated with hormones into the European market. Space does not permit a detailed examination of the details of these cases, but I would like to offer a few observations that might provide some perspective on key issues. Although the tuna case was heard as a GATT dispute and there are important procedural differences between how cases were handled under the GATT compared to the WTO, there is a similar logic to the findings of the tuna case that will be familiar to students of the shrimp case. The essence of the GATT panel finding was that the US import restrictions on tuna from Mexico were applied in a manner that put Mexican producers at a disadvantage relative to their US counterparts and there was no compelling case to support this discriminatory practice on environmental grounds. It would be unfair and inaccurate to conclude that the GATT and more recently the WTO have refused and will continue to refuse to permit environmental exceptions to the non-discrimination principle. The case for an exception needs to establish that discrimination is necessary and appropriate for the attainment of the environmental aim in question.

The US position on the gasoline case is perplexing. This would seem to be a case where the US policy objective could have been achieved in a non-discriminatory manner and to be safe from an adverse WTO panel ruling if the USA had opted to impose a product standard on domestic and imported gasoline. This case did not confront the reluctance of the GATT or the WTO to allow countries to impose process standards on imports. It did involve a concern that gasoline with certain objective physical characteristics would have an adverse effect on urban air quality in the USA, at least during certain times during the year. The fact that the USA did not apply a transparent and uniform product standard on domestic and foreign refiners gave the WTO panel a predictable, and I would argue unavoidable, basis for finding against the US measure. But this outcome can hardly be blamed on structural defects in the WTO agreement in general or on the procedures used to mediate disputes between members.

The EU beef hormone case has been examined by Kerr and Hobbs (2000) and by Roberts (1998), and I have little to add to that insightful literature, expect for one point. This case is frequently mentioned as an indication of the failure of the WTO, and especially the WTO agreement on Sanitary and Phytosanitary measures (SPS agreement) to affirm the so-called precautionary principle.

According to some proponents of this principle, WTO members should be able to take measures, including placing restrictions on imports, before definitive scientific evidence has been produced to document a certain risk. According to some critics of the WTO, this option was denied to the EU by the panel.

Examination of the 800 pages of the panel reports, however, indicates that this criticism is misplaced. Throughout the case, the EU steadfastly refused to exercise its clear option under the SPS agreement to impose import restrictions on a temporary basis while it pursued more complete scientific information on the human health risks associated with beef that had been treated with the controversial hormones. It insisted that it be granted the right to impose permanent measures and proceeded to develop a legally skillful but ultimately unsuccessful tactical case in support of this position. Once again, it would be unfair and inaccurate to conclude, as some critics have concluded, that the SPS agreement is by design incompatible with the precautionary principle.

ENVIRONMENTAL POLICY AND TRADE LIBERALISATION AFTER SEATTLE

The polarised rhetoric of some of the environmental critics of the WTO leading up to the Seattle meeting indicated that the organisation was incapable of successfully handling environmental trade disputes. On its face, the record is worrisome. US efforts to protect dolphins were ruled non-compliant with GATT membership obligations. US efforts to protect sea turtles were found to violate WTO rules. EU bans on beef imports were found to violate WTO obligations. Some commentators have invoked a 'three strikes and you are out' rule against the WTO and its predecessor, the GATT. According to this view, it is not possible to develop effective national measures to protect environmental values or human health that will not be struck down by a future WTO Panel.

My view is that it is premature to conclude that it is impossible to develop WTO-proof environmental policies. There are important policy development lessons to be learned from the existing short list of cases that have been addressed in multilateral trade disputes. Rather than conclude that the WTO is incapable of resolving trade disputes over national environmental policies or over trade provisions in international environmental agreements, it is my view that the findings of the Panel, clarified by the findings of the Appellate Body, provide constructive guidance to global efforts to improve natural resource and environmental stewardship. I would like to suggest the following checklist[3] that can be applied to the efforts of national governments:

---

[3] Ervin (1999) has also proposed a 7-item Code of Good Process for designing agricultural environmental programmes that are consistent with WTO obligations. His list includes:
- specify clear environmental objectives for programmes;
- clarify property rights in environmental resources to establish applicability of payments, charges and subsidies;
- prefer the least trade-distorting agri-environmental management instrument;
- establish scientific linkage of the environmental objective with the programme instrument;
- implement monitoring and evaluation programmes to document policy/programme efficacy;
- apply equal treatment (for domestic products and imports) if applicable; and
- ensure the transparency of agri-environmental measures.

- Devise environmental initiatives in light of international treaty obligations, including but not exclusively limited to obligations under the WTO.
- Show a good faith effort to develop an international cooperative consensus based agreement to address the environmental issue in question.
- Allow for flexibility in the means used to achieve a desired environmental outcome to accommodate differences in ecological and social conditions in other countries.
- Treat firms in other countries in a similar fashion to firms in your own country and treat other countries equally.
- Avoid using the most potent trade remedy available (typically an import embargo) when a less potent remedy will do.
- In applying a trade remedy, do so in a transparent, fair and reasonable manner. Recognise that the burdens of compliance with environmental measures are often greater the shorter the period of time allowed before compliance is necessary.

The general conclusion of many critics of the GATT's and the WTO's findings in environmental trade cases is that the standards that I have just listed are impossible to meet. I disagree. In the shrimp and sea turtle case, in the beef hormone case, in the US gasoline case and in the tuna dolphin dispute, policy was made with what appears to be flagrant disregard for these principles. It not justifiable to conclude that WTO proofing of environmental trade measures is impossible based on the experience we have had with such measures so far. A more reasonable interpretation of recent experience is that much progress can be made through the development of more institutionally aware environmental policies.

REFERENCES

Anderson, K. and Strutt, A. (1996) On measuring the environmental impact of agricultural trade liberalization. In: Bredahl, M., Ballenger, N., Dunmore, J. and Roe, T. (eds) *Agriculture, Trade and the Environment: Discovering and Measuring the Critical Linkages*. Westview Press, Boulder, pp. 151-172.

Anderson, T. and Leal, D. (1991) *Free Market Environmentalism*. Westview Press, Boulder.

Audley, J. (1997) *Green Politics and Global Trade: NAFTA and the Future of Environmental Politics*. Georgetown University Press, Washington D.C.

Bhagwati, J. and Hudec, R. (eds) (1996) *Fair Trade and Harmonization*. MIT Press, Cambridge.

Bredahl, M. and Holleran, E. (1997) Technical Regulations and Food Safety in NAFTA. In: Loyns, A., Knutson, R., Meilke, K. and Sumner, D. (eds) *Harmonization/Convergence/Compatibility in Agriculture and Agri-Food Policy: Canada, United States and Mexico*. University of Manitoba, University of Guelph, Texas A&M University and University of California, Winnipeg, Guelph, College Station and Davis, pp. 71-87.

Brubaker, E. (1995) *Property Rights in the Defence of Nature*. Earthscan, Toronto.

Casella, A. (1996) Free trade and evolving standards. In: Bhagwati, J. and Hudec, R. (eds) *Fair Trade and Harmonization*. MIT Press, Cambridge, pp. 119-156.

De Bruyn, S., Van den Bergh, J. and Opschoor, J. (1998) Economic growth and emissions: reconsidering the empirical basis of environmental Kuznets' curves. *Ecological Economics*, 25, 161-175.

Ervin, D. (1999) Toward GATT-Proofing environmental programmes for agriculture. *Journal of World Trade*, 33(2), 63-82.

Ervin, D. and Fox, G. (1998) Environmental policy considerations in the grain-livestock subsectors in Canada, Mexico and the United States. In: Loyns, A., Knutson, R. and Meilke, K. and Sumner, D. (eds) *Economic Harmonization in the Canadian/U.S./Mexican Grain-Livestock Subsector*. Texas A&M University and University of Guelph, College Station and Guelph, Section 4, pp. 275-304.

Esty, D. (1994) *Greening the GATT: Trade, Environment and the Future*. Institute for International Economics, Washington D.C.

Friends of the Earth England, Wales and Northern Ireland (1998) *Position Statement on the Multilateral Agreement on Investment (MAI)*. London.

Friends of the Earth England, Wales and Northern Ireland (1999a) *The World Trade System: How it Works and What's Wrong with It*. London.

Friends of the Earth England, Wales and Northern Ireland (1999b) *Seattle and the WTO: A Briefing*. London.

Friends of the Earth International (1999a) *The World Trade System: Winners and Losers*, Amsterdam.

Friends of the Earth International (1999b) *The World Trade System: An Activist's Guide-Forests*, Amsterdam.

Friends of the Earth International (1999c) *The World Trade System: An Activist's Guide*, Amsterdam.

Friends of the Earth International (1999d) *The World Trade System: An Activist's Guide-Food and Food Security*, Amsterdam.

Friends of the Earth International (1999e) *The World Trade System: An Activist's Guide-The WTO and Finance*, Amsterdam.

Fugazzotto, P. and Behera, C. (1999) *Dead Turtles: Good for the Global Economy?* Sea Turtle Restoration Project, Earth Island Institute, Forest Knolls, California.

Fugazzotto, P. and Steiner, T. (1998) *Slain by Trade: The Attack of the World Trade Organization on Sea Turtles and the US Endangered Species Act*. Sea Turtle Restoration Project, Earth Island Institute, Forest Knolls, California.

Gabel, L. (1994) *The Environmental Effects of Trade in the Transportation Sector*. Organization for Economic Co-operation and Development, Paris.

General Agreement on Tariffs and Trade (GATT) (1991) United States - Restrictions on Imports of Tuna. Report of the GATT Panel, Geneva.

Goklany, I. (1999) *Clearing the Air: The Real Story on the War on Air Pollution*. Cato Institute, Washington, D.C.

Goldman, P. and Fawcett-Long, J. (1999) *Farms and Food Safety at Risk: The World Trade Organization's Threat to Our Food System*. Earthjustice Environmental Defense Fund, San Francisco.

Goldman, P. and Scott, J. (1999) *Our Forests at Risk: The World Trade Organizations Threat to Forest Protection*. Earthjustice Environmental Defense Fund, San Francisco.

Goldman, P. and Wagner, M. (1999) *Trading Away Public Health: The World Trade Organization Obstacles to Effective Toxics Control*. Earthjustice Environmental Defense Fund, San Francisco.

Grossman, G. and Krueger, A. (1991) *Environmental Impacts of a North American Free Trade Agreement*. Discussion Paper No. 158, Woodrow Wilson School, Princeton University, Princeton.

Johnson, P. and Beaulieu, A. (1996) *The Environment and NAFTA: Understanding and Implementing New Continental Law*. Island Press, Washington D.C.

Josling, T. (1997) Policy dynamics in North American agriculture: definitions of and pressures for harmonization, convergence and compatibility in policies and programmes affecting the agri-food sector. In: Loyns, A., Knutson, R., Meilke, K. and Sumner, D. (eds) *Harmonization/Convergence/Compatibility in Agriculture and Agri-Food Policy: Canada, United States and Mexico*. University of Manitoba, University of Guelph, Texas A&M University and University of California, Winnipeg, Guelph, College Station and Davis, pp. 7-20.

Kerr, W. (2000) Is it time to re-think the WTO? A return to basics. *The Estey Centre Journal of International Law and Trade Policy*, 1(2), 99-107.

Kerr, W. and Hobbs, J. (2000) The WTO and the dispute over beef produced using growth hormones. CATRN Paper 2000-07, University of Saskatchewan, Saskatoon.

Kirton, J., Cavanagh, R., Fernandez de Castro, R., Foley, D., Fox, G., Hoyt, E, Moscarella, J., Nadal, A., Ramirez, R., Reardon, C., Runge, C. and Graber, D. (1999) *Assessing Environmental Effects of the North American Free Trade Agreement (NAFTA): An Analytical Framework (Phase II) and Issue Studies*. Commission for Environmental Cooperation, Montreal.

Klevorick, A. (1996) Reflections on the race to the bottom. In: Bhagwati, J. and Hudec, R. (eds) *Fair Trade and Harmonization*. MIT Press, Cambridge, pp. 459-468.

Leebron, D. (1996) Lying down with procrustes: An analysis of harmonization claims. In: Bhagwati, J. and Hudec, R. (eds) *Fair Trade and Harmonization*. MIT Press, Cambridge, pp. 41-118.

Levinson, A. (1996) Enviromental regulations and industry location: International and domestic evidence. In: Bhagwati, J. and Hudec, R. (eds) *Fair Trade and Harmonization*. MIT Press, Cambridge, pp. 429-458.

Lindsey, P. and Bohman, M. (1997) Environmental policy harmonization. In: Loyns, A., Knutson, R., Meilke, K. and Sumner, D. (eds) *Harmonization/Convergence/ Compatibility in Agriculture and Agri-Food Policy: Canada, United States and Mexico*. University of Manitoba, University of Guelph, Texas A&M University and University of California, Winnipeg, Guelph, College Station and Davis, pp. 93-112.

Loyns, A., Knutson, R., Meilke, K. and Sumner, D. (eds) (1997) *Harmonization/ Convergence/Compatibility in Agriculture and Agri-Food Policy: Canada, United States and Mexico*. University of Manitoba, University of Guelph, Texas A&M University and University of California, Winnipeg, Guelph, College Station and Davis.

Loyns, A., Knutson, R. and Meilke, K. and Sumner, D. (eds) (1998) *Economic Harmonization in the Canadian/U.S./Mexican Grain-Livestock Subsector*. Texas A&M University and University of Guelph, College Station and Guelph.

Loyns, A., Knutson, R., Meilke, K. and Yunez-Naude, A. (eds) (2000) *Policy Harmonization and Adjustment in the North American Agricultural and Food Industry*. Texas A&M University, the University of Guelph and El Colegio de Mexico, College Station, Guelph and Mexico City.

Mabey, N. and McNally, R. (1999) *From Trade Liberalization to Sustainable Development: A Critique of the OECD Paper*. World Wildlife Fund, Gland, Switzerland.

Perrin, M. (1998a) *Initiating an Environmental Assessment of Trade Liberalization in the WTO*. Volume II, World Wildlife Fund, Gland, Switzerland.

Perrin, M. (1998b) *Developing a Methodology for the Environmental Assessment of Trade Liberalization Agreements*. Discussion Paper, World Wildlife Fund, Gland, Switzerland.

Roberts, D. (1998) Preliminary assessment of the effects of the WTO agreement on Sanitary and Phytosanitary Trade Regulations. *Journal of International Economic Law*, 1(3), 377-405.

Runge, C. (1994) *Freer Trade, Protected Environment: Balancing Trade Liberalization and Environmental Interests*. Council on Foreign Relations, New York.

Runge, C. and Fox, G. (1999) Feedlot production of cattle in the United States and Canada: some environmental implications of the North American Free Trade Agreement. In: Kirton, J., Cavanagh, R., Fernandez de Castro, R., Foley, D., Fox, G., Hoyt, E., Moscarella, J., Nadal, A., Ramirez, R., Reardon, C., Runge, F. and Graber D. *Assessing Environmental Effects of the North American Free Trade Agreement (NAFTA): An Analytical Framework (Phase II) and Issue Studies.* Commission for Environmental Cooperation, Montreal, pp. 183-258.

Sampson, G. (2000) *Trade, Environment and the WTO: the Post Seattle Agenda.* Policy Essay No. 27, Overseas Development Council, Johns Hopkins University Press, Baltimore.

Selden, T. and Song, D. (1994) Environmental quality and development: is there a Kuznets' curve for air pollution? *Journal of Environmental Economics and Management*, 27, 147-162.

Stillwell, M., Caldwell, J., Godfrey, C. and Arden-Clarke, C. (undated) *Dispute Settlement in the WTO: A Crisis for Sustainable Development.* A World Wildlife Fund, Centre for International Environmental Law, Oxfam-GB and Community Nutrition Institute (US) Discussion Paper, World Wildlife Fund, Gland, Switzerland.

Van Putten, M. (1999) *What's Trade Got To Do With It? A Guide to Trade Policy and Saving the Environment.* National Wildlife Federation, Washington, D.C.

Wagner, M. and Goldman, P. (1999) *The Case for Rethinking the WTO: The Full Story Behind the WTOs Environment and Health Cases.* Earthjustice Environmental Defense Fund, San Francisco.

Wolf, C. (1979) A theory of non-market failure. *Journal of Law and Economics*, 22(1), 107-139.

World Trade Organization (1996) United States - Standards for Reformulated and Conventional Gasoline. Report of the Panel, WT/DS2/R, Geneva.

World Trade Organization (1997a) EC Measures Concerning Meat and Meat Products (Hormones) Complaint by the United States. Report of the Panel, WT/DS26/R/USA, Geneva.

World Trade Organization (1997b) EC Measures Concerning Meat and Meat Products (Hormones) Complaint by Canada. Report of the Panel, WT/DS48/R/CAN, Geneva.

World Trade Organization (1998a) EC Measures Concerning Meat and Meat Products (Hormones). Report of the Appellate Body, WT/DS26/AB/R and WT/DS48/AB/R, Geneva.

World Trade Organization (1998b) United States - Import Prohibition of Certain Shrimp and Shrimp Products. Report of the Panel, WT/DS58/R, Geneva.

World Trade Organization (1998c) United States - Import Prohibition of Certain Shrimp and Shrimp Products. Report of the Appellate Body, WT/DS58/AB/R, Geneva.

Index